100

MOST IMPORTANT
SCIENCE
IDEAS

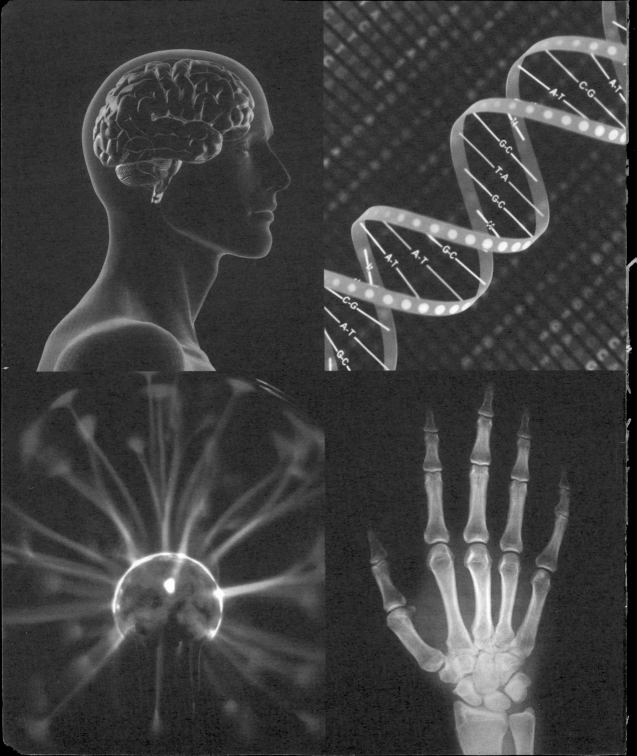

100

MOST IMPORTANT
SCIENCE
IDEAS

KEY CONCEPTS IN GENETICS, PHYSICS AND MATHEMATICS

MARK HENDERSON • JOANNE BAKER • TONY CRILLY

FIREFLY BOOKS

CONTENTS

INTRODUCTION

Science is a perpetual source of wonder. From Archimedes' eureka moment in his bathtub, to the splitting of the atom and the mapping of the human genome, its practice has transformed humanity's knowledge and understanding of life and the universe, if not quite everything.

Its great discoveries underpin all manner of technologies that have changed how we live. Science has brought us medicines and vaccines, computers and mobile phones, cars and airplanes, and modern crops that can just about feed a global population of 6.5 billion and rising. It has allowed us to investigate when and how the universe began, and the laws of mathematics and physics that govern its ways. It has traced how life on Earth evolved, and the common genetic heritage that connects every organism known to have existed. And it continues daily to generate breathtaking new discoveries, which often pose fresh questions that nobody had previously thought to ask.

Science is so broad in its scope that it is impossible for any one person to understand it all. Even professional scientists must specialize, and develop expertise in a narrow field: most molecular biologists will know little of particle physics, or vice-versa. But what unites all aspects of science is the scientific method, a uniquely rigorous approach to thinking that makes it possible to establish reliable knowledge. Scientists collect data, which they use to formulate hypotheses, and then they test these to destruction. They try to prove their most brilliant ideas wrong — and in doing so, they progressively grasp more about what seems to be right.

Discoveries build one on another, so that our understanding becomes ever more refined. Newton's theories of physics were improved upon by Einstein, and Einstein's insights will not be the last word. Darwin's theory of evolution and Mendel's genetics led Crick and Watson to the structure of DNA, and thence to the slow unraveling of the code of life — a process that is still incomplete. The great mathematical challenge of Fermat's last theorem has been resolved — but the Riemann hypothesis remains unproven. There is always another question to ask.

This book aims to introduce 100 of the most important ideas that science has put forward — the choicest fruits of its search for evidence. The great diversity of scientific achievement, of course, means that such a list could never purport to be exhaustive: there are thousands of concepts that could make deserving claims to a place. The three authors have thus each selected one of the many branches of science to explore in some depth, with a series of clear and concise essays that explain the fundamentals you really need to know.

The fields included in this volume are important, groundbreaking, and increasingly front-of-mind: genetics, physics and mathematics. Each has had a profound influence on our way of life, which continues to grow with advancing understanding. And the history of each discipline also illustrates something of the rigorous method by which science makes new discoveries.

The insights into the genetic code that have emerged in the past 50 years have changed the way we understand life on Earth. DNA has shed light on history, confirming the fact of evolution and tracing how the first humans left Africa to populate the world. It has provided new forensic tools for convicting the guilty and exonerating the innocent, and it explains how our individuality is forged by nature and nurture. Genetic medicine and stem cell research are promising a revolution in healthcare. And cloning, genetic engineering and designer babies have stirred ethical controversies that were once the stuff of science fiction.

The laws of physics explain why we see our reflections when we look in the mirror, why roads freeze over on winter nights, and why the Moon orbits the Earth while the Earth orbits the Sun. An understanding of electromagnetism has allowed us to harness it to light our homes, and to send words and pictures across continents. Vast atom-smashers like the Large Hadron Collider, at the CERN laboratory near Geneva, have made it possible to study the dawn of space and time, and to map the building blocks of matter. Then there is the weird world of quantum mechanics, in which particles start to behave very strangely indeed.

Neither genetics nor physics could have advanced as far or as fast, however, without mathematics. It is the cleanest and most fundamental of the sciences, which provides all the others with critical tools for the job. Mathematical principles explain the patterns in which genes are passed from generation to generation. The laws of physics are underpinned by equations. Mathematics can also reveal great beauty, as in the theory of prime numbers and the Fibonacci sequence — the inspiration for *The Da Vinci Code*.

The marvels of science offer limitless engagement and entertainment for curious minds. This book is where they should begin.

GENETICS

Life is all around us. Since the first living things emerged, almost 4 billion years ago, the Earth has been home to a cornucopia of organisms, memorably encapsulated by Charles Darwin as its "endless forms most beautiful and wonderful." Its biodiversity encompasses well over a million species known to science, from the largest blue whale to the tiniest bacterium, and perhaps ten times as many probably remain to be discovered. Countless more have been and gone, flourishing for thousands and millions of years before fading into extinction.

Save perhaps for some of the very first sparks of life, all its variety is founded on the same building blocks. Life, whether as complex as a human being or as simple as a single-celled microbe, is written in one alphabet: the letters of DNA. All the living things that have existed on our planet are bound by this common thread, and share their origins in an entity that scientists call LUCA — the last universal common ancestor. The study of genetics, at its heart, is thus the study of life. It explains both how all its terrestrial forms are linked, and how they are different.

The chapters that follow will explore in depth how humanity has come to understand its genetic code, and some of the opportunities and challenges this knowledge has presented. But we

1802 Jean-Baptiste Lamarck sets out hypothesis of inheritance of acquired characteristics

1859 Charles Darwin publishes *On the Origin of Species*, which sets out the theory of evolution by natural selection

1865 Gregor Mendel presents his laws of inheritance to Natural History Society of Brünn

1869 Friedrich Miescher discovers DNA, from pus-soaked bandages

1900 Rediscovery of Mendel's ideas by Hugo de Vries, Erich von Tschermak and Carl Correns

will start here with a few of the fundamental concepts that place life, and thus genetics, in its proper context.

The variety of life: Linnaean classification

Though attempts to break down the multiplicity of living things into distinct categories dates back at least to Aristotle, the great philosopher of ancient Greece, it came of age as a science in the 18th century with the work of Carl Linnaeus. The Swedish biologist is today considered the father of taxonomy, or biological classification, and the "binomial" system he devised lives on in the name of every species, from *Escherichia coli* to *Homo sapiens*.

Under the Linnaean system, which has been updated to reflect improved scientific understanding, life is classified in a hierarchy. As you move down this hierarchy, organisms become more closely related, and share more characteristics. At its head are the domains, followed by phyla, classes, orders, families, genera and species.

There are three domains: the bacteria are a group of single-celled organisms, which make up much of the world's biomass. The archaea are also single-celled, but have a separate evolutionary history. Then there are the eukaryotes: organisms, whether single- or multi-celled, in which the cell's genetic material is enclosed in a nucleus.

Each domain is then subdivided. The next level down is the kingdom: among eukaryotes, there are four, of animals, plants, fungi and protists (protist organisms are single-celled, or multi-celled but without specialized tissues, and there is disagreement whether protists comprise a true kingdom). After that comes the phylum:

1910 T.H. Morgan's experiments with fruit flies show chromosomal basis of inheritance

1927 Hermann Muller shows X-rays can cause genetic mutations

1941 George Beadle and Edward Tatum show that genes make proteins

1944 Oswald Avery, Colin McLeod and Maclyn McCarty show DNA carries genetic information

1953 Francis Crick and James Watson identify double-helix structure of DNA

9

humans belong to the chordates, which include all animals with a backbone and a few invertebrates. The mammals are a class — others include the reptiles, amphibians and birds — and primates are an order.

The human family, in Linnaean terms, is hominidae, the great apes. Chimpanzees, gorillas and orangutans — our closest living cousins in evolutionary terms — are members too. All the other animals in our genus, *Homo*, are extinct, such as *Homo erectus* and *Homo neanderthalis*, or Neanderthal man. Our species is *Homo sapiens*, and every other species' scientific name takes the same binomial form: the genus, followed by a cognomen specific to the species (and subspecies). The horse is *Equus caballus*, the giant redwood is *Sequoiadendron giganteum*, and the bacterium that causes stomach ulcers is *Helicobacter pylori*.

The structure of life: the cell

All the forms of life that appear in the Linnaean hierarchy share a characteristic besides descent from a common ancestor. Whether large or microscopic, simple or complex, they are built from cells. The name, which comes from the Latin *cellula*, meaning small room, was coined in the 17th century by Robert Hooke, to describe the tiny tessellating blocks he saw when looking at a slice of cork through an early microscope. Cells are the bricks of biology, the most basic units that can be considered alive. Viruses, it is true, are not made up of cells, but as they cannot reproduce without hijacking a host cell, they are not strictly classified as living.

1960 Jacques Monod identifies role of messenger RNA in translating DNA into proteins

1961 Marshall Nirenberg discovers first triplet DNA code for an amino acid

1975 Fred Sanger develops chain-termination sequencing, the first efficient method of reading genomes

1976 Richard Dawkins publishes *The Selfish Gene*, which popularizes the gene-centered view of evolution

1981 Martin Evans isolates embryonic stem cells in mice

Bacteria, archaea and some eukaryotes have just a single cell. More complex organisms can have many more: an adult human has about 100 trillion, including more than 200 specialized varieties such as neurons and lymphocytes. All cells, however, share a few unifying features. They are self-contained within a porous membrane. They take in nutrients through this membrane, which they metabolize in chemical reactions to create energy and specialized proteins and enzymes. And they can reproduce, by copying the genetic instructions written in their DNA.

The most important difference is between prokaryotic and eukaryotic cells. In prokaryotes — bacteria and archaea — the cell's DNA floats loose in its cytoplasm fluid, together with the ribosomes that convert it into protein. It is enclosed within a cell envelope, comprising a plasma membrane and, usually, a cell wall, and it has a tail-like appendage known as the flagellum, which it uses to get about.

Eukaryotic cells are typically at least ten times bigger than prokaryotes, and their defining quality is that their DNA is wrapped up and contained in a solid nucleus: the term *"eukaryote"* means "true nucleus" in Greek. A secondary difference is that these cells contain structures called organelles in their cytoplasm. The most important organelles are the mitochondria, which serve as energy-generating power plants, and, in plants, the chloroplasts — the crucibles of photosynthesis. These organelles were once independent prokaryotes that lived symbiotically within eukaryotic cells: they proved so useful that they were eventually absorbed completely.

1984 Alec Jeffreys develops genetic fingerprinting. First murder conviction using DNA evidence follows in 1988

1985 Creation of the first genetically modified crop, a tobacco strain

1989 Creation of the first "knockout" mice, in which genes have been switched off to investigate disease

1990 Human Genome Project launched

1990 First successful use of gene therapy to treat a human disease

The drivers of life: photosynthesis and respiration

Whether organisms are comprised of a single cell or many, the overwhelming majority would never have existed were it not for photosynthesis. While the question of how life emerged remains an open one, it is photosynthesis that keeps the largest part of it going. It is the means by which energy from the Sun powers life on Earth.

Not every life form, of course, can harness this energy directly: photosynthesis is the preserve of one kingdom of eukaryotes — the plants — and of some bacteria. Yet almost every species depends upon it: those that cannot do it themselves eat other species that can.

Photosynthesis uses energy from sunlight to convert carbon dioxide and water into carbohydrate — sugars such as glucose that can be used immediately as fuel, or laid down in storage. The necessary light is absorbed by protein structures known as photosynthetic reaction centers, which contain pigments — the most important is chlorophyll, which gives plant leaves their green hue.

Photosynthesis is also critical to most life on Earth for another reason: its by-product is oxygen, and it is thus the ultimate source of the part of the air we breathe. It was only after the first photosynthetic organisms evolved, 3.5 billion to 3 billion years ago, that the atmosphere began to fill up with oxygen, allowing the emergence of more complex life forms.

The advent of oxygen was important because it improves the efficiency of respiration, the process by which sugars are ultimately converted into the energy used by a plant — or by an animal that has eaten a plant. Respiration converts sugars into adenosine triphosphate, commonly known as ATP, a chemical carrier of energy

1990 Birth of Natalie and Danielle Edwards, the first babies screened as embryos to be free from a genetic disease

1993 Discovery of Huntington's disease mutation

1996 Birth of Dolly the sheep, the first mammal cloned from an adult cell

1998 Celera launches private effort to sequence the human genome

1998 Jamie Thomson isolates human embryonic stem cells

that can be used to power a cell's many specialized metabolic processes. As aerobic respiration in a cell's mitochondria, which requires oxygen, can make almost 20 times more ATP from the same quantity of glucose than anaerobic respiration, oxygen allowed for the evolution of larger and more sophisticated bodies. Without photosynthesis, and the aerobic respiration it enables, you would not exist to be reading this book.

The code of life: DNA

Genetics is a young science. The laws of inheritance were first formulated only in 1865, by Gregor Mendel, and knowledge of the double-helix structure of DNA is barely 50 years old. Yet in recent years a great wealth of discoveries about the genetic code that holds all life's instructions has started to transform our understanding of the subject. We know not only how species vary in form, but also how their genomes are related to one another. We have begun to identify which genes build cells, and drive biological processes like photosynthesis and respiration. And as well as helping to explain life, DNA is changing the way we live, providing tools with which to improve healthcare, forensic science and agriculture.

The chapters that follow highlight the history of genetics, its most important principles, and some of the technologies it has led to and the challenges they raise. It would of course have been possible to write many more. We are fortunate to live at a time when genetics has begun to reveal more about the workings of life than has ever been known. This is an introduction to that knowledge.

2001 Human Genome Project and Celera publish first drafts of humanity's genetic code

2006 Discovery of a new kind of genetic variation: widespread copy-number variation

2007 Publication of first wave of genome-wide association studies, identifying common genes that contribute to common diseases, and launch of first direct-to-consumer genetic screening

2007 Consortium shows that junk DNA and RNA are much more important to biology than had been thought

01 The theory of evolution

Charles Darwin: "There is grandeur in this view of life . . . that . . . from so simple a beginning endless forms most beautiful and most wonderful have been, and are being, evolved."

"Nothing in biology," wrote the geneticist Theodosius Dobzhansky, "makes sense except in the light of evolution." It is a truth that applies particularly strongly to its author's specialist field. Though Charles Darwin had no concept of genes or chromosomes, those concepts and all the others that will be described in this book have their ultimate origins in the genius of his insights into life on Earth.

Darwin's theory of natural selection holds that while individual organisms inherit characteristics from their parents, they do so with small and unpredictable alterations. Those changes that promote survival and breeding will multiply through a population over time, whereas those that have negative effects will gradually disappear.

As is often the case with truly great ideas, evolution by natural selection has a beautiful simplicity that, once grasped, immediately becomes compelling. When the biologist Thomas Henry Huxley first heard the hypothesis presented, he remarked: "How extremely stupid not to have thought of that!" Once a skeptic, he became evolution's most vociferous champion, earning the nickname "Darwin's bulldog" (see box).

The argument from design

For centuries before Darwin, natural philosophers had sought to explain the extraordinary variety of life on Earth. The traditional solution, of course, was supernatural: life, in all its diversity, was created by a god, and the traits that fit a particular organism to an ecological niche were a function of the creator's grand plan.

This "argument from design" dates back at least to the Roman orator Cicero, but it is most commonly associated with William Paley, an English clergyman. In an 1802 treatise, he likened the intricacy of life to a watch found on a heath, the very existence of which presupposes the existence of a watchmaker. It rapidly became scientific orthodoxy — even Darwin was much taken with it early in his career.

As was already clear to the philosopher David Hume in the 18th century, however, the argument from design begs the question: who designed the designer? The absence of an obvious naturalistic explanation for a phenomenon is a poor reason to look no further. Those who make it, from Paley to today's rebranded "intelligent design" creationists, are essentially saying: "I don't understand, so God must have done it." As a way of thinking, it is no substitute for science.

Acquired characteristics

While Paley was invoking the watchmaker, Jean-Baptiste Lamarck took a more intellectually curious approach to the problem. Organisms, he suggested, were descended one from another, with differences emerging by means of subtle modifications in each generation. His was the first theory of evolution.

Lamarck's evolutionary driver was the inheritance of acquired characteristics: anatomical changes caused by the environment would be passed on to offspring. The son of a blacksmith would inherit the strong muscles his father built up in the forge. Giraffes stretch their necks to reach higher branches, elongating the necks of subsequent generations of calves.

The theory is often lampooned today, not least because of its revival in the 1930s by Stalin's favorite biologist, Trofim Lysenko. His insistence that wheat could be trained to resist cold snaps caused millions of deaths from famine in the Soviet Union. Lamarck's ideas are

Darwin's bulldog

T.H. Huxley won his nickname, "Darwin's bulldog," during the 1860 meeting of the British Association for the Advancement of Science, when he defended Darwin's theory against the argument for design advocated by Samuel Wilberforce, the Bishop of Oxford. Though no verbatim account exists, Wilberforce began to mock his rival, asking whether he claimed descent from an ape through his mother or father. Huxley is said to have replied: "I would rather be descended from an ape than from a cultivated man who used his gifts of culture and eloquence in the service of prejudice and falsehood."

sometimes even described as heresy. Yet though he was wrong about the details of evolution, his broader thinking was astute. He correctly ascertained that biological characteristics are inherited — a perception of vast importance. He was mistaken only about the means.

On the Origin of Species

The real means were soon elucidated by Darwin. In the early 1830s, he had sailed aboard the marine survey ship HMS *Beagle*, as naturalist and "gentleman companion" to its captain Robert FitzRoy, on a voyage that enabled Darwin to make detailed observations of the flora and fauna of South America. He found particular inspiration in the Galapagos Islands, west of Ecuador, each of which was home to subtly different species of finch. Their similarities and differences led him to consider whether these species might be related, and had become adapted over time to the environment of each island.

In this, Darwin's assessment differed little from Lamarck's. What set his hypothesis apart was its mechanism. The economist Robert Malthus (1766–1834) had described how populations that increase in size would compete for resources, and Darwin now applied this principle to biology. Chance variations that helped an organism to compete for food and mates would help it to survive, and to pass those traits to its offspring. Variations with negative effects, however, would die out over time as their carriers lost out to others better adapted to their surroundings. Changes were not caused by the environment, but selected by it.

Only a theory

Creationists like to dismiss evolution as "only a theory," as if this gives their alternative scientific parity. This reflects their overwhelming misunderstanding of science, which does not use the term "theory" in its common sense of a hunch. Rather, it means a hypothesis that is confirmed by all available data. Evolution more than meets this definition — it is supported by evidence from genetics, paleontology, anatomy, zoology, botany, geology, embryology and many other fields. If the theory were wrong, almost everything we know about biology would have to be reassessed. It is like the theory of gravity — not an idea we can take or leave, but the best explanation currently available for an observed set of facts.

The implications of this natural selection were brutal. It worked towards no goal or purpose, and gave no special consideration to human life. What mattered, in Herbert Spencer's famous phrase, was "the survival of the fittest."

Darwin first sketched out his ideas as early as 1842, but he did not publish them for another 17 years, fearing the derision that had been heaped upon treatises such as *Vestiges of the Natural History of Creation*, an 1844 pamphlet which argued that species can morph into new ones. In 1858, however, two years after starting to write up his theory, he received a letter from Alfred Russel Wallace, a younger naturalist who had developed similar notions. After presenting jointly with Wallace to the Linnean Society of London, Darwin rushed *On the Origin of Species* into print in 1859.

> **The theory of evolution by cumulative natural selection is the only theory we know of that is in principle capable of explaining the existence of organized complexity.**
> **Richard Dawkins**

Clerical naturalists, including Darwin's old tutors Adam Sedgwick and John Stevens Henslow, were outraged by the new theory. Another critic was Robert FitzRoy, who considered himself betrayed by an old friend, who had abused his kindness to promote views akin to atheism. But Darwin's theory found favor with a younger generation of intellectuals, who recognized both the theory's importance and its utility in undermining a scientific establishment that was still under heavy church influence.

The theory has been updated since 1859, not least by Darwin himself: in *The Descent of Man* (1871), he described how mating preferences can drive evolution as well as the environment, adding sexual selection to the scientific lexicon. But the central principle that all species are interrelated, and emerge one from another by way of random changes that are passed on if helpful to survival or breeding, has become the glue that holds biology together. It is also the foundation stone of genetics.

02 The laws of inheritance

For all Charles Darwin's brilliance, his theory still lacked something critical at its core: it had no way of accounting for the individual variations that were supposed to be passed on from one generation to the next. Darwin himself favored "pangenesis," the idea that the characteristics of each parent merge in the offspring, but he was as wrong in this as was Lamarck about acquired characteristics. If only he had read a paper by a contemporary, a Moravian monk by the name of Gregor Mendel.

In 1856, the same year in which Darwin started work on *On the Origin of Species*, Mendel began a remarkable series of experiments in the garden of the Augustinian monastery of St. Thomas in Brünn, now Brno in the Czech Republic. Over the next seven years, he was to breed more than 29,000 pea plants, with results that would make him known — when the rest of the world finally took notice — as the founder of modern genetics.

> What will doubtless rank as one of the great discoveries in biology, and in the study of heredity perhaps the greatest, was made by Gregor Mendel, an Austrian monk, in the garden of his cloister, some 40 years ago.
> William Castle

Mendel's experiments

Botanists had long known that certain plants "breed true" — that is, characteristics such as height or color are reliably transmitted to the next generation. Mendel exploited this in his experiments on variation, by taking seven true-breeding pea traits, or phenotypes, and cross-breeding the plants that bore them to create hybrids. Pea strains that

Mendelian Inheritance in Man

The Online Mendelian Inheritance in Man database includes more than 12,000 human genes that are thought to be passed on according to Mendel's laws, with dominant and recessive alleles. Of these, at the time of writing, 387 variable genes have been sequenced and linked to a specific phenotype, including diseases such as Tay–Sachs or Huntington's, and more neutral traits such as eye color. Several thousand other phenotypes are known to follow a Mendelian inheritance pattern, but the parts of the genome responsible have yet to be identified or mapped. About 1 percent of births are affected by Mendelian disorders, which result from variation in a single gene.

always produced round seeds, for example, were crossed with strains with wrinkled seeds; purple flowers with white; tall stems with short. In the next generation, known to geneticists as F_1, only one of the traits would remain — the progeny always had round seeds, purple flowers or tall stems. Parental characteristics did not blend, as pangenesis suggested, but one characteristic invariably seemed to dominate.

Next, Mendel took each hybrid and used it to fertilize itself. In this F_2 generation, the trait that seemed to have been erased suddenly came back. Around 75 percent of the peas had round seeds, with the remaining 25 percent coming out wrinkled. In all seven of his samples, this same ratio of 3:1 emerged. His results fit the pattern so well,

---

——FIRST EXPERIMENT——

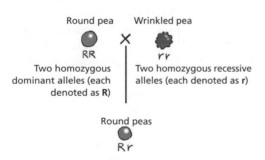

In the F_1 generation, all offspring are heterozygous, with one allele of each kind. The peas are round as the round allele, **R**, is dominant

SECOND EXPERIMENT: ——
USES OFFSPRING OF FIRST EXPERIMENT

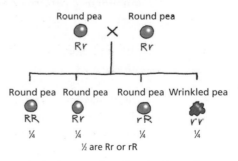

In the F_2 generation, the ratio of round (dominant) to wrinkled (recessive) peas is 3:1

indeed, that some later scientists have suspected fraud. The principles he discovered are now too well attested for that, but it is quite possible that Mendel realized the implications of this ratio early on, and stopped experimenting just when the numbers added up nicely.

What was happening, Mendel recognized, was that these phenotypes were being transmitted by paired "factors" — we would now call them genes — some of which are dominant and others recessive. The parent plants bred true because they carried two dominant genes for round seeds or two recessive genes for wrinkled — in the language of genetics, they are homozygous. When they were crossed, the progeny were heterozygous — they inherited one gene of each kind. The dominant gene won out, and all the seeds were round.

In the F_2 generation, there were three possibilities. A quarter, on average, would have two round-seed genes, and thus round seeds. Half would have one gene of each sort, producing round seeds because that gene was dominant. And another quarter would inherit two wrinkled-seed genes, producing wrinkled seeds: such recessive genes can generate a phenotype only when no dominant gene is present.

Mendel's laws

Mendel used these results to draw up two general laws of inheritance (to avoid confusion, I will use the language of modern genetics here and not his own). His first principle, the law of segregation, holds that genes come in alternative varieties known as "alleles", which influence phenotypes such as seed shape (or eye color in people). Two alleles governing each phenotypic trait are inherited, one from each parent. If different alleles are inherited, one is dominant and expressed, while the other is recessive and silent.

His second principle was the law of independent assortment: the inheritance pattern of one trait does not influence the inheritance pattern of another. The genes that encode seed shape, for example, are separate from the ones that encode seed color and will not affect it. Each Mendelian trait will be passed on in 3:1 ratios, according to the dominance pattern of the genes involved.

Neither of Mendel's laws is quite correct. Some phenotypes are linked and often inherited together — as are blue eyes and blond hair among Icelanders — and not all traits follow the simple patterns of dominance found in the monk's peas. But they were a good first effort. Genes found on different chromosomes are indeed inherited separately in line with the second law, and there are plenty of diseases that fit with the first. These are known as Mendelian disorders — conditions

Complex dominance

Not all traits that are governed by single genes follow quite the inheritance pattern that Mendel discovered. Some genes are incompletely dominant, meaning that when an organism is heterozygous, with one copy of each allele, the resulting phenotype is intermediate. Carnations with two alleles encoding red color are red; those with two white alleles are white; and those with one of each are pink. Genes can also be co-dominant, meaning that heterozygotes express both traits. Human blood groups work like this: while the O allele is recessive, the A and B alleles are co-dominant. So both the A and B alleles are dominant to O, but a person who inherits one A and one B will have type AB blood.

such as Huntington's disease, which always occurs in people who have one copy of a dominant mutated gene; or cystic fibrosis, caused by a recessive mutation that is dangerous only when two copies are inherited, one from each parent.

Rejection, ignorance and rediscovery

Mendel presented his paper on inheritance to the Natural History Society of Brünn in 1865, and it was published the following year. But if Darwin's opus was a sensation, Mendel's was barely read, and those who did read it missed its significance. It appeared, indeed, in a volume in which Darwin annotated both the preceding and following articles, but he left the work that was ultimately to underpin his theory unmarked. In 1868, Mendel was elected abbot and his research ceased,

> ❝Mendelism supplied the missing parts of the structure erected by Darwin.❞
> Ronald Fisher

though he remained certain of its significance. Shortly before his death in 1884 he is said to have remarked: "My scientific work has brought me a great deal of satisfaction, and I am convinced that it will be appreciated before long by the whole world."

He was right. In the 20th century, Hugo de Vries, Carl Correns and Erich von Tschermak each developed similar theories of inheritance to Mendel's, and acknowledged the monk's priority. A new science was born.

03 Genes and chromosomes

C.H. Waddington: "Morgan's theory of the chromosome represents a great leap of imagination comparable with Galileo or Newton."

When T.H. Morgan (1866–1945) began experimenting with fruit flies in 1908, he accepted neither Darwin nor Mendel. Although he was satisfied that some form of biological evolution had taken place, he doubted natural selection and Mendelian heredity as the means. His results, however, were to convince him that both theories were in fact correct, and revealed the cellular architecture that allows traits to be conveyed from one generation to the next.

Morgan proved not only that phenotypes are inherited in the fashion that Mendel proposed, but also that the units of heredity reside on chromosomes. These structures in the cell's nucleus, of which humans have 23 pairs, had first been discovered in the 1840s, but their function remained uncertain. In 1902, biologist Theodor Boveri and geneticist Walter Sutton independently proposed that chromosomes might hold the material of inheritance, to great controversy. Though Morgan was among the skeptics, his fruit flies put the argument to rest. He provided the physical evidence that cemented the Mendelian revolution.

The field of study that it threw open now had a name. Mendel had called the codes for heritable traits "factors," but in 1889, before his role in rediscovering the monk's work, Hugo de Vries had used "pangen" to describe "the smallest particle [representing] one hereditary characteristic." In 1909, Wilhelm Johannsen produced a more elegant contraction — the gene — along with the term "genotype" to signify an organism's genetic makeup, and "phenotype" to signify the features that genes produce. William Bateson, an English biologist, rolled all this together into a new science: genetics.

The threads of life

Chromosomes, we now know, are threads composed of chromatin — a combination of DNA and protein — that sit in the cell's nucleus and hold the vast majority of its genetic information (a little lies elsewhere, in mitochondria and chloroplasts). They are usually depicted as sticks that are pinched in the middle, but they actually take on this form only during cell division. For most of the time, they are long, loose strings, like fabric necklaces. Genes are like patches of color woven into the design.

The number of chromosomes differs from organism to organism, and they almost invariably come in pairs: individuals inherit one copy from their mother and one from their father. Only in reproductive cells called gametes — in animals, the eggs and sperm — is just a single set present. Ordinary paired chromosomes are known as "autosomes", of which humans have 22 pairs, and most animals also have sex chromosomes that can differ between males and females. In humans, people who inherit two X chromosomes are female, while those who have one X and one Y are male.

In the 1880s, the advent of dyes that can stain chromatin allowed the embryologist and cytologist Edouard van Beneden to observe that each cell's maternal and paternal chromosomes remain separate throughout cell division, a discovery that led Boveri and Sutton to suggest a role in Mendelian inheritance. If genes were held on discrete chromosomes that came from each parent, that could explain how recessive traits could be preserved to re-emerge in later generations.

The fly

Boveri and Sutton were proved right by one of their biggest critics — Morgan. His instrument was the humble fruit fly, *Drosophila melanogaster* — the Latin name means "black-bellied dew-lover." Females can lay 800 eggs a day, and its fast reproductive cycle, which can produce a new

Chromosomal disorders

Inherited diseases are not always caused by mutations in specific genes; they can also be caused by chromosomal abnormalities, or aneuploidies. One example is Down's syndrome, caused when people inherit three copies of chromosome 21 instead of the usual two. This extra chromosome causes learning disabilities, a characteristic physical appearance and an increased risk of heart disorders and early-onset dementia.

Aneuploidies of other chromosomes are almost invariably fatal before birth. They are often responsible for miscarriage and infertility, and it is becoming possible to screen in-vitro fertilization (IVF) embryos for these faults to improve couples' chances of a successful pregnancy.

Humans and other animals

Humans have 23 pairs of chromosomes — the 22 autosomes, plus the sex chromosomes X and Y. Until 1955, however, it was widely agreed that we had 24 pairs, in common with our closest animal relatives, the chimpanzees and other great apes. This was overturned when Albert Levan and Joe-Hin Tjio used new microscopy techniques to reveal 23 pairs. Closer examination of the human chromosome 2 shows that it was formed by the fusion of two smaller chromosomes that still exist in chimpanzees. This merger was one of the evolutionary events that made us human.

generation every two weeks, allowed Morgan's lab to cross-breed millions of the insects to examine patterns of inheritance.

Drosophila usually has red eyes, but in 1910 Morgan found a single white-eyed male. When he mated the mutant with an ordinary red-eyed female, their progeny (the F_1 generation) were all red-eyed. These flies were then mated with one another, to produce the F_2 generation in which Mendel's recessive traits reappeared. The white-eyed phenotype came back — but only in about half of the males, and in none of the females. This result seemed to be linked to sex.

> **Morgan's findings about genes and their location on chromosomes helped transform biology into an experimental science.**
> Eric Kandel

In humans, sex is determined by the X and Y chromosomes — females are XX and males XY. Eggs always contain an X, while sperm can bear an X or a Y. As the X chromosome affects the sex of fruit flies in similar fashion, Morgan realized that his results could be explained if the mutant gene that turned eyes white was recessive, and was carried on the X chromosome.

In the F_1 generation, all the flies were red-eyed as they inherited an X chromosome from a red-eyed female, and thus had a dominant red-eyed gene. The females were all carriers of the recessive gene, but it was not expressed. None of the males had it at all.

In the F_2 generation, all the females were red-eyed as they received an X chromosome with a dominant gene from a red-eyed father — even if their mothers were carriers and passed on a mutant X, they would not develop white eyes as the trait is recessive. Among the F_2 males, however, the half that received a mutant X from their mothers were white-eyed: they had no second X chromosome to cancel out the effects of the recessive gene.

Morgan had hit on a critical principle. Many human diseases, such as hemophilia and Duchenne muscular dystrophy, follow this sex-linked pattern of inheritance: the rogue genes responsible lie on the X chromosome, hence the diseases are developed almost exclusively by males.

Genetic linkage

As Morgan studied *Drosophila* still further, his team found dozens more traits that seemed to be carried on chromosomes. Sex-linked mutations were the simplest to spot, but it soon became possible to map genes to the autosomes, too. Genes that lie on the same chromosome tend to be inherited together. By studying how often certain fly traits are co-inherited, Morgan's "drosophilists" were able to show that certain genes lie on the same chromosome, and even to calculate their relative distance from one another. The closer the genes, the more likely they are to be passed on together. This concept, called genetic linkage, is still a key tool for finding genes that cause disease.

Morgan had been wrong about Mendel, wrong about Boveri and Sutton, and indeed wrong about Darwin. But he was not pig-headed in his wrongness. Instead, he used experimental data to overcome it and develop a fundamental idea. His conversion is a perfect illustration of one of science's great strengths. Unlike in politics, when the facts change in science it's OK to change your mind.

___FIRST EXPERIMENT___

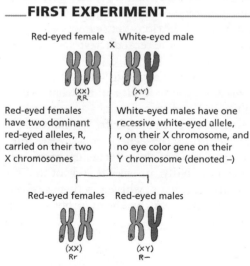

Red-eyed female   x   White-eyed male

(XX)
RR

(XY)
r−

Red-eyed females have two dominant red-eyed alleles, R, carried on their two X chromosomes

White-eyed males have one recessive white-eyed allele, r, on their X chromosome, and no eye color gene on their Y chromosome (denoted −)

Red-eyed females   Red-eyed males

(XX)
Rr

(XY)
R−

In the F₁ generation, all flies are red-eyed, as they all have one copy of the dominant red-eyed allele, R

SECOND EXPERIMENT:___

USES OFFSPRING OF FIRST EXPERIMENT

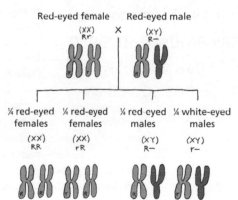

Red-eyed female   x   Red-eyed male

(XX)
Rr

(XY)
R−

¼ red-eyed females   ¼ red-eyed females   ¼ red-eyed males   ¼ white-eyed males

(XX)
RR

(XX)
rR

(XY)
R−

(XY)
r−

In the F₂ generation, all females are red-eyed as they have at least one dominant X-linked red-eyed allele, R. Half the males have the dominant allele, R, and are red-eyed, but half have the recessive allele, r, and are white-eyed

04 **Sex**

Sex is one of the great problems of life. That isn't just because of the amount of time we spend thinking about it, but also because it is an evolutionary and genetic puzzle.

Many organisms — most, indeed, given that bacteria make up a high proportion of the world's biomass — are perfectly happy procreating by themselves. Why, then, does asexual reproduction not apply across the board? It is good enough for most of the cells in the human body — the somatic cells that make up organs like the liver and kidneys divide as if they were asexual microorganisms. The only exceptions are our germ cells, which make the sperm and eggs (gametes) that ultimately make new humans.

Asexual reproduction allows any organism to duplicate its entire genome into its offspring, give or take a few random copying errors. Sex, however, means only half a population can bear young, reducing the reproduction rate, and it requires both sexes to spend time and energy finding partners. Only half a parent's genes get into its sons or daughters. All these things should be bad in terms of natural selection. Yet sex not only persists, it thrives — it is the reproductive system used by most visible life.

> " Sex is the queen of problems in evolutionary biology. Perhaps no other natural phenomenon has aroused so much interest; certainly none has sowed as much confusion. "
>
> Graham Bell

Its survival against the apparent odds is explained by what happens at a genetic level, and by what that means for evolution. Random mutations are not the only fuel for natural selection and drift. Sex causes variation too, by shuffling the genetic pack every time it happens. This process, known as "crossing-over" or "recombination", repeatedly throws the code of life together in new arrangements, which can be passed on to future generations. Any that prove particularly advantageous will be favored, just like beneficial mutations.

Meiosis and mitosis

This opportunity for variation emerges from a special method of cell division that is unique to sex. The overwhelming majority of the human body's cells are diploid, with a full count of 46 chromosomes, arranged in two sets of 23 pairs. When these somatic cells divide, as the body grows or heals, they copy their genomes completely by a process called mitosis. All the pairs of chromosomes are duplicated, and the two sets are drawn apart as the cell splits, so that one set ends up in each of its daughters. The result is two new diploid cells, each with 46 chromosomes identical to those of their parent.

Mitosis, in its essentials, is asexual reproduction. And the one place in the body where it doesn't apply is the parts that are specialized for sex. In germ cells, eggs and sperm are made by another method of cell division — meiosis. During meiosis, the diploid precursor cells of gametes duplicate their DNA, then share it out equally between four daughter cells with 23 chromosomes each. In men, these become sperm, and in women, one becomes an egg while the other three are discarded as "polar bodies."

Cells of this type are known as "haploid" — they have just one copy of each chromosome, instead of the pairs found in diploid somatic cells. When the two types of gamete fuse after sex to generate an embryo, the full complement of 46 is restored, with one copy of each chromosome furnished by each parent.

Crossing-over

This fusion of genetic material from two individuals provides variation, by creating different combinations of chromosomes. But it is not the only way in which sex adds value: the actual composition of the chromosomes that go into each sperm and egg is also unique.

When paired chromosomes line up during meiosis, they swap genetic material between themselves. The two strands of DNA — one originally inherited from an individual's mother and one from its father — intertwine, and then break at the points where they are twisted together. These segments then fuse with their neighbors, so that genes "cross over" between the chromosomes. The result is a gamete with an entirely new, jumbled chromosome that is an amalgam of paternal and maternal genes.

This crossing-over means that while every gamete gets a copy of every gene, the combination of alleles it carries will be unique. A man's sperm will not have chromosomes that come wholly from his mother or

RECOMBINATION

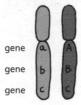

Chromosomes inherited from mother and father line up during meiosis

Chromosomes cross over

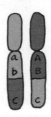

Chromosomes exchange segments of DNA to produce new configuration

Genetic relatedness

Recombination explains how much of your DNA you share with your family, and why you differ genetically from your siblings. Half of your genetic material comes equally from your mother and father, because you were conceived from gametes produced by each. But although you might think that 50 percent of your DNA is common to your brothers and sisters, too, that is true only on average. The randomness of recombination means that it is theoretically possible, though statistically extremely unlikely, that you have inherited a wholly different set of alleles to your siblings.

from his father, like the ones in his somatic cells. It will have new ones that contain chunks of genetic material from each of his parents. Genes are thus forever being assembled in slightly different arrangements, and recombination can occasionally even merge them to create new genes. Some permutations and gene mergers may be beneficial, while others are damaging. It is another source of heritable variation on which natural selection can act.

> ❝ The 'sexual' method of reading a book would be to buy two copies, rip the pages out, and make a new copy by combining half the pages from one and half from the other, tossing a coin at each page to decide which original to take the page from and which to throw away. ❞
> **Mark Ridley**

Recombination also allows scientists to map where genes are placed on chromosomes, using the concept of linkage that was introduced in Chapter 3. As T.H. Morgan understood, genes that lie close together on chromosomes tend to be inherited together too, and crossing-over is the reason. Genes are not swapped individually between chromosomes, but as parts of larger blocks. If two genes are contained in the same block or "haplotype," they will be linked — individuals who get one will tend to get the other as well.

A reason for sex

In species that reproduce sexually, meiosis and recombination give every individual a genotype of its own, and this extra variation can be adaptive. In asexual reproduction, mutations are invariably transmitted to offspring, even if they are deleterious. This leads to an effect known as "Muller's ratchet" (see box), by which genomes tend to deteriorate in quality over time. Sex, through crossing-over, allows offspring to differ from their parents. Half of them will miss out on rogue genes that would have been passed on asexually, giving the species an advantage.

The genetic variety that sex engenders, too, means that germs and parasites find it harder to spread through entire populations at once. Such diversity makes it more likely that some individuals will have a degree of genetic resistance, so that some will always survive new epidemics, and breed future generations with some immunity. Sexual variety gives the species that enjoy it a head start in life.

Muller's ratchet

When an organism reproduces asexually, its entire genome will be copied into its offspring. Hermann Muller realized that this has a big drawback: if a copying error causes a deleterious mutation, it will always be passed on to all that individual's descendants. The same thing will happen each time new mutations arise, so that the organism's overall genetic quality deteriorates over time. Muller likened the process to a ratchet, with teeth that allow movement in one direction only.

Sex and recombination circumvent Muller's ratchet, as they mean that not every mutation in a parent is passed to its children. Many asexual organisms, such as bacteria, have evolved other means of swapping genes to avoid negative effects.

05 Genes, proteins and DNA

It is rather distressing to pass urine that turns pitch black on exposure to the air, yet the condition that causes this, alkaptonuria, was little studied for centuries as it is largely harmless. In the 1890s, it caught the attention of Archibald Garrod, an English physician. When Mendel's ideas were rediscovered soon afterwards, Garrod noticed that this disorder followed a Mendelian pattern of inheritance. He identified not only one of the first diseases confirmed to have a genetic origin, but also a rule of the field as a whole: that genes work by producing proteins.

Though alkaptonuria is rare, affecting about one in 200,000 people, Garrod noted it was much more common following marriages between first cousins, and that in susceptible families the ratio of unaffected to affected children was almost exactly three to one. This, he realized, was precisely what would be expected if alkaptonuria was caused by a recessive gene, and not by infection as was commonly assumed.

> **"**Once the central and unique role of proteins is admitted, there seems little point in genes doing anything else.**"**
> Francis Crick

Garrod's knowledge of biochemistry, too, led him to propose a function for that gene. What blackens the urine in alkaptonuria is a substance called homogentisic acid, which the body usually breaks down. Garrod suspected, correctly, that patients with the condition lacked an enzyme (a protein that catalyzes chemical reactions) that was critical to its elimination. The result was that the chemical was excreted in urine, turning it black.

One gene, one protein

Garrod deduced from these observations that the function of genes was to make proteins. Many other medical problems might be caused by similar "inborn errors of metabolism," as he called them in the title of a 1909 book. It was a hugely significant insight, providing a means by which genes and genetic mutations influence biology. Perhaps because of the relative obscurity of the diseases he studied, however, Garrod's theories, like Mendel's, went unremarked for decades.

They also lacked direct evidence. That was to be supplied in the 1940s by George Beadle — another pupil of T.H. Morgan's — and American geneticist Edward Tatum. Beadle's work with fruit flies had suggested that eye color might be fixed by chemical reactions under the control of genes, but the organism was too complex to prove the theory experimentally. Beadle and Tatum turned instead to a simple bread mold called *Neurospora crassa*, which was irradiated to generate mutations.

When the mutants were crossed with normal mold, some of their offspring multiplied freely, but others would divide only when a specific amino acid, arginine, was added to the growth medium. These molds had inherited a mutation in the gene for an enzyme critical to arginine production. Unless the essential amino acid was provided from elsewhere, the yeast could not grow.

This suggested a simple rule: every gene contains the instructions for making a particular enzyme, and that enzyme then goes to work in cells. Even though the rule has since been modified — some genes are capable of making more than one enzyme, or smaller components of proteins — it is essentially correct. Genes do not guide cellular chemistry directly, but by proxy through the proteins they make, or fail to make because of mutations.

Life on Mars?

If primitive life is ever discovered on Mars — or anywhere else for that matter — the first question scientists will ask will be: "is it based on DNA?" The genetic instructions of every terrestrial organism are written in DNA (the exception is some RNA viruses, and they cannot reproduce without a DNA-based host). That offers overwhelming evidence that they are all ultimately descended from a common ancestor.

If extraterrestrial life uses DNA as well, the same implication holds true. Perhaps Mars was seeded with life by microorganisms carried from Earth on a meteorite. Or perhaps the reverse is true — and we are all really Martians.

The alphabet of DNA

Each DNA molecule is made up of phosphates and sugars,
which provide its structural architecture, and nitrogen-rich chemicals
known as "nucleotides" or bases, which encode genetic information. The
bases come in four varieties — adenine (A), cytosine (C), guanine (G) and
thymine (T) — and together, these provide the letters in which the genetic
code is written.

The bases can be further subdivided into two classes: adenine and
guanine are larger structures called purines, and cytosine and thymine are
smaller pyramidines. Each purine has a complementary pyramidine, to
which it will bind — A binds to T, and C to G. Mutations also tend to
substitute a purine for a purine, or a pyramidine for a pyramidine —
A will usually mutate to G, and C to T.

It is an insight that has had profound implications for medicine:
while altering the defective genes that cause disease is difficult, some
genetic conditions can be treated by the more straightforward means of
replacing the missing protein. Hemophiliacs, for example, can be given
the blood-clotting enzyme that their bodies are genetically incapable of
producing themselves.

Enter DNA

The discovery that genes carry the code for making proteins challenged
conventional wisdom about their construction, as it had been widely
thought that genes were proteins. If proteins were actually the products
of genes, the chemical basis of heredity had to lie elsewhere. It was to
be found in a mysterious substance that had first been purified from
pus-soaked bandages by Swiss scientist Friedrich Miescher as long ago
as 1869: deoxyribonucleic acid, or DNA.

DNA was known to exist in almost every kind of cell, yet though
Miescher had suspected it might play a part in inheritance, this
function remained purely speculative until Oswald Avery, Maclyn
McCarty and Colin MacLeod began an important series of experiments
in 1928. Avery's team was intrigued by a bacterium that causes
pneumonia, which exists in two forms that are either lethal or harmless.
When the scientists injected mice with both live harmless bacteria, and
lethal ones that had been inactivated, they were surprised to see the
rodents fall ill and die. The harmless germs had somehow acquired the
virulence of the inert ones.

To find what they called the "transforming factor," the scientists experimented with more than 20 gallons of bacteria for more than a decade. They treated these colonies with enzyme after enzyme that knock out particular chemicals, to test various candidates that might be conveying lethal instructions from germ to germ. Only when an enzyme that breaks down DNA was tried did the transformation stop: DNA was the messenger. Further evidence for its role came in 1952 from Alfred Hershey and Martha Chase, who tagged DNA with radiation to show that it is the genetic material of a phage — a kind of virus that attacks bacteria.

DNA is not just the stuff of life for bacteria and phages: it writes the genetic recipes for every living thing on Earth. The only exception is certain viruses that use its chemical cousin ribonucleic acid (RNA) instead — and as these cannot reproduce on their own, there is some debate about whether they can really be considered to be alive.

The DNA code is written in only four "letters" known as nucleotides or bases (see box). Yet this simple alphabet is sufficient to make organisms as different as humans and herring, frogs and ferns. It builds both the genes that produce proteins and the genetic switches that turn them on and off, and it is self-replicating, so the whole code can be copied every time a cell divides. It is life's software, containing the information needed to build and run a body.

06 The double helix

When Francis Crick sat down to lunch with James Watson at the Eagle pub in Cambridge, England, on February 28, 1953, and announced that the pair had "found the secret of life," other drinkers could have been forgiven a little skepticism. Crick was a 36-year-old physicist who had yet to finish his PhD. His American collaborator was just 24. They had also been expressly forbidden from studying the problem they now claimed to have solved: the structure of the DNA molecule that had been recognized for almost a decade as the conveyor of heredity. Even Watson, not a man known for circumspection, was somewhat disconcerted by his friend's boldness, as he was still concerned that their answer might be wrong.

He need not have worried. Their discovery that DNA is wound into a double helix ranks as one of the most important scientific achievements of the 20th century, on a par with Einstein's relativity and the splitting of the atom. While early genetics had shown clearly that genes drive inheritance, it had had little to say about the chemical processes involved. Crick and Watson changed that, demonstrating how genes actually work. They opened a new era of molecular biology, in which genetic activity could be tracked, charted and ultimately even changed.

> "It seems likely that most if not all the genetic information in any organism is carried by nucleic acid — usually by DNA, although certain small viruses use RNA as their genetic material."
> Francis Crick

The idea of the double helix also pointed firmly to the route by which the code of life is copied as cells divide, with each strand providing a template from which genetic instructions can be duplicated. As Crick and Watson put it in the short paper they published in *Nature* that April: "It has not escaped our notice that the specific pairing we have postulated immediately suggests a possible copying mechanism for the genetic material."

The search for the structure

The significance of DNA to heredity was widely suspected by the early 1950s, and several teams were seeking to resolve the molecule's structure. In the U.S., Linus Pauling had already shown that many proteins were coiled into a spring-like helix, and proposed, incorrectly, a triple helix for DNA. At King's College London, meanwhile, Rosalind Franklin and Maurice Wilkins were studying DNA using X-ray diffraction, which analyzes how molecules scatter radiation for clues to their form.

In Cambridge, Crick and Watson were meant to be using the same tool for different purposes — Crick's target was protein structure, and Watson's a tobacco virus — but they found DNA more interesting. For a period, however, they were told not to study it by Laurence Bragg, their laboratory head, who felt it would be distracting and ill-mannered for them to stray into the same territory as King's.

They continued to work on the problem, at first surreptitiously and eventually with Bragg's approval, and solved it by combining others'

The dark lady of DNA

Rosalind Franklin's role in the discovery of the double helix remains a source of great controversy. The importance of her x-ray images is beyond dispute, and observers like Brenda Maddox, her biographer, have argued that she was a victim of sexism who has never been given the credit she deserves.

Crick, Wilkins and, particularly, Watson, certainly failed to acknowledge her contribution properly at the time, but there is some merit in their counterargument that while Franklin's work was pivotal, she never properly grasped its significance. She was also excluded from the Nobel Prize for Medicine that the trio shared in 1962 for a perfectly innocent reason: she died of ovarian cancer in 1958, and Nobel Prizes are never awarded posthumously.

Linus Pauling

In the race to identify the structure of DNA, the smart money was not on Watson and Crick but on Linus Pauling — the brilliant American chemist who had already made key discoveries about protein structure and chemical bonding. Pauling was the first to suggest a helical structure for the DNA molecule, even if it was wrong in several details, and he might well have beaten the Cambridge team to the punch had it not been for his political activism.

In 1952, he was accused of having communist sympathies, and his passport was suspended. He was thus forced to abandon a trip to the UK, and never got to see the images taken by Franklin that helped Watson and Crick to solve the problem.

work with their own through luck, brilliance and deviousness. Their first stroke of fortune came from a 1952 visit to the UK by Erwin Chargaff, whose experiments in the U.S. had shown that the four bases of DNA always occur in the same ratios — cells have equal amounts of the base pairs adenine (A) and thymine (T), and cytosine (C) and guanine (G). His lectures led Crick and Watson to understand that DNA bases come in pairs, with the letter A always linked to T and C linked to G. A critical piece of the double helix was in place.

A second vital clue came from Franklin's research. In 1952, she had taken an x-ray image of the DNA molecule, known as "Photo 51," which Wilkins had shown to Watson without her knowledge. Crick, too, had learned of her results from Max Perutz, his thesis supervisor, who had reviewed the King's work for the Medical Research Council. The pair realized that the picture's significance had passed their rivals by, and that in combination with Chargaff's ratios it suggested a potential structure for DNA.

They were then able to turn this insight into results because, unlike Franklin, they did not confine their investigations to the laboratory. While the x-ray image was crucial, Crick and Watson understood its meaning by lower-tech means, playing with tin and cardboard models of DNA's components to test possible structures by trial and error. Photo 51 worked like the key to a jigsaw puzzle, indicating a framework into which all the pieces might fit. And the framework — the double helix — worked perfectly.

How the helix works

The DNA molecule is composed of two linked chains of bases. Each base is joined to its natural partner — A to T and C to G — by a hydrogen bond, and held at the other end by a sugar and phosphate backbone. This pairing system means that two DNA strands coil around each other in a double helix, like a twisted rope ladder. Each strand is the mirror image of the other — where one has an A, its partner will always have a T, and vice versa. If the first strand reads ACGTTACCGTC, the other will read TGCAATGGCAG.

This structure betrays its function. The sequence of the DNA bases encodes genetic information twice over, making it wonderfully easy to copy. When a cell divides, an enzyme breaks the hydrogen bonds that connect the base pairs, unzipping the double helix down the middle into its two constituent strands. These can then serve as templates for replication. A second enzyme called DNA polymerase tacks new bases onto the letters of each strand, matching As to Ts and Cs to Gs. The result is two new double strands of DNA, to provide the genetic software for two daughter cells.

Like so many great ideas in genetics, the double helix is elegantly simple. Yet it immediately explained how the code of life is copied, and it cleared the way for further discoveries about how that code influences biology. It was the harbinger of a new genetic age in which it was to become possible to use DNA to diagnose disease, to develop drugs, to catch criminals and even to modify life. If the structure proved simple, the same cannot be said for its consequences.

DOUBLE HELIX

REPLICATION

1 Double helix unzips during cell division
2 Each independent strand of DNA acts as a template for a complementary strand to be created, adding As to Ts, Cs to Gs, etc.
3 Two new double-stranded DNA molecules are created, one of which migrates into each new cell

07 **Cracking the code of life**

The double helix explained how genes are copied, and thus how genetic information is reliably passed on from cell to cell and from generation to generation. It also suggested that mutations in DNA's letters would be inherited, allowing for Darwin's descent with modification. What the structure did not elucidate, however, was how genes perform their other vital task besides copying themselves: the synthesis of the proteins that drive biology.

The code of life was clearly written in an alphabet of four letters — the A, C, G and T of DNA — which composed instructions for producing the 20 amino acids that make up proteins. But until it could be cracked, the code was meaningless. Biology had no Rosetta Stone, no key for deciphering the messages encrypted in DNA.

66 It now seems very likely that many of the 64 triplets, possibly most of them, may code one amino acid or another, and that in general several distinct triplets may code one amino acid. 99
Francis Crick

That key was to be provided by some perceptive theorizing by Francis Crick, followed by experiments led by American biochemist Marshall Nirenberg and French biologist Jacques Monod that found evidence to match. Little more than a decade after its discovery, written into the double helix, the genetic code had been broken and molecular biology had an organizing principle.

The adaptor molecule: messenger RNA
Crick and Watson had identified the double helix by assembling existing evidence and interpreting it correctly. Crick's next stroke

The central dogma

Another of Francis Crick's important contributions was what he called the "central dogma" of biology — that genetic information generally travels through a one-way system. DNA can copy itself into DNA or transcribe itself into mRNA, and mRNA can make protein, but it isn't possible to reverse the process.

There are three exceptions to this rule. Some viruses can replicate themselves by copying RNA directly into RNA, or perform "reverse transcription" from RNA into DNA. It is also possible to translate DNA directly into protein, but only in the laboratory. The information contained in proteins, however, can never be converted into RNA, DNA or even other proteins. The redundancy of the genetic code makes this impossible.

of genius, however, was much more speculative, emerging well in advance of any experimental data. It was the notion that DNA might be translated into amino acids by means of an "adaptor molecule" — an interpreter that carries orders from genes to cellular protein factories.

By 1960, Crick's intuition had been proved right. At the Pasteur Institute in Paris, Monod's team used bacteria and bacteriophage viruses that prey on them to show that DNA indeed produces an adaptor molecule, which is made out of a close chemical relative called ribonucleic acid (RNA).

RNA is similar to DNA, but has a few structural differences. The most important is that instead of the base thymine, it uses a similar nucleotide called uracil (U). It is also more unstable, and thus

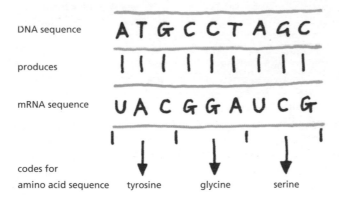

DNA sequence	A T G C C T A G C
produces										
mRNA sequence	U A C G G A U C G
codes for amino acid sequence	tyrosine glycine serine

shorter lived in the cell, and it forms many different types of molecules with specialized functions. Crick's adaptor molecule is a form known as messenger RNA (mRNA), a single-stranded molecule into which genes are "transcribed." This mRNA is used to make protein, in a process known as translation.

As in replication, the double helix is unzipped, and then one strand is read to produce a mirror-image strand of mRNA. In this transcription, Cs in the genes will become Gs in the mRNA, Ts will become As, Gs will become Cs, and As will become Us — uracil replacing thymine in the RNA molecule. These genetic signals then migrate from the cell nucleus to protein-making structures called ribosomes, which add amino acids into chains one at a time in the order specified by the code.

Another kind of RNA, transfer RNA, collects amino acids and threads them onto the growing protein string. The instability of mRNA means that, as in *Mission: Impossible*, the messages self-destruct when they have done their job. There is no danger that they will stick around to create rogue proteins when they are no longer needed.

Triplets

How, though, do the ribosomes know which amino acids have been ordered? And how do they know where protein chains should start and finish? The answer lies in the sequence of the bases in genes, by which passages of DNA and mRNA specify particular amino acids. The code — first proposed by Crick — is extremely simple, based on combinations of just three DNA letters, or "triplets."

Exons and introns

Not all the DNA in a gene is actually used to make protein. The parts that matter are known as exons. These are interspersed with stretches of non-coding DNA called introns, which have no bearing on the protein recipe that the gene carries. While all the DNA is copied into mRNA, the introns are then edited out by special enzymes, and the exons are spliced together to place an order for a protein.

A good analogy is a film screened on television. The scenes you want to watch are the exons, but these are broken up by commercial breaks — the introns — which aren't part of the story. And if you have recorded the whole thing, you can fast-forward through the commercials to watch it all at once — much as a ribosome reads a spliced string of exons.

The meaning of these triplets began to emerge with the work of Nirenberg, who in 1961 mixed ribosomes from *E. coli* bacteria with amino acids and single RNA bases. When he added pure uracil, the result was long protein-like chains of the amino acid phenylalanine. The first triplet had been deciphered — mRNA bearing the message "UUU" means "put a phenylalanine molecule on the protein string." Within five years, the meaning of all 64 combinations of the four bases had been established. The code had been cracked.

As there are 64 possible triplets or "codons," yet only 20 amino acids, some amino acids are specified by more than one triplet. Phenylalanine, for example, is made not only by the codon UUU, but also by the codon UUC. There are six ways of making each of the amino acids leucine, serine and arginine. Only two of the 20 are specified by unique codons: tryptophan (UGG) and methionine (AUG). AUG is also the "start codon," telling ribosomes to begin adding amino acids, meaning that most proteins start with methionine. And there are three "stop codons" — UAA, UAG or UGA tell the ribosome: "this protein is now complete."

" It seems virtually certain that a single chain of RNA can act as messenger RNA. "
Francis Crick

This system is not as simple as it might be. Crick, indeed, first proposed a more elegant code of 20 possible triplets — one for each amino acid. What nature's version lacks in style, though, it makes up for in substance, for its redundancy has considerable advantages. The fact that the most important amino acids can be produced by multiple codons creates resistance to mutation. Glycine, for example, can be written GGA, GGC, GGG or GGU. If the final base mutates, the product remains the same.

This leaves less scope for catastrophic copying errors that could compromise a whole organism. About a quarter of all possible mutations, indeed, are "synonymous" in this way, and the workings of natural selection mean that a still larger proportion of those that survive — about 75 percent — have no effect on a protein's function. The genetic code is a "Goldilocks" language — the amount of variation it allows is neither too much or too little, but just right for evolution.

08 Genetic engineering

A good code isn't just there to be deciphered and read. It can be used for creative purposes, too. If the code of life could be understood, it could potentially be altered and manipulated.

Hermann Muller had realized, when exposing his fruit flies to radiation in the 1920s, that deliberately inducing mutations might allow humanity to direct evolution in desirable ways. The double helix and the cracking of the genetic code meant this might be achieved with precision: instead of waiting for a random x-ray mutation with a useful function, perhaps chromosomes and genes could be edited with specific functions in mind. Genetic engineering was now on the agenda.

It is one thing to imagine genetic engineering, however, and quite another to perform it. An engineer of any kind is nothing without tools, and DNA triplets can hardly be cut and pasted with scissors and glue. What turned genetic engineering from science fiction into reality was the discovery in the 1970s of molecular "scissors and glue," a series of enzymes that could be used for writing and copying genes, cutting them out, and splicing them into a genome. Scientists would now be able to play God, creating new combinations of DNA that had never before existed in nature.

Molecular scissors

The first tools to be discovered were a class of proteins called restriction enzymes, which are sometimes nicknamed molecular scissors. Bacteria use these chemicals to "restrict" bacteriophage viruses that infect them, by recognizing particular sequences of their invaders' DNA and chopping them up at those points.

The process, first described by Swiss microbiologist Werner Arber in the 1950s, had obvious potential in genetics. If such enzymes target specific stretches of DNA, they could be used to cleave it into fragments at specific spots. In 1972, American microbiologist Hamilton Smith identified a restriction enzyme produced by the bacterium *Hemophilus influenzae* that did just that, attacking a phage at the same stretch of six base pairs every time.

More than 3,000 restriction enzymes are now known, each specialized to a particular DNA sequence. They are fundamental to genetic engineering, allowing scientists to splice genes and parts of genes. If a gene is known to start with one DNA sequence and to finish with another, the two restriction enzymes specific to those sequences can be used to cut it out.

Recombinant DNA

Once restriction enzymes have been used to cleave DNA, enzymes called ligases can be used to stick it back together. If restriction enzymes are molecular scissors, ligases are molecular solder or glue. Various cut fragments can thus be joined to one another, or stitched into the genome of another organism. The result is known as "recombinant DNA" — a sequence that has been pieced together by recombining segments in the laboratory.

Recombinant DNA was first created in the 1970s by American biochemist Paul Berg, who stuck together parts of a monkey virus called SV40 and a bacteriophage. His original plan was to insert this genetically modified virus into *E. coli* bacteria to allow it to replicate, but he paused. SV40 is harmless to humans, but what if genetic engineering were to change that? SV40 was known to promote tumor growth in mice, and *E. coli* inhabit the human gut. If bacteria carrying the recombinant virus were to escape, they might infect people and churn out carcinogenic SV40 proteins.

Reverse transcriptase

Another enzyme that has proved important to genetic research is reverse transcriptase, which was discovered by David Baltimore and Howard Temin in 1970. It is used by retroviruses such as HIV to transcribe their RNA code into DNA, and then insert it into cells so they can replicate. Many drugs for treating HIV and other viruses work by inhibiting reverse transcriptase.

The enzyme also makes it possible to turn messenger RNA into DNA in the laboratory. This can be a valuable tool for gene-hunting, allowing scientists to find transcribed mRNA messages, and use these to infer the DNA sequences from which they are derived.

The potential biohazard led Berg to suspend this stage of his experiments, and to call for a moratorium on allowing recombinant DNA to replicate until the risks could be properly assessed. He resumed only in 1976, after the Asilomar conference had drawn up strict safety protocols for future research (see box). Similar issues have dogged genetic engineering ever since: though thousands of recombinant products have been used safely in the past three decades, many critics still argue for a precautionary approach.

The first genetically modified organisms

Less squeamish in outlook — or more reckless, depending on your point of view — were Herbert Boyer, of the University of California at San Francisco, and Stanley Cohen, of Stanford University. When they teamed up, Boyer was studying restriction enzymes, while Cohen was investigating plasmids — circular packages of DNA found in bacteria, which they sometimes swap between themselves as a defense mechanism against antibiotics or phages. Boyer and Cohen used the new tools of genetic engineering to add a gene that confers antibiotic resistance to a plasmid, and then inserted it into *E. coli*. The bacteria became antibiotic-resistant. They were the first true genetically modified organisms (GMOs).

The first application of recombinant DNA was in the laboratory, to "clone" interesting genes by cutting them out and splicing them into plasmids. When placed into bacteria, these plasmids would then replicate, making multiple copies of the genes for scientists to study. A variation on this procedure was used to clone the segments of the human genetic code that were mapped by the Human Genome Project (see Chapter 9).

The Asilomar conference

In February 1975, Paul Berg assembled 140 scientists, doctors and lawyers at the Asilomar State Beach conference center in California, to discuss the ethical questions raised by genetic engineering. It established a number of biosafety principles, designed to prevent the accidental release of a recombinant organism that could infect humans or animals. The key recommendation was that when human or animal viruses were being studied, bacterial hosts that cannot survive outside the laboratory should be used. That way there was little chance of unwittingly unleashing a "superbug" on the world.

Still more exciting — and more lucrative — was the medical potential. Boyer saw that if human genes were engineered into plasmids, it would be possible to induce bacteria to make human proteins that could be used in therapy. In 1976, he set up a company called Genentech to commercialize the technology, with backing from Robert Swanson, a venture capitalist.

The company's first success was a recombinant version of insulin (the hormone essential to sugar metabolism that is lacking in patients with type 1 diabetes) which had previously been obtained from pigs. Boyer created this by placing the human insulin gene into *E. coli* via a plasmid. The bacteria that took up the plasmid became insulin factories, producing vast quantities of the hormone suitable for medical use.

> " The concern of some that moving DNA between species would breach customary breeding barriers and have profound effects on natural evolutionary processes has substantially disappeared as the science revealed that such exchanges occur in nature. "
> Paul Berg

A similar approach is now used to create scores of drugs and other commercial products, many of which have significant advantages over the alternatives. Human growth hormone for treating dwarfism, for example, was once extracted from the pituitary glands of cadavers, and contamination infected many recipients with Creutzfeldt–Jakob disease — the human equivalent of mad cow disease. The recombinant version has no such risks. There is even a recombinant form of rennet for making vegetarian cheese — the real thing comes from cows' stomachs. Muller had been spot on: genes could be molded to our purposes.

09 The human genome

As DNA sequencing began to reveal human genes in the 1980s, an even bigger prize started to edge into view. If science could learn so much about biology and disease by mapping a few short sections of DNA, how much more might be revealed by reading the entire genetic code of our species?

While sequencing had to be conducted by hand, a project to decipher the whole human genome had looked like fantasy. But with the advent of automated techniques, influential figures started to argue that it might be both possible and worthwhile. In 1986, Renato Dulbecco, a Nobel laureate, called on the U.S. government to support such an effort to underpin cancer research. In Britain, Sydney Brenner — a future Nobel prizewinner — was urging the European Union to do the same thing.

> **The only reasonable way of dealing with the human genome sequence is to say that it belongs to us all — it is the common heritage of humankind.**
> John Sulston

The U.S. Department of Energy, which had been tasked with investigating the effects of radiation on DNA, soon took up the baton. "Knowledge of the human genome is as necessary to the continuing progress of medicine and other health sciences as knowledge of human anatomy has been for the present state of medicine," it declared in a 1986 report. But other scientists and institutions, including the U.S. National Institutes of Health, were more skeptical. Some considered the task too ambitious and expensive. Others thought it would divert both intellectual and financial capital away from more achievable genetic research.

The Human Genome Project

By the end of the decade, the case had been made. The international Human Genome Project, funded by governments and charities, was launched in 1990, under the leadership of James Watson. Its goal was to read every one of the 3 billion base pairs in which humanity's genetic instructions are written, which its architects envisaged would take 15 years and $3 billion — a dollar per DNA letter.

The project was so grand in scope that the last thing that it expected was competition. Yet in 1998, when the public consortium had finished just 3 percent of the code, a private-sector challenge emerged. Craig Venter, the geneticist who had identified more genes than any other, struck a $300 million deal with the main manufacturer of DNA sequencing machines to produce his own version of the genome.

Armed with a new technique he had developed, called whole-genome shotgun sequencing, Venter's company, Celera, promised to finish in just two to three years, well before the public project's planned completion date. Watson had won the first great contest of the genetic era, to find the structure of DNA. Now another race was on, which was to become one of the most bitter rivalries in modern science.

Different techniques, different philosophies

The human genome is much too big to be read in one go. It would have to be split into sections that sequencing machines could manage, and the rival teams took different approaches to this problem. The public project first divided it into large chunks of 150,000 base pairs, cloned thousands of copies in bacteria, and then mapped these clones' positions on their chromosomes. Each clone was then broken down still further into random

Whose genome?

Both the Human Genome Project and Celera used genetic material from several donors: DNA was extracted from blood taken from women, and from sperm provided by men. Celera's genome used five individuals — two white men and three women of African-American, Chinese and Hispanic backgrounds. The male donors, it later emerged, were Craig Venter and Hamilton Smith. The public project used DNA from two males and two females. Though they remain anonymous, one is known to have been a man from Buffalo, New York, code-named RP11, whose sample was used most often because it was of the best quality.

fragments, sequenced and reassembled by matching the fragments' overlapping ends. The sequenced clones would then be mapped back to their chromosomal positions to provide the complete code.

The technique was thorough, but extremely slow. Celera, which took its name from the Latin for "speed," thought it could do better by skipping the mapping stage, and assembling the whole genome at once from small fragments. This "shotgun" method had already been used to sequence bacteria and viruses, but many experts doubted it would work on the human genome, which is larger by a factor of 500 or more. Venter, however, proved its utility by sequencing the genome of an old friend to geneticists — the fruit fly — and moved on to the human code.

Had the two groups been divided by nothing more than professional approach, relations might have remained cordial. But they differed in world view, too. The Human Genome Project saw the genetic code as the universal property of humankind, and placed all results into a public database, GenBank, as soon as they were ready. Celera, however, was there to make money.

While Venter had forced his financial backers to accept that he would publish his data openly, he was also running a business. Celera hoped to sell access to a powerful genetic database, complete with software that companies could use to find new genes and develop new drugs. University researchers could see the data for free, but would have to pay royalties on any commercial products they might develop as a result.

This was anathema to scientists like John Sulston, who led the British contribution to the public project. They saw Venter as a kind of

Data release: a double-edged sword?

By publishing its new data on a daily basis, the Human Genome Project hoped to make it impossible for Celera to patent the entire genetic code. The strategy worked, but at a price. Celera was free to download the fruits of its rival's labors to refine its own sequence, and other biotechnology companies could and did do the same thing. As Craig Venter has pointed out, this probably led to more genetic patenting, not less, as businesses pored over the public results and filed claims on the most interesting-looking genes.

genetic pirate, who was trying to appropriate something that belonged to everybody for his investors and himself. Though Venter always insisted the genome itself was unpatentable, the fear was that Celera was seeking to privatize it. The Human Genome Project stepped up the pace of its efforts, hoping to block any bid for ownership by putting data into the public domain before its rival could lay a claim.

An agreed draw

Venter finished first, but the public project was so close behind that he agreed to a draw. Critical to the uneasy truce were two interventions by Bill Clinton, the U.S. president. In April 2000, Clinton announced that he thought the genome should be public property, and his statement sent biotech stock prices — including Celera's — into freefall. Mortified by this unintended consequence, he then determined to make amends by bringing the two parties together. He negotiated a joint announcement at the White House between the two camps, at which he formally acknowledged Venter's contribution.

> **We are learning the language in which God created life.**
> Bill Clinton

Celera was true to its word and published, and its value-added database proved so useful that most public science institutions and pharmaceutical companies subscribed. The public project's gear-change, however, precluded any possibility of the genome being patented. In 2004, after Venter had fallen out with his backers and resigned from Celera, his reference genome was even added to GenBank, free from any access restrictions. The genome war was over, and the fierce rivalry had served humanity well. The stimulus of competition meant the genome had been sequenced far more quickly than anyone had thought possible a decade before.

10 Lessons of the genome

When the acrimonious race to read the human genome reached its final furlong, the competing parties could at least agree on one thing. The "book of man" was going to contain an awful lot of genes.

The fruit fly, Craig Venter had shown, had around 13,500 genes. John Sulston's project to sequence *Caenorhabditis elegans*, a microscopic nematode worm, had revealed about 19,000. Human life was so complex, it was reasoned, that many more genes than that would be needed to write the instructions. The consensus figure was around 100,000, and one biotech company even claimed to have characterized 300,000 human genes.

The publication of the two draft genome sequences in 2001 was to deliver quite a surprise. Analysis suggested that it only contained between 30,000 and 40,000 genes, and the tally has fallen steadily ever since. At the time of writing, the latest count is about 21,500 — slightly more than the zebrafish, and slightly fewer than the mouse. There is little correlation between the biological complexity of an organism and its number of protein-coding genes.

GENE SPLICING

1 Gene's sequence

TACGCA   AACCT   GATCGA   TCCG   TACCAG
[EXON 1] [INTRON 1] [EXON 2] [INTRON2] [EXON3]

2 Whole gene transcribed into mRNA

[EXON 1] [INTRON 1] [EXON 2] [INTRON2] [EXON 3]
AUGCGU   UUGGA   CUAGCU   AGGC   AUGGUC

One gene, many proteins

Since the experiments of George Beadle and Edward Tatum in the 1940s established that genes make proteins, the notion that one gene codes for one protein had become a mantra of molecular biology. Yet there are hundreds of thousands of human proteins, but only tens of thousands of human genes. The mantra was wrong. Both genes and proteins, it turns out, are more versatile than had been assumed.

Single genes can in fact contain the recipes for many different proteins — in part because of their structure. As explained in Chapter 7, only the sections of genes known as "exons" actually carry instructions for protein synthesis. Information from non-coding introns is removed from messenger RNA, and the exons are stitched together, before proteins are made.

These exons can be spliced in many different ways, and this "alternative splicing" means a single gene can specify multiple proteins. Some genes, too, make only chunks of proteins, which can then be joined together in different orders to produce a wider variety of enzymes. Proteins can also be modified by cells after they have been produced. The result of all these processes is a protein population or "proteome" that is much more diverse than the human gene count would suggest.

The surprisingly low number of human genes also indicates that "junk DNA" — the 97 to 98 percent of the genome that does not code for protein — might be more important than had been imagined. Some non-coding regions produce different cellular messengers made out of specialized forms of RNA (see Chapter 32). These work as switches that turn gene activity on or off and up or down, or that direct splicing to change which protein a gene makes. Much junk DNA, indeed, is now thought to be anything but junk (see Chapter 29). Some of it is critical to regulating how genes are expressed, and is as physiologically significant as genes themselves.

3 Gene splicing removes introns, which do not carry protein-coding information

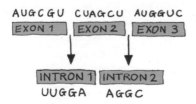

4 Exons, which carry protein-coding information, are translated into amino acids, which are then strung together into proteins

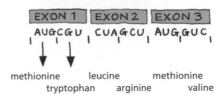

Variation between species

When the human genome was compared with those of other species, it became clear that very few human genes are truly unique: most have a counterpart in other organisms. About 99 percent are shared with chimpanzees, and about 97.5 percent with mice. Natural selection does not reward change for change's sake, and thus genes that work well tend to be "conserved" by evolution. A very similar code, making a very similar protein, will do the same job in related species. Both humans and pigs, for example, share a similar insulin gene: that is why pig insulin could be used to treat diabetics. Evolution does not often drop genes or create entirely new ones so, in retrospect, perhaps it is not so odd that most mammals have been found to have comparable gene counts.

What often happens instead is that a few genes are co-opted to perform new functions as evolution progresses. Many acquire slight mutations that are peculiar to a particular species, which allow them to do new things. A human gene called FOXP2, for example, has a counterpart in both mice and chimpanzees, but the human version differs from the chimp version in two places, and from the mouse version in three. These tiny changes may have played a role in the evolution of speech: people with defective FOXP2 genes suffer from language impairment.

Many of the differences between humans and other animals have emerged not because we have new genes unknown in our relatives, but because some of our shared genes have been altered, so they work

How finished is the genome?

Most people tend to think the sequencing of the human genome was completed either in 2000, when the feat was announced at a White House press conference, or in 2001, when the rival groups first published their data. All that had then been produced, however, were working drafts that were riddled with holes: almost 20 percent of the code had yet to be sequenced. Even in the supposedly "finished" version released in 2003, about 1 percent of the protein-coding regions was missing, along with higher proportions of non-coding junk DNA. Efforts to fill in the gaps are continuing, and the sequences of certain sections — the centromeres at the middle of chromosomes, and the telomeres at their ends — remain unmapped. They contain so much repetitive DNA that standard technology has struggled to read them.

There is no human genome

When we talk about the human genome, we are in a sense discussing a fictional entity. The only people who share every letter of the genetic code are identical twins, and every one of us is otherwise unique. What the human genome sequence provides is an average, a reference point against which all of our individual genetic variations can be compared. It tells us where the important genes that we do share lie, making it simpler to investigate what they do. This means that when scientists find SNPs that seem to be linked to a disease, it is possible to trace them back to the genes in which they occur, revealing clues to their effects.

in subtly different ways. Others are thought to reflect changes in the regulatory regions of junk DNA, and in the RNA messages that they send.

Variation between people

People, of course, are genetically even more similar to one another than are humans and chimpanzees. By standard measures, 99.9 percent of the genome sequence is universal, shared by every person on the planet. We also have the same genes as one another, except in rare cases in which one or more have been deleted entirely. The 0.1 percent of DNA that is not shared, however, provides plenty of scope for variation: with 3 billion base pairs in the genome, that still leaves 3 million places in which individuals' DNA can differ.

This kind of variation involves random substitution of one DNA letter for another. The places where this occurs are known as "single nucleotide polymorphisms" or SNPs — pronounced "snips." Many SNPs have no effect at all: as discussed in Chapter 7, the genetic code has redundancy, so some mutations do not change the amino acid sequence of proteins. Others, however, make a material difference to the protein that a gene will produce, or alter the way junk DNA controls gene expression.

These SNPs are one of the primary ways in which genetics make individuals different. Some have trivial effects, altering characteristics such as hair or eye color. Others are more insidious, either directly causing disease or altering metabolism in ways that make people more vulnerable to particular conditions. They are responsible for much of the variety of human life.

11 Genetic determinism

When his version of the human genome sequence was published in February 2001, Craig Venter attended a biotechnology conference in Lyon, France. In his keynote speech, he extolled it as a landmark in human understanding, not only because of what it explains about the significance of genetics, but for what it says about the field's limits as well.

As it contained so few genes, Venter said, the genome put paid to any notion that the behavior, character and physiology of individual human beings are wholly determined by their genetic constitutions. "We simply do not have enough genes for this idea of biological determinism to be right," he said. "The wonderful diversity of the human species is not hard-wired in our genetic code. Our environments are critical."

> **It would be quite practicable to produce a highly-gifted race of men by judicious marriages during several consecutive generations.**
>
> **Francis Galton**

Venter's logic was somewhat flawed — John Sulston, indeed, has accused him of making a "bogus philosophical point." It is quite true that the 30,000 to 40,000 genes that the genome was then thought to contain are utterly insufficient to hard-wire every human trait. But his implication — that three times as many genes would have been able to do this — is wrong. Both genetic *and* environmental factors are important in explaining the human condition, and the genome did not initially shed much light on the relative importance of each.

Venter's intention to discredit genetic determinism, however, is in itself worth noting. For since the inception of the science genetics has commonly been misinterpreted to imply predestination, and the idea that people are prisoners of their genes has had dreadful social and scientific consequences.

Social Darwinism

When Charles Darwin published *On The Origin of Species* in 1859, he avoided discussion of what evolution had to say about human behavior, but it was not long before contemporaries tried to apply his theories to society. Figures like Herbert Spencer, a philosopher who coined the phrase "survival of the fittest," reasoned that human societies could learn from nature, and improve themselves by marginalizing and discarding their weakest members. These "social Darwinists" argued that interventions to help the poor and the sickly might be noble in intention, but that they would ultimately weaken the human race by subverting natural selection.

Other thinkers appropriated Darwin to support their own notions of biological determinism. Cesare Lombroso (1836–1909) and Paul Broca (1824–80) contended that criminals, the mentally ill and the intellectually challenged were physiologically different from ordinary law-abiding citizens, and that their poor character was inherited and immutable. The now-discredited pseudosciences of phrenology and craniology, which hold that certain physical features and skull shapes reflect moral or mental degeneracy, were widely used to support such views.

Evolutionary theory was also employed to champion racism, with the argument that certain ethnic groups, particularly those with dark skin, represented more primitive forms of humanity that owed savagery to their less evolved status. Robert Knox, a Scottish anatomist, developed a particularly repellent anthropological theory, which contended that mankind was a genus, and that the different human

Galton the polymath

Francis Galton is usually remembered today for eugenics, but many of his other achievements were based on much sounder science, and proved more lasting. His experiments on rabbits showed that traits are not passed on by mixing parental characteristics, as Darwin had thought likely, and foreshadowed Mendelian genetics. He effectively founded modern statistics, introducing the principle of regression to the mean, by which abnormal results tend to be followed by a return to the average. He also helped to develop the forensic science of fingerprinting, and to advance meteorology: Galton produced the first weather map. His one bad idea has largely obscured his many good ones.

races were species of greater or lesser sophistication, which could be scientifically classified in order of superiority. White Anglo-Saxons, of course, stood at the apex of his ethnic hierarchy.

Eugenics

Darwin rejected social theories that drew on his biology, influenced in part by his own family's history of medical infirmities: two of his ten children died in infancy, and he was particularly devastated by the death of his daughter Annie at the age of ten. His cousin Francis Galton, however, was to take up such thinking with fervor. A polymath of formidable intellect, Galton drew the conclusion from his research into human inheritance that the species could be improved by selective breeding, like any other species. He was the founder of eugenics.

This philosophy, which takes its name from the Greek words for "good breeding," at first had the goal of producing a gifted elite caste by encouraging "eugenic marriages" between people of good health and high intelligence. It soon took on a more sinister form, with advocates seeking to discourage or even to prevent reproduction among those deemed to come from lesser genetic stock. At worst, they promoted the forcible sterilization of "imbeciles," the disabled, the insane and others considered to be genetically unfit.

> **And for the rest — those swarms of black and brown and yellow people who do not come into the new needs of efficiency? Well, the world is a world, not a charitable institution, and I take it they will have to go.**
> H.G. Wells

In the late 19th and early 20th centuries both positive and negative eugenics were widely seen as progressive and scientific. Some of the movement's most enthusiastic supporters were socialists such as H.G. Wells and Beatrice and Sidney Webb, who saw it as a means of improving the genetic quality — and hence the social prospects — of the working classes.

Despite its British origins, eugenic measures were never incorporated into UK law (see box), but many other countries adopted them enthusiastically. In the U.S., many states passed eugenic marriage

Britain's eugenics bill

In 1912, the UK's Liberal government introduced the Mental Deficiencies Bill, backed by no less a champion than Winston Churchill. It would have imposed penalties on people who married partners thought to be intellectually subnormal, and it was drafted so it could later be amended to approve compulsory sterilization. The campaign against it was led by Josiah Wedgwood, a Liberal MP who, like Galton, was related to Darwin. Wedgwood attacked both the shaky scientific principles on which the bill was based, and its assault on individual liberty, and won enough support to secure its withdrawal. It was the closest Britain came to eugenic legislation.

laws that banned the "feeble-minded" or even epileptics from marrying, and 64,000 people had been forcibly sterilized by the time the practice was finally outlawed in the 1970s. Nazi Germany went still further, progressing from 400,000 forced sterilizations in the name of "racial hygiene" to euthanasia of the disabled, and ultimately to the Holocaust.

Multiple misunderstandings

Leaving aside these appalling infringements of human liberty, the sort of biological determinism that drove the eugenics movement rested on a huge scientific misunderstanding. While genes have an important influence on many aspects of human health and behavior, and many diseases and mental disorders are inherited, most of the traits and conditions that the eugenicists sought to affect are not governed by genetics alone. Venter was right in the broadest sense: genes by and large do not programme human behavior and health, but exert an altogether subtler influence.

The crimes inspired by a warped misinterpretation of inheritance, however, have had a lasting impact on a whole field of study. Past abuses of genetics have left many people so suspicious of any suggestion that genes play a part in forming human character or behavior, that even investigating such effects is often seen as politically incorrect. That is no more scientific than the mistaken theories of Galton or Knox.

12 Selfish genes

To many people, the "bible" of genetic determinism was published in 1976 by Richard Dawkins, then a little-known Oxford University zoologist. Though *The Selfish Gene* contained little original research, drawing heavily on other scientists like George Williams, William Hamilton and John Maynard Smith, it can justifiably be claimed as one of the most influential works in modern biology. It remains the outstanding account of the gene-centered view of evolution.

The Selfish Gene's argument is that many traditional accounts of evolution and genetics have got a fundamental principle the wrong way around. Organisms do not use genes to reproduce; rather, it is genes that exploit organisms to replicate and pass themselves on to another generation. The gene is the basic unit of natural selection. Evolution is best understood as acting on these self-copying packages of information, and not the creatures, plants or bacteria that carry them.

> "We are survival machines . . . robot vehicles blindly programmed to preserve the selfish molecules known as genes. This is a truth which still fills me with astonishment."
> **Richard Dawkins**

At one level, this is banal — since the modern evolutionary synthesis, it has been accepted that genetic variation is the raw material that allows evolution to take place. At another level, however, it is highly provocative. It suggests that the phenotypes that genes create have no inherent value: though these might improve the survival and reproduction of individuals, groups and species, they are not ultimately selected for this purpose. Such benefits are the incidental means by which genes secure their future. *The Selfish Gene* is the strongest possible interpretation of natural selection's amoral nature, suggesting there are few facets of behavior or physiology in which genetic influences might not be found.

Memes

Perhaps the most truly original idea in *The Selfish Gene* is that cultural phenomena can be subject to a form of natural selection, in similar fashion to genes. Dawkins coined the term "meme" to describe a unit of cultural information — such as a religion, song or anecdote — which is passed from person to person and which competes for popularity. Like genes, memes can mutate when people copy them incorrectly. Advantageous mutations, which make a meme more memorable, tend to thrive, while those that ruin its meaning die out. The concept is highly controversial: some philosophers think it elegant, but others find the analogy too neat, and lacking evidence in its support.

Survival machines

The short life spans of all living things mean that individuals are here today, gone tomorrow. Their genes, however, are functionally immortal — at least for as long as they can continue to duplicate themselves, and live again in another body. They do this by building "survival machines" — Dawkins' elegant phrase for roses, amoebae, tigers and people, which ferry genes from one generation to the next.

The genes that thrive and succeed in making the most copies of themselves are the ones that build survival machines that are best adapted to their environments. Thus genes often have beneficial functions in the organisms that carry them: they instruct cells to produce adrenaline to aid flight from predators, insulin to metabolize sugar, or dopamine to run the brain. But these adaptations are nothing more than a by-product of Darwinian selection's action at a genetic level, where it rewards those genes that copy themselves most often.

This is what Dawkins meant by invoking his brilliant metaphor: to an outside observer, genes look as if they are behaving selfishly. Organisms breathe, feed and behave in certain ways because it suits the interests of their genes that they do so. It is a paradigm that explains many known phenomena in biology and medicine — including the matter of why we grow infirm as we get older and eventually die. From a gene's perspective, there is no point in building survival machines that last a lot longer than their purpose, which is to live long enough to breed and raise young, so that their genes can prosper all over again.

The naturalistic fallacy

A common misconception about Dawkins, and the evolutionary psychologists he has helped to inspire, is that selfish gene theory seeks to justify a dubious morality. This argument falls into an intellectual trap called the naturalistic fallacy. That something is natural does not make it right. If genes can promote violence or rape, to help them to propagate, that provides no justification for such crimes, as Dawkins makes perfectly clear. Indeed, we need to study such influences if we want to prevent them. "Let us understand what our own selfish genes are up to, because we may then at least have a chance to upset their designs, something that no other species has ever aspired to do," he says.

A misunderstood metaphor

His choice of language, however, left Dawkins open to misinterpretation — often wilful — from critics who felt his theory was overly bleak, reductionist and deterministic. Genes, of course, are not conscious and have no intentions: they are not selfish in the way people can be. As Mary Midgley, a philosopher, said in a famous review: "Genes cannot be selfish or unselfish, any more than atoms can be jealous, elephants abstract or biscuits teleological." This line of argument, however, was a classic assault on a straw man. Dawkins had made it perfectly clear that genes are not actually selfish, but that they work in ways that make them appear so. The entire point of his hypothesis is that evolution is motiveless.

Another implication that is often wrongly drawn from the book is that if genes work selfishly, individuals must behave in this manner too. Yet as Dawkins had again explained, selfish genes do not necessarily generate selfish people. In fact, they offer a wealth of potential evolutionary explanations for altruism. Within families, in which many genes are shared, individuals have an obvious genetic motivation for helping others. Mathematical biologists have also used game theory to show that selfish genes can thrive by making organisms co-operate to their greater shared benefit — a concept known as "reciprocal altruism."

Neither does selfish gene theory imply that organisms can be explained solely in terms of their genes, as critics such as Midgley seem to think. The gene-centered view of evolution is a reductionist theory, but it is not deterministic: it does not exclude environmental inputs. Individuals' phenotypes, Dawkins says, are always a product of both genes and their surroundings. That, indeed, is one of the main reasons why evolution does not act on phenotypes, which always differ between individuals and are thus destroyed by death, but does act on the longer-lasting and much less mutable genes.

Evolutionary psychology

One effect of *The Selfish Gene* was to inspire a generation of biologists to think afresh about how genes affect human life, helping to shape not only our bodies but also our minds. The gene-centered view fed a burgeoning understanding that people are animals, that the brain is an evolved organ, and that its propensities have not escaped the influence of selfish genes working to promote their own survival.

> *The Selfish Gene* brought about a silent and almost immediate revolution in biology. The explanations made so much sense, the fundamental arguments were so clearly stated and derived completely from first principles, that it is hard to see after reading the book how the world could ever have been any different.
>
> Alan Grafen

This has been particularly important to the development of the new fields of evolutionary psychology and sociobiology, which seek to explain aspects of our species' behavior in terms of Darwinian adaptation. Scientists like Leda Cosmides, John Tooby, David Buss and Steven Pinker have argued persuasively that many phenomena that occur throughout different human societies — such as aggression, co-operation, gossip and typical male and female attitudes to sex and risk — are shared because they have evolved. These traits are found everywhere because, at least in times and places past, they helped humans to survive and thrive, ensuring that plenty of copies of the genes that influence them spread through the gene pool. Selfish genes have helped to make people who they are.

13 Nature versus nurture

The monster Caliban, according to his master Prospero, was "a devil, a pure devil, on whose nature nurture can never stick." Yet only a few decades before Shakespeare wrote *The Tempest*, St. Ignatius Loyola had founded the Jesuit order, with its famous maxim: "Give me the child until he is seven, and I will show you the man." The debate over the relative contributions of inheritance and experience to the human condition has deep historical roots.

It became one of the most politically charged questions of the genetic age. On one side stood those who sought and saw genetic explanations for human psychology, on the other, those who believed it to be molded by culture. There was not supposed to be much common ground. Sarah Blaffer Hrdy, an evolutionary psychologist, has even joked that perhaps we are genetically programmed to set nature against nurture.

The blank slate

The doctrine of the blank slate, which argues that humans share few innate character traits and instead develop them through experience and learning, is usually traced to 17th-century philosopher John Locke. It grew popular in the Enlightenment, fitting the mood of challenge to the authority of monarchy and aristocracy: if human capacities were not innate but learned, there was little to justify hereditary rule. For Locke, the blank slate was a statement of individual freedom.

It was later to become strongly associated with the political left. Though many early socialists had been enthusiasts for eugenics, later generations grew suspicious of genetics, from the way it was used to justify oppression of disadvantaged racial and social groups, most brutally in Nazi Germany. Liberal opinion turned decidedly against the concept of a biological human nature, which was increasingly seen as a

> " Once [social scientists] staked themselves to the lazy argument that racism, sexism, war and political inequality were logically unsound or factually incorrect because there was no such thing as human nature (as opposed to morally despicable, regardless of the details of human nature), every discovery about human nature was, by their own reasoning, tantamount to saying that racism, war and political inequality were not so bad after all. "
>
> Steven Pinker

tool with which male and bourgeois elites could rationalize their hegemony.

Those who argue that genes influence personality and abilities still struggle to shake off this association with racism and far-right politics. The rise of the post-modern belief that all things are socially constructed has also fed a view that it is politically incorrect even to investigate genetic influences. When E.O. Wilson suggested in the 1970s that human behavior was evolved like that of other animals, his lectures were picketed and students doused him with water. The left-wing biologists Steven Rose, Leon Kamin and Richard Lewontin responded with a book entitled *Not In Our Genes*, which accused Wilson of a crude determinism designed to legitimize the status quo. "Its adherents claim, first, that the details of present and past social arrangements are the inevitable manifestations of the specific action of genes," they said.

Nature via nurture

The rival camps claiming nature and nurture as the main influence on human behavior, however, are not pitched so far apart as is usually assumed. Partisans of both have often caricatured one another's position, and many of their disagreements are really about emphasis. Few, if any, members of the "nature school" are true genetic determinists who believe every human trait can be directly mapped to triplets of DNA. Equally, while strong cultural determinism is more common, most critics of genetic theories argue that the importance of

The genetics of aptitude

There is often an element of chicken-and-egg about the interaction of nature and nurture. Let's take sporting ability as an example. If a boy has inherited genes that give him strong fast-twitch muscles and a good lung capacity, he is quite likely to find himself a quicker sprinter than many of his peers. As a result, he's likely to enjoy sport, to attract the attention of his school's track coach, to make the 100 yards team, and to get training that adds further to his speed. He seeks out an environment that suits his genes.

Something similar probably applies to other abilities, such as intelligence and music. Genes might not influence intelligence per se, so much as create an aptitude for learning, so that a child concentrates in class and spends their free time in the library.

genes is exaggerated rather than non-existent. The great controversy, indeed, is now starting to give way to consensus, as improved understanding of how genes actually work makes it clear that the two forces are often impossible to separate.

Twin studies

Much of the critical evidence has emerged through the study of twins. Identical twins share all their DNA, while fraternal twins share only half — they are no more closely related on a genetic level than are ordinary siblings. Both kinds of twins, however, share a womb, a family and a cultural environment. Comparisons between the two types can thus tease out the extent to which inheritance is important.

Across a wide range of traits, including IQ, personality indicators such as extroversion and neuroticism, and even homosexuality, religiosity and political conservatism, identical twins are more similar to one another than are fraternal pairs. This indicates that genes must affect these aspects of personality.

The concordance between identical twins, however, is rarely 100 percent — their IQ scores, for example, tend to be around 70 percent similar, compared to around 50 percent for non-identical pairs. By definition, inheritance therefore cannot be the only factor involved: if it was, identical twins would always turn out the same. For most human qualities, neither the extreme-nurture nor the extreme-nature hypothesis can be correct.

The Dunedin Cohort study

Even more striking evidence has come from a recent series of studies led by Avshalom Caspi and Terrie Moffitt. These scientists have been following up a cohort of children born in 1972-73 in Dunedin, New Zealand, recording details of their life experiences and testing their DNA. The results have demolished the nature-nurture dichotomy.

> ❝The argument about intelligence has been about nature versus nurture for at least a century. We're finding that nature and nurture work together.❞
> Terrie Moffitt

First, Moffitt and Caspi studied a gene called MAOA, which has two variants or alleles. Boys with one allele are more likely to behave antisocially and get into trouble with the law — but only if they were also maltreated as children. When raised in well-adjusted families, those with the "risky" allele are fine. It is not a gene "for" criminality, and no determinism — genetic or environmental — is involved. A genetic variant must be activated by an environmental influence to do any potential harm.

The serotonin transporter gene, 5HTT, also has two alleles, and is known to be involved in mood. Moffitt and Caspi found that people with one allele were 2.5 times more likely to develop clinical depression than those with the other — but again, only under particular circumstances. The risk only affects people who also experience stressful life events such as unemployment, divorce or bereavement — and even then, it is a matter of raised risk, not determinism. When their environments are happy, their genotypes make no difference.

All this goes to show the sterility of the nature–nurture debate. The question should not be which is the dominant influence, but how they work together. Nature works through nurture, and nurture through nature, to shape our personalities, aptitudes, health and behavior.

14 Genetic diseases

Michael Rutter: "Most ordinary people, and even many medics, still think in terms of 'genes for' particular conditions. Yet genes with large effects like that are very much the exception, not the rule."

When genes make the news, it is more often than not in the context of disease. Headlines regularly proclaim the discovery of "Alzheimer's genes," "breast cancer genes," even "obesity genes." We know that the gene for Huntington's disease lies on chromosome 4, and that the gene for sickle-cell anaemia lies on chromosome 11. Embryos can be tested for the cystic fibrosis gene or the hemophilia gene, so that only healthy ones are transferred to the womb.

You could thus be forgiven for assuming that the main function of many genes is to cause disease. Yet as the science writer Matt Ridley has pointed out, this is as misleading as it is to define the heart by heart attacks, or the pancreas by diabetes. In truth, there are no such things as genes "for" particular diseases. The "Huntington's gene" is not carried only by people with the devastating neurological disease. We all have it. What is different about Huntington's patients is that they carry a version with a destructive mutation. If you like, they have a "patho-gene."

Many of the genes that are commonly described as being "for" this disorder or that are not even deterministic. The genes BRCA1 and BRCA2, for instance, are so closely identified with breast cancer that they are named after the disease. Women who carry mutated copies have a very high lifetime risk of breast tumors, of up to 80 percent. But by definition, that means that at least 20 percent of carriers will not get breast cancer. Genes like these are known as incompletely penetrant — they influence disease, but they do not inevitably cause it.

Simple and complex inheritance

Some mutations, of course, do have an inevitability about them. If you inherit too many repeats of the triplet CAG in a particular gene, you will get Huntington's disease. The number of repeats can even tell you the age at which you are likely to start to experience tremors, mood swings and neurological damage, leading to death. With 40 repeats, you'll stay healthy, on average, until the age of 59, but with 50 repeats, you'll fall ill in your late 20s.

Huntington's is one of the very few examples in which absolute genetic determinism holds true. People can escape from these mutations only if science develops a treatment, or if something else kills them first. More than 200 such conditions are known, and they are in general passed on through normal Mendelian inheritance. There is a simple match between genotype and phenotype, between mutation and disease.

Some are autosomal (carried on non-sex chromosomes) and dominant, meaning that inheriting one copy is enough to cause disease: examples include Huntington's and hereditary non-polyposis colon cancer. Others, such as cystic fibrosis and sickle-cell anaemia, are autosomal recessive. Only people who are "homozygous," with two copies of the defective allele, will suffer, while "heterozygous" carriers with just one copy experience no ill-effects. Still more are carried on the X chromosome, as with hemophilia and Duchenne muscular dystrophy, and most commonly affect boys.

Most diseases that are affected by genetics, however, are not so simple. The common conditions that are the chief causes of ill-health and death in the developed world, such as heart disease, diabetes and most cancers, are influenced by inheritance, but there is no one-to-one relationship between a particular mutation and a disease.

Sometimes, as with the BRCA genes, a defective gene has a very large effect, but not an inevitable one. More often, dozens of genes, each with small individual effects, combine to make people more susceptible to a disease. On their own, such genetic variants are virtually harmless.

Autism

Even when medical conditions are heavily influenced by inheritance, it can still be infuriatingly difficult to trace the genes that are responsible. Autism, for example, is known from twin and family studies to be highly heritable, indicating that genes are strongly involved. Despite years of research, however, no genes that definitively predispose to this developmental disorder have yet been found.

This suggests one of two things. Either there are no "autism genes," but the chances of developing the condition are raised or lowered by dozens or even hundreds of normal genetic variants, each of which has only a small effect by itself; or it is affected by very rare spontaneous mutations, which are unique to individuals or their families. Autism is explored further in Chapter 19.

Put together, they explain why some families struggle with high blood pressure, while others tend to develop cancer.

Why do disease genes survive?

As patho-genes like the ones that cause Huntington's and cystic fibrosis are so harmful, you might expect them to have been weeded out by evolution. Natural selection is brutal towards alleles that confer even the slightest survival disadvantage, and these alleles have catastrophic effects. How have they managed to hang on to their places in the human gene pool?

> " We all have the Wolff–Hirschhorn gene, except, ironically, people who have Wolff–Hirschhorn syndrome. "
> Matt Ridley

Sometimes, the answer is pure bad luck. A spontaneous mutation in the egg or sperm from which a person is conceived can occasionally prove catastrophic, if it occurs in a vital place. Diseases caused by extra genetic repeats, such as Huntington's and fragile X syndrome (which can cause mental impairment), are especially likely to be triggered in this way. It often takes only a small error to transform an acceptable number of repeats into a damaging one.

Other deleterious mutations survive because they cause damage only late in life, long after the carrier has had time to breed and raise a family. Many genetically influenced cancers, and Huntington's again, are good examples, as most patients live well into their 50s before symptoms strike. In such circumstances, natural selection does not apply. People with these defects have just as many offspring as those without them.

For recessive genetic disorders, another factor can be involved. Often, these have flourished because people who carry just one copy of a mutated gene have some kind of advantage. A single copy of the defect that causes sickle-cell anemia, for example, confers a degree of resistance to malaria. The evolutionary benefits of being heterozygous can outweigh the costs of conceiving some homozygous children with a crippling illness. The sickle-cell mutation is most common in regions where malaria is endemic, and where the genetic trade-off is worthwhile.

Superbugs

A disease does not have to be the result of an inherited mutation to be influenced by genetics. Bacteria, viruses and other pathogens are based on DNA and RNA no less than we are, and are forever evolving to evade our immune systems and the drugs with which we bolster these natural defences, and thus reproduce themselves more effectively.

A great example is the phenomenon of hospital "superbugs." Bacteria multiply so rapidly that their genomes rarely stand still for long: each of the billions of cell divisions that a colony will experience each day presents an opportunity for mutation — and some of these mutations will confer a degree of antibiotic resistance. Natural selection means that if an antibiotic is then used in treatment, a few bacteria will survive, and then divide to seed a new colony with resistant progeny. The result has been the emergence of strains such as MRSA, which stands for methicillin-resistant *Staphylococcus aureus*, and which resists the entire penicillin family of antibiotics.

With complex conditions such as heart disease, to which many genes contribute, the picture is different again. The variants that add slightly to risk are not best thought of as disease genes at all. These are common variations, with multiple effects — in the jargon, they are pleiotropic, from the Greek for "many influences." These influences can be positive as well as negative, which explains why they have spread so widely through the gene pool.

Genes are not for disease, and even rogue genes do not underlie most widespread diseases. Rather, these are influenced by the action of perfectly normal genes, working in concert with the environment.

15 Gene hunting

Mark McCarthy, University of Oxford: "We are mostly finding that for any given disease there are zero or at best one or two genes with large effects. Then there is a sprinkling of genes, perhaps five to ten, with modest effects of 10 to 20 percent, and there may be many hundreds with even smaller effects."

In the late 1970s, Nancy Wexler, the daughter of an American doctor, set out in search of the genetic mutation that causes Huntington's disease. Wexler's mother and uncles had the condition, and she knew there was a 50 percent chance she had inherited it. Finding the defect, she reasoned, would allow people in her position to discover whether they had been given a genetic death sentence. It might also lead to a treatment.

On learning of an extended family with a high incidence of Huntington's in Venezuela, Wexler traveled to Lake Maracaibo in 1979 to collect blood from more than 500 people. She then sent the samples for genetic analysis by her collaborator, Jim Gusella. His team began to compare DNA taken from people with and without Huntington's, and by 1983 he had narrowed down the search to the short arm of chromosome 4. It took another decade, however, to identify an actual gene, which makes a protein called Huntingtin.

> We have now entered a new era of large-scale genetics unthinkable even a few years ago.
>
> **Peter Donnelly**

The discovery, in 1993, was one of the great early successes of disease genetics, but it was the result of an extremely laborious process. The project took 14 years to deliver — and though it has led to a test (which Wexler decided not to take), it has yet to produce a therapy.

The Huntington's mutation, too, was among the genome's low-hanging fruit. It has a catastrophic effect, and it is autosomal dominant, inherited in simple Mendelian fashion. These factors meant it would be one of the easier genes to find. Other genes that influence disease in subtler fashion, however, were going to be very difficult to pin down.

Linkage analysis

The Huntingtin gene was identified using a technique called linkage analysis, which was until recently the most effective way of detecting how genetic variations influence disease. It relies on the way genes that rest close together on chromosomes tend to be inherited together, because of the recombination effect discussed in Chapter 4.

First, scientists select a number of single nucleotide polymorphisms (SNPs) — DNA sequences that are known to vary in spelling by one letter — as markers spaced at appropriate intervals through the genetic code. The next step is to search for these markers in people from families in which an inherited disease like Huntington's occurs. If a marker is always found in people with the disease, but not among those who are healthy, it must lie close to the mutation responsible, which can then be identified and sequenced. As family members share so much of their DNA, it is usually necessary to look only at a couple of hundred markers, in a few dozen people, to get results.

This technique, however, is readily applicable only to fairly rare disorders caused by mutations with large effects, as with Huntington's or BRCA1 (see Chapter 14). To find subtler influences on disease, many more people have to be screened. The numbers required make it essential to look beyond families, to less closely related people who share less of their DNA. That, in turn, means that hundreds of thousands of genetic markers have to be scanned, to get a statistical relationship strong enough to reveal a gene. Until recently, that was so expensive and time-consuming as to be functionally impossible.

Genome-wide association

It has now become practical with the advent of two new tools that have transformed disease genetics. The first is the micro-array or "gene chip" (see box), which can screen a person's DNA for a million genetic

Gene chips

Research projects like the CCC would not have been possible without the development of DNA micro-arrays or "gene chips." These hold a collection of up to a million microscopic spots of DNA, each in the configuration of a particular SNP. When the DNA that you wish to test is exposed to this chip, any sequences that are present will bind to the corresponding spot. They can be used to screen for hundreds of thousands of genetic markers at once, revealing which SNPs are carried by the person who is being tested.

The 1,000 genomes project

One of the next stages in the hunt for genes that affect our health is an international effort to map the entire genomes of more than 1,000 people, which has been made affordable by new sequencing technology. This should allow scientists to find and catalogue every single genetic variant that is carried by at least one person in 100. It will effectively work as an index to the genome. When a marker SNP suggests that a part of the genome is linked to a disease, geneticists will immediately be able to look up all the reasonably common variants in its chromosomal neighborhood, to investigate which, if any, are responsible for the effect.

variations at once. The second is the HapMap, a chart completed in 2005 that shows which segments of the genome, called haplotypes, tend to be inherited together.

The new technique, called genome-wide association, starts with the HapMap, from which scientists select 500,000 SNPs as markers for every haplotype block. Gene chips are then used to look for these markers among thousands of people with a particular disease — say type 2 diabetes — and a similar number of healthy controls. Any markers that are significantly more common in either group are then investigated further, to pinpoint parts of the genome that are associated with a higher or lower risk.

The beauty of this method is that it can reveal completely unexpected results. If a variant raises the risk of a condition by more than about 20 percent, genome-wide association will find it, even if this effect had never been suspected. A variant in a gene called FTO, for example, causes fused toes in mice. In 2007, one of the first large genome-wide association studies, by the Wellcome Trust Case Control Consortium (CCC), found that in people it predisposes slightly to obesity.

Early in 2007, science knew of hardly any common genetic variants that influenced disease. By the spring of 2008, the tally stood at more than 100, as genome-wide association studies began to bear fruit. The CCC has found genes linked to heart disease, rheumatoid arthritis, Crohn's disease, bipolar disorder and both kinds of diabetes, as well as obesity and height. Other teams have found new variants that affect breast cancer, prostate cancer, heart attacks and multiple sclerosis. Exciting new data is being published all the time. Even cautious voices are speaking openly of a step-change in humanity's ability to read and understand the genome.

Each of these variants has a small effect on its own, raising risk by between 10 and 70 percent. When considered alongside other variants, however, their combined effects can be large. They are also extremely common — those identified by the CCC are carried by between 5 and 40 percent of the Caucasian population. As the diseases they influence are common, they clearly affect hundreds of millions of lives.

Genetics is suddenly being taken to a different level. It was once a science limited to finding mutations with devastating effects, but for very few people. It is now tracing variants with a more limited impact, but on common medical conditions. You might call it the democratization of the genome.

16 **Cancer**

Despite the fact that most common diseases are the result of complex interactions between inheritance and our surroundings, the products of nature via nurture, there is one that always features genetics at its core. Rather, it is not one disease, but a group of more than 200 — the cancers. Brain and breast tumors, carcinomas of the lung and liver, melanomas of the skin and leukemias of the blood share a common characteristic. They are ultimately diseases of our genes.

That may come as something of a surprise, given that cancer is often thought of as an environmental disease. Whether it is sunbeds and melanoma, the human papilloma virus and cervical cancer, asbestos and mesothelioma, or smoking and any cancer you choose, there is overwhelming evidence that environmental influences can contribute, often decisively, to the formation of tumors. All these carcinogens that can seriously damage your health, however, do so in essentially the same way. They damage DNA.

> " It would surprise me enormously if in 20 years the treatment of cancer had not been transformed. "
> Mike Stratton, Cancer Genome Project leader, 2000

Cancer is the result of genetic failure. Every time a cell divides, it must successfully copy its DNA. It is estimated that 100 million million cell divisions take place during an average human lifetime. Every one has the potential to introduce an error into a daughter cell's genetic code that can turn it cancerous.

In healthy tissue, cell division is a controlled process, ordered by genetic signals that ensure it happens only when it is supposed to. Cancer develops when it starts to run out of control. In every case, the trigger is a copying mistake during cell division, often at a single DNA letter. Many mistakes of this sort are harmless, doing nothing to alter the genome's function, but when mutations strike in the wrong place, the results can be disastrous.

Oncogenes and tumor suppressors

The genetic errors that start cancer can be inherited or acquired from exposure to carcinogenic chemicals or radiation. To launch the destructive career of a tumor, though, they need to affect two broad categories of gene. The first class are the oncogenes — genes which, when defective, give cells new properties that turn them malignant. The second are the tumor suppressors — the genome's policemen, whose job is to spot oncogene mutations and tell malignant cells to kill themselves.

Most cells that acquire oncogene mutations are shut down by their tumor suppressors, and commit suicide by a process called apoptosis. A cell with mutations in both kinds of gene, however, can escape this programed death and become cancerous, though sequential damage to many different genes is usually required. The cell will divide unchecked, passing its mutant genetic legacy to its progeny, which proliferate to create rogue tissue that can eventually metastasize through the body, damage organs, and kill.

Many of the oncogenes that drive cancer are implicated in tumors found in very different parts of the body. Mutations in the BRAF gene, for instance, are common in both malignant melanomas, in which they are often caused by ultraviolet light, and in colon cancer. The same tumor suppressors are often damaged too — the p53 gene is mutated in almost half of all human cancers. Most inherited mutations that

Telomeres

Another genetic clue to cancer comes from stretches of repetitive DNA at the end of each chromosome called telomeres, which protect against the loss of genetic information. Without them, some important genes would be disrupted every time a cell divides, because DNA cannot normally copy itself all the way to the end of the chromosome. The telomeres absorb this damage, shortening a little with each cell division, and when they are lost completely, the cell usually dies. Telomere loss is one of the main causes of aging.

One of the reasons why cancer cells grow out of control is that most can copy their telomeres, because of mutations that allow them to make an enzyme called telomerase. This helps them to divide unchecked, but it also suggests a medical plan of attack. Several telomerase-inhibitor drugs have begun clinical trials.

contribute to cancer affect tumor suppressors as well — both BRCA1 and BRCA2 have this role. These defects greatly raise the lifetime risk of cancer by reducing by one the number of genetic hits a cell must take to set it on the path to malignancy.

Genetic therapy

To treat cancer, it is necessary to root out the genetically abnormal cells that cause it, either by killing them with drugs or radiation, or removing them with surgery. All these methods can be pretty brutal: operations such as mastectomies can be disfiguring, while chemotherapy and radiotherapy respectively poison and burn healthy tissue as well as the tumors they are designed to cure. Their side-effects are legion.

These blunt instruments, however, are now being supplemented with smarter weapons, and genetics provides the guidance system. If it is possible to characterize the precise genetic mutations that are driving a particular cancer, it is also possible to target these with drugs. A prime example is Herceptin, a drug prescribed to women whose breast tumors have mutations in the gene for a receptor called HER-2. The drug binds to this receptor, killing the cancer. It can halve the relapse rate — but only among those patients whose cancers are genetically susceptible to it. In others, it has no effect. Had it been tested in the general population, rather than in a targeted group, it would never have made it through clinical trials.

> **I imagine that in the future, machines that read the genetic signatures of patients' cancers will be more important than their oncologists.**
> **Richard Marais,**
> **Institute of Cancer Research**

This is the future of cancer treatment, which a project called the International Cancer Genome Consortium should help to realize. This $1 billion initiative aims to identify all the mutations that drive 50 common types of cancer, so that doctors can pinpoint the precise genetic factors that are responsible for the growth and spread of their patients' tumors. Cancers could then be treated not so much according to where they occur in the body, but on the basis of the genetic makeup of their rogue cells. We may soon think not of bowel or stomach cancers, but of BRAF-positive or p53-positive tumors.

Mike Stratton, of the Wellcome Trust Sanger Institute, a leader of the consortium, is already seeking to develop therapeutic strategies based on this approach. His team is currently investigating how 1,000 cancer cell lines, each with known mutations, respond to

400 different drugs. The goal is to determine whether some of these agents are effective against tumors with a particular DNA profile, but not against others.

Another benefit of cancer genomics should be to make chemotherapy kinder, through drugs that home in on DNA targets that are found in cancer cells, but not in healthy tissue. It may also be possible to avoid damage to a patient's reproductive cells: these are particularly vulnerable to existing cancer therapies, which often cause sterility as a consequence.

The cancer paradox

While both life expectancy and its quality have improved significantly in Western countries in the past century, cancer rates are going up. Between 1979 and 2003, the UK incidence rose by 8 percent in men, and by 26 percent in women. This is sometimes blamed on pollution or other environmental factors, but its principal cause is actually the success of modern medicine.

Antibiotics, sanitation, better nutrition and other improvements in public health mean that fewer people are dying young of infectious diseases — but longer lives allow more DNA damage to accumulate, to the point at which tumors can form. The genetic nature of this disease explains the apparent paradox of medicine. As it defeats other foes, more of us will live long enough to develop cancer. The challenge, which genetics will help to meet, is to turn it from a fatal disease into a chronic one.

17 Behavioral genetics

Certain behaviors and personality traits are well known to run in families. People with religious parents are more likely to be churchgoers, while those who grew up in left-wing households are more likely to vote that way when they reach 18. We all know people, too, whose character quirks remind us of their close relatives — nervous daughters of nervous mothers, and fathers and sons who share a passion for fishing or fast cars.

Folk wisdom tends to attribute all this to upbringing: to the way a child's outlook on life is molded by that of its parents, whether through deliberate indoctrination or passive exposure to their tastes. That conclusion, however, is too simple. Children of course share a home environment with their mothers and fathers, which can greatly affect personal development, but that is not all they share. They also inherit half their DNA from each parent, and the science of behavioral genetics has shown that this can be equally important, if not more so.

> **"**It would be unwise to assume that genetics will not be able to assist in determining degrees of blame, even if the 'all-or-nothing' question of responsibility is not affected by genetic factors themselves.**"**
> Nuffield Council on Bioethics

Natural experiments

The relative contributions of nature and nurture are fiendishly hard to separate when studying families, as either could account for shared traits, from spirituality to spitefulness. As it is unethical to separate children from their parents in controlled experiments, such research must rely on natural experiments instead.

As we saw in Chapter 13, identical twins share both a home environment and all their DNA, while fraternal sets share the same home but only half their genes. Comparisons between the two are therefore illuminating; for traits affected by genetics, identical pairs will look more alike. Adoption studies are also useful. For characteristics that are strongly heritable, adopted children should conform more closely to their birth families than to their adoptive ones.

Such studies have shown that genetics does not just affect physical attributes such as height and obesity. Many aspects of mental, psychological and personal development are at least partially heritable, too. The list includes intelligence, antisocial behavior, risk-taking, religiosity, political views and all the "big five" personality traits — neuroticism, introversion/extroversion, agreeableness, conscientiousness and openness to experience. There is even evidence to suggest that a woman's ability to have an orgasm may be influenced by her genes.

Heritability

These effects can be quantified, using statistical techniques to calculate heritability quotients. These are expressed as a percentage or decimal, and they are easily misunderstood. When behavioral geneticists say that a trait, say thrill-seeking, is 60 percent heritable, it does not mean that any given person can attribute 60 percent of his aptitude for bungee-jumping to his genes. Neither does it mean that of 100 people who like extreme sports, 60 have inherited this passion while 40 have learned it.

Height

One pitfall of behavioral genetics is well illustrated by a non-behavioral trait in which genes are certainly involved — height. It is estimated that about 90 percent of the differences between individuals' heights reflect genetic variation, and 20 genes that are involved have been identified. Though environmental effects like nutrition matter, the influence of genetics is strong.

Nobody of sound mind, however, would suggest that height should be assessed by genetic tests. You will always get more accurate answers by measuring people. The same is true of all sorts of heritable characteristics, such as personality, intelligence or aggression. When a phenotype can be reliably evaluated, the genotype that contributed to it is often irrelevant in the real world.

The true implication is much subtler: it is that 60 percent of the differences we find between different people's attitudes to risk are down to inherited variation.

To call a trait heritable is thus meaningful only at a population level; it says nothing about precisely how genetics has affected a particular person. In some, genes will be the most important factor, while in others it will be formative experiences. Heritability quotients reflect an average. Unless the value is zero (as for the language you speak) or one (as for Huntington's disease), both nature and nurture are always involved.

It is a misconception that findings of heritability imply genetic determinism. If anything, the reverse is true: most heritability quotients for behavior and personality hover between 0.3 and 0.7, leaving a large role for environmental influences.

Ethical conundrums

Most of the time, this sort of research is benign. If we learn how far genetics is involved in reading disabilities, or antisocial behavior, it may be possible to identify the genes — or the environmental factors — that play a part, and develop drugs or social programs to intervene. But knowledge about genetic effects on behavior can also open more difficult ethical territory.

In 1991, Stephen Mobley robbed a branch of Domino's Pizza in Oakwood, Georgia, and shot the manager, John Collins, dead. He was convicted of murder and sentenced to death, but his lawyers then appealed on innovative grounds. Their client came from a long line of violent criminals, and carried a genetic mutation that had been linked to similar behavior in a Dutch family. Mobley's genes made him do it, they argued, and his sentence should be commuted as a result.

Intelligence

Perhaps the most vexed question in behavioral genetics concerns the inheritance of intelligence. Twin studies suggest that between 50 and 70 percent of variance in general intelligence is explained by genetics, but the search for genes that underlie this has so far drawn a blank. One of the probable reasons is that there is no "gene for" intelligence. At least half of the 21,500 or so human genes are expressed in the brain, and any of these could plausibly affect mental development.

The appeal was thrown out, and Mobley was executed in 2005. Most scientists think his claim was specious, as the link between his mutation and violence is far from robust. If a number of genes are reliably shown to influence aggression or psychopathy, however, then future cases may not be quite so clear-cut.

Genetic tests are unlikely to provide a defence: genes may predispose people to patterns of behavior, but they do not inevitably cause them. Some people argue, though, that such information could be considered in mitigation, in similar fashion to psychiatric illness. Britain's Nuffield Council on Bioethics recently suggested that genetics could helpfully "assist in determining degrees of blame."

> Behavioral genetics cannot deal with highly complex behaviors, and certainly not generic ones like good and evil. There is no data of genes that predispose toward good or evil, and any such data would be so weak as to apply [only] to a minority of cases.
> **Philip Zimbardo, Stanford University psychologist**

Other possibilities are more sinister. Genetic profiling might be used to identify those whose genes make them more likely to become offenders, in the manner of the movie *Minority Report*. Similar techniques might be employed in schools, to select genetically gifted pupils for special tuition, or to assess job applicants for inherited aptitude for particular roles.

Applications such as these, however, would rest on a misunderstanding: behavioral genetics is probabilistic, not deterministic, and applies to populations, not individuals. To use its findings to prejudge people would therefore be a gross infringement of liberty. The way people behave is the result of a complex interaction between their genes and their experiences, and the balance may differ for every one of us. Individual capacities are best assessed by considering people as they are, not by trying to predict what their genes say they should be.

18 Genetic history

Chris Stringer, Natural History Museum, London: "All men and women are Africans under their skin. It's as simple as that."

When Charles Darwin wrote *The Descent of Man* in 1871, scientific racism was in its heyday. People of white European origin had come to dominate the globe, and this was widely thought to reflect their superior biological status. Many intellectuals considered humanity to be not one species but several, and Darwin's ideas led some to conclude that dark-skinned people had been left behind by evolution. The idea that we are all Africans would have struck fashionable opinion as preposterous. That, however, is precisely what Darwin suggested in his second great book. As our closest animal cousins, the chimpanzees and gorillas, are native to Africa, he reasoned that the same is probably true of our own species, *Homo sapiens*.

It was a prescient insight. Within 50 years, fossil discoveries had started to point towards an African origin for humanity, and this has now been overwhelmingly confirmed by genetic research. DNA has not only shown that people are all closely related, and far more alike than different. It has allowed us to trace the history of our own species and others, and even to identify some of the biological idiosyncrasies that make us human.

Out of Africa

Many of the most significant fossils belonging to ancient human relatives, and all of those that are more than about 2 million years old, have been discovered in Africa. Fossils like Lucy, the celebrated *Australopithecus afarensis* specimen unearthed by Donald Johanson in Ethiopia in 1974, have left little doubt that the lineages of humans and chimps separated south of the Sahara.

The more recent evolutionary history of *Homo sapiens*, however, has been less clear. Other human species, such as *Homo erectus* and Neanderthal man, had spread beyond Africa well before anatomically modern humans appeared around 160,000 years ago, and two hypotheses have competed to explain the origin of our species.

The "out of Africa" theory contends that we evolved just once, in Africa, and then migrated to displace our relatives on other continents.

The multi-regional view claims instead that pre-existing populations of proto-humans evolved separately, or at least interbred with traveling *Homo sapiens* bands, to give rise to modern races.

The fossils have always favored "out of Africa," but genetics has supplied clinching evidence. Two types of human DNA have proved particularly instructive. While most chromosomes are constantly shuffled by recombination, this process does not apply to the genes held by the male Y chromosome, and by the mitochondria, which are transmitted from mothers to their offspring. Both are inherited intact, and vary only because of spontaneous mutations.

As such mutations occur at a fixed rate, the DNA of living people can be used to reconstruct their ancestry. And the evolution of Y chromosomal and mitochondrial DNA has progressed exactly as the out of Africa theory predicts. It even charts the routes by which *Homo sapiens* populated the globe.

Further evidence comes from genetic diversity. The "out of Africa" theory suggests that several thousand people were living in the continent about 70,000 years ago, when a small group crossed the Red Sea. The descendants of these migrants then went on to populate the rest of the world. Non-Africans, who all originate from this small founding group, should therefore be less genetically diverse than Africans, who sprang from a population that was larger and more varied to begin with.

Are we Neanderthals?

A great controversy in human evolution has been the Neanderthals' place on our family tree — did these ancient inhabitants of Europe die out when *Homo sapiens* reached the continent, or were they at least partially assimilated through interbreeding?

Sufficient genetic material has now been recovered from Neanderthal fossils to allow the species' genome to be sequenced, and the results have settled the debate. Modern humans do not appear to have any Neanderthal DNA. If any of our ancestors mated with Neanderthals, their offspring did not survive to contribute to the human genome as it exists today.

Another surprising insight from the Neanderthal genome is that the species had the same version of the FOXP2 gene that modern humans have today. This could mean that they were capable of language, and that they were not the grunting brutes that popular culture often portrays them to have been.

Whither human evolution?

If humans are themselves the results of past evolutionary forks, could *Homo sapiens* itself evolve into different species? In 2007, research led by American anthropologist Henry Harpending suggested that the answer might be yes. Genetic differences between population groups, he found, had widened over the past 10,000 years. Left to continue, this could potentially result in two or more new species.

Harpending's study, however, looked at the pre-industrial world, when ethnic groups were usually separated by distances too great to travel. Now that air transport and globalization have broken down many geographical boundaries, most evolutionary biologists think a fresh human speciation event is highly unlikely.

Once again, this is just the pattern that DNA reveals. Human genetic diversity is much greater within Africans than between Africans and any other ethnic group — or even between other ethnic groups that appear closely related. On a genetic level, a Finn will resemble some Africans more than he resembles a Swede. The variability of human DNA declines with increasing distance from the motherland — Australian Aborigines and Native Americans are the least diverse populations of all. Genetic reconstruction techniques are so good that we even know roughly how many people — about 150 — left Africa in that critical first wave.

What makes us human?

Similar methods can be used to chart the evolutionary history of any species and to establish genetic relationships between them. Molecular evidence, for example, shows that the closest living relatives of whales and dolphins are hippos. DNA proves the fact of evolution as surely as the fossil record. Genetic comparisons are also capable of pinpointing some of the evolutionary events that may have been important in the development of particular species. In our own case, they have highlighted at least a few of the genes that appear to make us human.

FOXP2, which we met in Chapter 10, is a prime example. Across mammals and birds, this gene is highly conserved — the sequence is almost exactly the same from species to species, which usually means it does something important. In mice and chimps, which last shared a common ancestor at least 75 million years ago, the FOXP2 protein differs by just a single amino acid.

Humans and chimps diverged much more recently — about 7 million years ago — yet our FOXP2 protein differs by two amino acids from the chimp version: twice as many mutations as separate chimps and mice have accumulated, in less than a tenth of the evolutionary time. This pattern suggests that natural selection is at work, preserving useful changes. In this case, it may be language: people with FOXP2 defects have major speech impairments. These mutations could be part of the explanation for a capacity that is unique to humans.

“You get less and less variation the further you go from Africa.”

Marcus Feldman,
Stanford University

Another stretch of DNA, called HAR1, shows signs of even stronger selection. It is 118 base pairs long, and in the 310 million years since chimps and chickens shared an ancestor, just two of these have changed. Human HAR1, however, is spelt differently from the chimp version in no fewer than 18 places. The rapid pace of its evolution has led scientists to speculate that it might be involved in brain size and intelligence — the most striking difference between humans and other animals. It may be one of the genes that makes us human.

19 Sex genes

Eve, according to the Bible, was fashioned from Adam's rib. Yet if genetics has surprised racists by revealing that the cradle of humanity is Africa, it has also surprised male chauvinists. DNA has revealed that the Book of Genesis got the story the wrong way around. By default, human beings are genetically programmed to be female.

In the movie *My Fair Lady*, Professor Henry Higgins famously asked: "Why can't a woman be more like a man?" But from a genetic perspective, the question is more interesting and revealing when posed in reverse.

Why can't a man be more like a woman?

The discovery of the underlying genetic reason for the differences between the sexes was made independently, and appropriately, by a woman and a man. In 1905, Nettie Stevens and Edmund Beecher Wilson noticed that male and female cells varied in chromosomal structure. While females had two copies of a large chromosome, the X, males had just one, along with another, much smaller chromosome, the Y. They had identified the system by which sex is determined in many animals, including humans: women have the chromosomal type XX, while men are XY.

> "In its brief moment of glory, [SRY] sends billions of babies on a masculine journey."
>
> Steve Jones

When meiosis separates pairs of chromosomes to create gametes with a single set, eggs always carry an X, while sperm can carry an X or a Y. X-bearing sperm will produce girls when they fertilize eggs; Y-bearing sperm, boys. For the first six weeks of gestation, both male and female embryos develop in identical fashion. They would continue to do so, to produce babies that look female, without the action of a single Y chromosome gene. A woman's extra X chromosome sends no extra signal that makes her into a woman. That is the way people will be, without the intervention of a gene called SRY.

The masculinity switch

SRY, which was discovered in 1990 by Robin Lovell-Badge and Peter Goodfellow, stands for sex-determining region Y, and it is the biological key to sex. Those people who have a working copy will grow a penis, testicles and beard, while those who lack one will develop a vagina, womb and breasts. It is perhaps the most influential single gene in the human body.

If it does not kick into action seven weeks into pregnancy, or if its instructions cannot be heard, the body will continue to develop along its default female path. Should an XY embryo's SRY gene be mutated and non-functional, or should other genetic faults make cells insensitive to the male hormones the gene directs the gonads to produce, that embryo will grow into a girl (though she will be infertile). On rare occasions, SRY can find its way into an X chromosome, by means of a kind of mutation called a translocation. It will not surprise you to learn that when that happens, XX individuals become male.

SRY works as a masculinity switch. After five weeks of pregnancy, all embryos start to develop unisex gonads, which have the potential to become testes or ovaries. Then, two weeks later, the SRY switch is flipped, or not. Once turned on, it tells these gonads to become testes. If it is absent, switched off, or silent, these gonads will later begin developing into ovaries.

After eight weeks, the newly sculpted testes start to make male hormones, and these androgens masculinize the body. Clusters of cells that would otherwise become the clitoris and labia form the penis and scrotum, and the sex organs are plumbed together using ducts that atrophy in females. SRY makes a man.

Sex selection

The chromosomal differences between men and women mean it is now possible to choose the sex of one's children. The most effective method is to create embryos through in-vitro fertilization, and then remove a single cell to check whether it has two X chromosomes or an X and a Y. Only embryos of the desired sex would then be transferred to the womb.

Another method, known as MicroSort, relies on the different sizes of the X and Y chromosomes. Sperm are treated with a fluorescent dye that stains DNA, and these are then passed under a laser. As the X is so much larger than the Y, X-bearing sperm will glow more intensely, and can be separated out. The technique is claimed to increase the chances of having a child of the desired sex to between 70 and 80 percent. It is currently permitted in the U.S., but not in the UK.

Sex differences

SRY is not only the root cause of the physiological differences between the sexes. It also plays a part in those behaviors that are more common among people with a Y chromosome, such as risk-taking and aggression. None of these is directly programmed by SRY, though some of the human Y chromosome's other 85 or so genes might be associated with traits that are more commonly observed among men. But they nonetheless fall under the influence of this powerful gene. The cascade of androgens that SRY sets off masculinizes minds as well as bodies.

This indirect genetic effect is probably at least as responsible for typically male personality traits as are culture and learning. Men's higher testosterone levels certainly make them more prone to violence and recklessness, and they may also affect personality.

Simon Baron-Cohen, of the University of Cambridge, has suggested that one example is the way that women tend to be better than men at empathizing — at identifying others' thoughts and emotions, and then acting appropriately. Men, on average, are better at systemizing — at building and understanding systems, like car engines, mathematical problems and soccer's offside law.

Baron-Cohen's work hints that this may be related to androgen exposure in the womb. In one study, his team examined pre-natal testosterone levels in 235 expectant mothers who had an amniocentesis test to check for fetal abnormalities, and then followed up the children

Masculinity and health

No common gene is more damaging to health than SRY. Women outlive men in every society, and one of the reasons is the hormonal profile that this male gene creates. High testosterone levels make men more likely to take risks that endanger their survival, whether through careless driving, aggressive behavior, or smoking and drug use. The female hormone estrogen also protects against cardiovascular disease, which is the biggest killer of both men and women. Alzheimer's is the only major disease that commonly affects both sexes for which women have a higher risk.

Being male is also a major risk factor for autism, which is four times more common among boys than girls. Simon Baron-Cohen has suggested that this could be related to excess pre-natal androgen exposure, which creates an "extreme male brain" that often excels at systemizing, but has little empathizing ability.

when they were born. Those exposed to more testosterone tended to look less at other people's faces, and to acquire stronger numerical and pattern-recognition skills.

This research is easily misunderstood. It in no way suggests that it is any better to be a "systemizer" or an "empathizer," or that either trait is associated with greater intelligence. Neither are all men created one way and all women another. It is only on average that more men will have the first brain type and more women the second, just as men are on average taller than women while some women are taller than some men.

These averages, however, are part of a growing understanding that men and women are not biologically identical in their thought processes and behavior, any more than in their reproductive systems. And the root cause of most of these differences is a single Y-chromosome gene.

20 The battle of the sexes

Some genetic mutations vary in their expression in succeeding generations depending upon whether they are passed on by the father or the mother, via a phenomenon known as imprinting.

In 1532, on a visit to the town of Dessau, Martin Luther was presented with a child that behaved so strangely that he doubted its humanity. "It did nothing but eat; in fact, it ate enough for any four peasants or threshers," the father of the Reformation noted. "It ate, shat, and pissed, and whenever someone touched it, it cried." Luther thought the diagnosis simple: the child was possessed by the Devil, and should be thrown into the river Molda to drown. "Such a changeling child is only a piece of flesh, a *massa carnis*, because it has no soul," he said.

Today, we would come to a different conclusion. From the symptoms described by Luther's chronicler, Johannes Mathesius, pediatricians would immediately suspect Prader–Willi syndrome. Far from having no soul, what it lacked was probably a genetic region called 15q11, which when missing causes the grotesque overeating, floppy muscles and learning difficulties that were described in 1956 by Andrea Prader and Heinrich Willi.

Luther's changeling must have inherited its 15q11 mutation from its father. We can say that with confidence because had the defect occurred in its maternal copy of chromosome 15, it would have developed an entirely different disease. In 1965, a British doctor named Harry Angelman reported three rare cases of what he described as "puppet children." They were small and thin, they moved stiffly and jerkily, they were severely mentally retarded, and they had an unusually happy demeanor. Incredibly, their condition is caused by precisely the same stretch of DNA as is Prader–Willi.

Imprinted genes

The genetic disease that a child with a 15q11 defect will get depends on which parent supplied the mutated chromosome. If it came from the mother, it will be Angelman syndrome, and if it came from the father, it will be Prader–Willi. The gene that underlies these disorders is imprinted — that is, it carries a biological marker that tells cells to express only the maternal or paternal copy. Imprinted genes can "remember" their parental history, through a process known as methylation that leaves some switched on and others switched off (see box).

> " To turn anthropomorphic, the father's genes do not trust the mother's genes to build a sufficiently invasive placenta, so they do the job themselves. "
> Matt Ridley

Dozens of imprinted genes are now known, and a large proportion of them are involved with the development of embryos. Imprinting seems to require that a viable embryo has genetic input from both a man and a woman. This is superficially obvious — conception occurs when a sperm fertilizes an egg, so of course both sexes are involved. After fertilization, however, the pronuclei of the two gametes do not fuse immediately, and the sperm pronucleus can be switched for one from an egg, or vice versa. Thus can scientists create embryos with two genetic fathers or mothers.

In principle, such embryos should develop normally — they have a full complement of chromosomes and the cellular architecture they need to grow. But they don't. Experiments with mouse embryos have

Methylation

Imprinting works because of a process called DNA methylation, by which the function of genes is altered by chemical modifications. It involves the addition of a chemical tag, known as a "methyl group", to the DNA base cytosine. This can turn down a gene's activity, or switch it off entirely. It is critical to ensuring that genes are expressed only at the right times in an organism's life cycle, and in the right kinds of tissues.

Most of these methyl tags are wiped away during the early stages of embryonic development. The main exceptions are the imprinted genes, which retain these marks to flag up their maternal or paternal origin.

shown not only that they fail and die, but that they fail in different ways according to the source of their genes.

When all the genetic material is female, the inner cell mass that will eventually become the fetus starts to form normally, but dies because it lacks a viable placenta. When the embryo has two genetic fathers, the placenta forms normally, but the inner cell mass is a shapeless mess — a *massa carnis*, as Luther might have called it. Both sexes are necessary: imprinted paternal genes are essential for a healthy placenta, while imprinted maternal genes are needed to organize the embryo.

A hostile takeover

The evolutionary reasons for this can be understood by thinking of the placenta as a fetal organ — an idea first proposed by Australian biologist David Haig in the early 1990s. It is a tool by which the fetus becomes a kind of parasite, effectively stealing from its mother the nutrients, oxygen and the other resources it needs to grow.

The interests of the fetus and its mother are therefore slightly different: while the fetus benefits by squeezing as much out of its

Artificial gametes

An exciting implication of stem cell research (a field described more fully in Chapter 24) is the prospect of growing artificial eggs or sperm, to allow men and women who make none to have their own genetic children. This possibility, however, has also led to speculation that sperm could be made from female cells, or eggs from male ones, to allow gay couples to conceive. It has even been suggested that the same person could produce both sets of gametes, in the ultimate form of self-love.

Imprinting, however, suggests that making working "male eggs" or "female sperm" will be very difficult. It would be necessary to ensure they carry all the right tags to denote maternal or paternal genes, the full range of which remains unknown. Sperm also require a Y chromosome, which female cells lack.

Imprinting issues are also thought to explain the developmental problems suffered by many animal clones. Another likely effect is the inability of mammals, unlike bees, lizards and sharks, to reproduce by parthenogenesis — a process by which eggs develop spontaneously into embryos without fertilization.

mother as it can without killing her, the mother tries to hold some resources back for the sake of her health. This leads to a uterine tug of love, which may be responsible for pregnancy complications such as pre-eclampsia and gestational diabetes.

This means that though half a fetus's genes come from its mother and half from its father, each set has different interests. Maternal genes will benefit from being less demanding, so the woman they came from has a good chance of breeding again. To paternal genes, however, the mother's future reproductive prospects do not much matter. They benefit from diverting as much of her investment as possible to the current fetus: she might conceive her next one with someone else. Imprinted genes from the father therefore create an aggressive placenta, which stages something of a hostile takeover of the mother's womb.

> ❝The phenomenon is called imprinting because the basic idea is that there is some imprint that is put on the DNA in the mother's ovary or in the father's testes which marks that DNA as being maternal or paternal, and influences its pattern of expression — what the gene does in the next generation in both male and female offspring.❞
> **David Haig**

Haig's hypothesis has been supported by subsequent discoveries about imprinted genes and embryonic development. The gene for an insulin-like growth factor called IGF2, for example, is active in building the placenta, but switched off in adults. It is also paternally imprinted. A gene that appears to counteract its effects, called H19, is imprinted too, but from the maternal side.

Imprinted genes are not widely found in egg-laying animals: as their embryos are not fed through a placenta, and so cannot influence the resources they get, there is no need for maternal and paternal genes to fight it out. The placenta is not just a highly efficient means of nourishing offspring. It is also the *casus belli* for a fierce battle of the sexes.

21 Genetic fingerprinting

In a few short years the science of genetic fingerprinting has revolutionized crime detection around the world and become a central tool of forensic science in many other contexts.

On August 2, 1986, the body of a 15-year-old girl named Dawn Ashworth was discovered in a wood near the English village of Narborough. She had been raped and strangled, in a very similar fashion to Lynda Mann, a 15-year-old from the same village who had been murdered three years previously. Richard Buckland, a 17-year-old who lived locally, was soon arrested. But while he confessed to the second murder, he would not admit to the first.

The police were convinced that the crimes were the work of the same man: the *modus operandi* was identical and semen of the same blood type had been found on both bodies. In search of evidence, officers approached Professor Alec Jeffreys, of Leicester University, a geneticist who had recently developed a method for identifying people from their DNA. He agreed to compare Buckland's DNA against the crime scene samples.

The results were shocking: the girls had been killed by the same man, but it could not have been Buckland. His DNA proved his confession had been false, and the case against him was dropped. The police began taking blood samples from more than 5,000 local men, but no matches were found until a man was overheard boasting that he had taken the test for a friend. The friend, Colin Pitchfork, was arrested, and his DNA was a perfect match. He confessed, and was jailed for life on January 23, 1988. Genetic fingerprinting had solved a murder for the first time.

The technique

The test that convicted Pitchfork relies on repetitive segments of junk DNA known as minisatellites, which are between 10 and 100 letters long. These feature the same core sequence — GGGCAGGAXG, where X can be any of the four bases. Minisatellites occur at more than 1,000 locations in the genome, and in each of these places they are repeated a random number of times.

> " There was a level of individual specificity that was light years beyond anything that had been seen before. It was a eureka moment. We could immediately see the potential for forensic investigations and paternity. "
>
> Alec Jeffreys

Jeffreys stumbled on their forensic potential by accident. While studying minisatellites for clues to the evolution of disease genes, he examined DNA samples taken from his lab technician, Vicky Wilson, and her parents. Though the number of minisatellite repeats showed a family resemblance, each profile was unique.

Jeffreys immediately realized the implications: as every person has a highly individual genetic fingerprint, these could be used to match suspects to blood or semen found at crime scenes. Other suggestions came from Jeffreys' wife: the technique could also prove whether would-be immigrants who claimed to be related to British citizens were telling the truth, or to confirm paternity of a child.

Use and abuse

DNA fingerprinting has transformed forensic science. It has convicted hundreds of thousands of criminals like Pitchfork — and, just as

Kary Mullis

The inventor of the polymerase chain reaction (PCR) is one of the more colorful Nobel laureates. He has spoken openly of his experiences with LSD, and his autobiography *Dancing Naked in the Mind Field* describes a 1985 incident in which he believes he encountered a talking, glowing raccoon. He has become a controversial figure for other reasons, too: he has supported the maverick claim that HIV does not cause Aids, and championed astrology. The importance of his contribution to molecular biology, however, is unquestioned. As PCR allows DNA to be amplified, it has greatly improved the sensitivity of genetic fingerprinting and genetic testing for disease.

Low copy number testing

Contamination is a particular issue for a forensic technique called low copy number testing, which can match genetic fingerprints to DNA from as few as five cells. It is very hard to prove, however, that these come from a guilty person and not an innocent third party.

When you pick up an object, your hands will always deposit a few cells, and pick up others left behind by other people who have handled it. Some of these cells can then be shed onto other surfaces you touch. A frequently touched object, like a door handle, could thus transfer innocent people's DNA onto the hands of a criminal, and thence to a crime scene.

When large biological samples such as semen stains are tested, this is not a problem. The criminal's cells will far outnumber those of third parties, which can be discounted. Tiny samples of just a few cells, however, present a problem: it is hard to be certain they have not been innocently transferred. In 2007, such concerns led to the collapse of the trial of Sean Hoey, who was accused of the 1998 Omagh bombing in Northern Ireland, which killed 29 people.

importantly, has exonerated innocent people like Buckland. Another forensic use has been in identifying bodies. In 1992, it proved that a man buried in Brazil under the name Wolfgang Gerhard was Josef Mengele, the fugitive Auschwitz doctor, and it was used after the 9/11 terrorist attacks to identify victims' body parts.

The actor Eddie Murphy, the film producer Steve Bing and the footballer Dwight Yorke are just three of the thousands of men whose disputed paternity has been confirmed by DNA. The technique even proved that the semen stain on Monica Lewinsky's infamous blue dress contained the "presidential DNA" of Bill Clinton.

The technology has advanced considerably since the Pitchfork case. A technique called the polymerase chain reaction (PCR), invented in 1983 by Kary Mullis (see box), was soon incorporated into forensic genetics. As this allows small quantities of DNA to be amplified, it means that as few as 150 cells can make up a readable sample: suspects can now be identified from mere traces of biological material. Microsatellite analysis has also been replaced by the use of smaller repetitive chunks of DNA called short tandem repeats, which are more likely to survive environmental exposure and amplify well with PCR.

Many countries now routinely store DNA from convicted criminals, and sometimes, as in the UK, even from people who are arrested but never charged. The UK database holds samples from about 4 million people — 6 percent of the population. As only one in a million people share the same genetic fingerprint, a match to a crime scene sample is often seen by lawyers and juries as conclusive proof. It has even been used by supporters of the death penalty to argue that miscarriages of justice are no longer possible.

Useful as genetic fingerprinting can be, however, its importance is often overstated. First, there is the "prosecutor's fallacy." If a profile is held by one person in a million, then in a country of 60 million like the UK, it will be shared by 60 people. Every crime scene sample thus has 60 potential origins that are all equally likely. Unless other evidence points to a suspect, a match means the chance someone is innocent is not one in a million, but 59 in 60.

Another problem is that genetic fingerprints merely place suspects at crime scenes: they provide circumstantial evidence, which may not indicate guilt. It is one thing if a suspect's DNA is found in semen recovered from a rape victim, but quite another if it is found at his local general store after a robbery. If he frequented that shop, his DNA may be present for entirely innocent reasons. Contamination is a further issue: it is possible for an innocent person's DNA to appear at a crime scene because he has opened the same door as the guilty party or shaken hands with him (see box).

Genetic fingerprinting has caught thousands of rapists and murderers, and there is no question that it serves the cause of justice. It is only a tool though, and it is by no means infallible.

22 GM crops

People have been genetically modifying plants for thousands of years. All the crops that are farmed around the world, from rice to cassava to apples, have genomes that differ markedly from their wild relatives, as a direct result of human intervention. Plants with sweeter fruit, bigger seeds or shatterproof stalks have been deliberately picked out for cultivation and selective breeding, creating the domesticated varieties we eat today. Agriculture has always been an unnatural business.

Hermann Muller realized in the 1920s that an understanding of genetics might be used to accelerate and direct this process. By bombarding crops with radiation, it was possible to induce hundreds of mutations, some of which would produce new strains with useful properties that might never have evolved naturally.

Then, in the 1970s, came a still more powerful tool: recombinant DNA technology, which allowed new genes to be spliced into organisms. Plant breeding no longer had to rely on a "hit-and-hope" strategy of inducing mutations and picking the ones that looked promising. Genes that confer desirable traits could be deliberately inserted into crops instead, using either a bacterial vector or a "gene gun" that fires new DNA into the genome on tiny particles of gold.

> " GM is a complex technology, not a homogeneous technology. It has to be considered on a case-by-case basis. "
> Sir David King, former UK government chief scientist

The potential

The first GM plant developed with this technology arrived in 1985 — tobacco engineered with a gene from *Bacillus thuringiensis* (Bt). This bacterium is toxic to many insects, and is used by organic farmers as a pesticide. The Bt tobacco made this insecticide itself, reducing the need for chemicals to be sprayed for pest control. Food crops took a little longer to develop, but the first — the Flavr Savr tomato, which had a longer shelf-life — hit the U.S. market in 1994. A similar product was launched in Europe in 1996, and tomato puree that was clearly labeled as "genetically altered" began to sell well.

GM animals

Genetic engineering is not just for plants: it can also be used to create GM animals. The overwhelming majority created so far are transgenic mice used for medical research, in which a particular gene has been knocked out so they act as better models of a human disease, such as cystic fibrosis or cancer.

A more recent application is nicknamed "pharming," and involves engineering animals so they produce useful products in their milk: goats, for example, have been modified to make spider-silk and blood-clotting agents. GM animals could also soon produce healthier meat: pigs have been modified so their flesh contains omega-3 fatty acids normally found in oily fish, though no GM meat is yet approved for human consumption.

More GM crops were soon rolled out by biotechnology companies like Monsanto: the first wave included Bt varieties of cotton and soya beans, and maize and oilseed rape resistant to herbicides. Both the industry and publicly funded scientists began to talk up the technology's potential for addressing food shortages and malnutrition in the developing world, with GM crops that can tolerate salty soil or drought, or that produce higher yields.

One of the most exciting prospects is golden rice, developed in 2000 by German scientist Ingo Potrykus. This is enhanced with a daffodil gene, which makes it produce the precursor of vitamin A. Up to 2 million people die each year, and 500,000 go blind, because their diets are deficient in this essential nutrient. As many live in countries where rice is the staple food, the technology offers a simple route to better health.

The backlash

GM crops have been embraced by farmers in many parts of the world. More than 100 million hectares are in cultivation, mainly in North and South America but increasingly in China, India and South Africa. More than half the world's soya is GM, and an estimated 75 percent of processed foods sold in the U.S. contain GM material.

The great exception, however, has been Europe. While the technology was accepted by American consumers with little fuss, it was launched in Europe at an inopportune moment. In the mid-1990s, it emerged that several dozen Britons had contracted a new form of

Food safety

Perhaps the biggest GM food safety scare blew up in 1998, when Arpad Pusztai, of the Rowett Research Institute, claimed to have found that potatoes modified with an insecticide called lectin were harmful to rats. The work received huge publicity, but Britain's Royal Society highlighted serious flaws in the research — such as the failure to use an appropriate control group. The finding is widely considered unreliable.

Another controversy surrounded the addition of a Brazil nut gene to GM soya, which inadvertently also led to the transfer of a nut allergen. The problem, though, was identified before the crop was marketed, and it was withdrawn. While this GM variety could have been harmful, the case demonstrates the thoroughness of safety testing, and says little about the technology as a whole.

Creutzfeld–Jakob disease, a fatal brain condition, from eating beef infected with mad cow disease or BSE — despite government assurances that there was no risk. This provoked a crisis of confidence in food safety, to which GM crops fell victim.

Though there has never been much evidence that genetic engineering raises any special safety issues, groups such as Greenpeace whipped up a wave of public hostility to these new "Frankenfoods." Scientists were accused of meddling with nature — and just as feeding animal carcasses to cows had caused BSE, it was felt this might have unpredictable consequences for human health.

Other concerns centered on environmental impacts. In theory, herbicide-tolerant varieties should be good for biodiversity, by reducing the need for chemical sprays, but many green activists worried that the reverse might happen in practice. If farmers knew they could douse weedkiller with impunity, what would stop them from using more of it?

Further fears emerged from a study that suggested the Bt toxin made by many GM crops could kill non-target insects, such as monarch butterflies. Organic farmers began to complain that GM pollen would contaminate their fields. Anti-GM activists started uprooting trial crops, and the public mood turned sour. Supermarkets cleared products from their shelves. While there has been no formal ban, no GM crops are grown in the UK at the time of writing, and just a single variety is licensed for commercial cultivation in the European Union.

Case by case

Some of these concerns are more legitimate than others. Food safety is probably a red herring — GM products have been eaten by millions of American consumers for over a decade without any adverse consequences, and the few studies that have suggested problems have been flawed. Environmental objections, however, may have more merit: the British field trials that were not ripped up found that herbicide-tolerant crops can hit biodiversity, depending on the spraying protocols that are used.

What this highlights is the folly of trying to consider GM crops as a whole. That a plant is genetically engineered says nothing about whether it can be safely eaten, or its likely impact on the environment. What matters is what the genes that are inserted do, and how the crop is then managed by the farmer. Some transgenic crops are likely to have ecological benefits when properly used, or to improve yields or produce more nutritious foods. Others may pose hazards, to the environment or to health. The technology has huge potential, but it is not a panacea. The only reasonable way to judge it is by looking case-by-case at the crops it is used to create.

23 Evo-devo

Seen through a microscope, the early embryos of all mammals look so similar as to be indistinguishable. Even trained eyes cannot tell whether a ball of a few cells will grow into a mouse, a cow or a human. All are formed in the same way, from the fusion of an egg and a sperm, each carrying half a genome's worth of chromosomes, and their development follows much the same pattern for the first weeks of life *in utero*.

From an evolutionary point of view, this is not so surprising. Humans and mice diverged only around 75 million years ago, and it makes sense that our early embryonic growth should proceed along similar pathways. People and fruit flies, however, are much more distantly related. We are vertebrates, they are not, and our last common ancestor — probably something known as a "roundish flatworm" — has been dead for well over 500 million years.

> **All complex animals — flies and flycatchers, butterflies and zebras and humans — share a common 'tool kit' of 'master genes' that govern the formation and patterning of their bodies.**
> Sean Carroll, California Institute of Technology

Yet the new science of evolutionary developmental biology, or "evo-devo" for short, has shown that on a genetic level, humans and flies look remarkably alike. Despite our manifold physiological differences, many of the genes that build our bodies are not just similar, but identical. The same passages of DNA determine the position of flies' compound eyes and humans' simple ones, and piece together the two species' body parts in the right order. They are universal software programs, which can run as happily on the hardware of *Drosophila melanogaster* as on *Homo sapiens*.

The developmental-genetic toolkit

Evo-devo combines genetics and embryology to determine the ancestral relationships between different organisms, and to establish how their DNA causes them to develop in their particular ways. It is the science of how genotype defines phenotype.

Its critical insights began to emerge in the early 1980s, when two German-based scientists, Janni Nüsslein-Volhard and Eric Wieschaus, used chemicals to generate random mutations in flies, then bred them and followed their offspring's development from embryos into adults. When a mutation had an interesting effect, such as giving an insect extra wings, or legs on its head, the scientists traced it back to the gene responsible. This allowed them to pinpoint the function of dozens of genes, and the points at which they tell the embryo to develop in certain ways.

A set of just 15 genes turned out to govern the layout of the early embryo. These include clusters that sit in a line on the same chromosome, called Hox genes (Hox is short for the homeobox, a 180-letter stretch of DNA that they all share). These set down the shape of the fly embryo, giving it a front and back, segments and sides, and they are placed on the chromosome in the same order in which they

Naming genes

While there is now a broad set of rules for naming genes, the scientists who made new genetic discoveries were long allowed to get away with calling them whatever they liked. Genetics is consequently rich in fanciful vocabulary. One of the first genes of the developmental toolkit to be identified is called hedgehog. It got the name because when this gene is missing or defective, fruit fly larvae grow hairs that make them look a little like the spiny animals. Mammals have a related gene that was named Sonic hedgehog, after the video game character, and fish have one called Tiggywinkle, after Beatrix Potter's prickly heroine.

The fruit fly has a mutation named Cleopatra, because it is lethal in combination with a gene called asp. Another mutation is Ken and Barbie: like the dolls, the flies that carry it have no external genitalia. Many key genes found by Nüsslein-Volhard and Wieschaus have kept their German names, such as *kruppel* (cripple), and *gurken* (cucumber). Sometimes, even creative minds get stuck. The gene called ring has nothing to do with shape or function: it stands for Really Interesting New Gene.

sculpt the body from head to abdomen. Hox genes tell the head to grow antennae, and the thorax to grow wings and legs. When they are mutated, they result in monsters, such as flies with legs where the antennae should be.

Though mice (and humans) have more Hox genes than flies, they do exactly the same job, directing the formation of body parts in the order in which they occur on chromosomes. They are the key elements of a "developmental-genetic toolkit" with which embryos are pulled into shape. These genes are so similar across species separated by hundreds of millions of years of evolution that it is even possible to transplant them from one animal to another, without losing function. Knock out a Hox gene in a fly, and replace it with the same gene from a mouse, and it is often impossible to tell that anything has happened. Human Hox genes do the trick just as well.

The Hox genes are just the most basic tools by which the body is made. Scores of others have been identified, which all perform a similar task in different species. The eyeless gene, for instance, is so called because flies that lack it grow no eyes. Knock it out, and replace it with its mouse equivalent, and the fly will grow perfectly normal eyes. This is particularly remarkable because insects have compound eyes, and mammals simple ones. The gene seems to say: "grow an eye of the type you would normally grow," with other genetic instructions specifying which type is appropriate to the species.

> "We didn't know it at the time, but we found out everything in life is so similar, that the same genes that work in flies are the ones that work in humans."
>
> Eric Wieschaus

Genetic switches

This understanding that the shape of very different species, with radically different body plans, is governed by a fairly small fundamental set of body-building genes, raises an obvious question. If we share these genes with flies and mice, why do we not have wings, antennae and segments, or whiskers and tails?

The answer seems to lie in "genetic switches" that turn genes on and off. Some of these are proteins known as "transcription factors":

these bind to sequences called promoters and enhancers, which surround genes and raise or lower their activity. Others are controlled by the 98 percent of the genome that is not involved in making proteins — the stretches of apparent nonsense often known as junk DNA. Much of this seems to play a critical role in telling genes when they should be active, and when they should keep quiet.

What the Hox genes, and the other tools in the kit, achieve is to set networks of these switches in motion in particular cells, according to their positions in the body. These networks, in turn, determine which genes go to work and which remain dormant. Every liver cell, pancreatic islet cell and dopamine neuron carries the same basic package of genetic software. Specialized programs from this software package, however, are activated in each cell type.

These changing patterns of gene expression also explain how the same genes can achieve such different results in different organisms. Species diversity is greatly influenced by the way the same genes are used in idiosyncratic ways.

This helps to solve the riddle of how so few human genes — about 21,500, as our genome sequence revealed — are sufficient to build such a sophisticated organism. The wonderful complexity of the human animal emerges only in part from genes that carry instructions for making proteins that are unique to our species. Evo-devo tells us that the intricate network of switches that conduct this genetic orchestra is just as important, if not more so.

24 **Stem cells**

In ancient Gaelic legend, Tir na Nog was the land of eternal youth, where sickness, aging and death did not exist. Ian Chalmers, of the University of Edinburgh, a Scot proud of his Celtic heritage, remembered the story in 2003, when he identified a gene with remarkable properties.

It is switched on only in the cells of early embryos, and it appears to be critical to their ability both to copy themselves indefinitely as if forever young, and to develop into any of the 220 or more cell types in the adult body. Chalmers named it Nanog, and it is one of the genetic keys to the unique properties of embryonic stem (ES) cells.

ES cells are the body's master cells, the raw material from which bones and brains, livers and lungs are grown. They occur only in young embryos, in which cells have yet to differentiate into the specialized tissues of the adult body. Because they are "pluripotent," with the ability to give rise to any of these tissues, they have great medical potential. They could generate replacements for cells that become diseased or damaged in conditions such as diabetes, Parkinson's and spinal paralysis, but they are also a source of controversy. As they must be harvested from embryos, some religious groups consider their use unethical.

> " Embryonic stem cells … are in effect, a human self-repair kit. "
>
> Christopher Reeve (d.2004), paralyzed actor and embryonic stem cell research advocate

The stem cell controversy

ES cells were first isolated from mice in 1981, by a Cambridge University team led by Martin Evans. Almost two decades later, in 1998, a group led by Jamie Thomson, of the University of Wisconsin, isolated human ES cells, raising hope that their versatility might be harnessed to treat diseases. If ES cells could be grown into dopamine neurons, which are lost in Parkinson's disease, they might be transplanted to treat it. For diabetes, they might be used to grow new beta cells that produce insulin.

Research into these cells has usually relied on embryos left over after in-vitro fertilization, though embryos are sometimes created specifically for this purpose. Such experiments have revealed how to grow these cells into self-perpetuating colonies or "lines," often using a bed of mouse cells to provide critical nutrients (though this technique is being phased out). Scientists are now investigating which genetic and chemical cues make ES cells pluripotent, and then tell them to develop into specialized tissue.

The work has been opposed by people who consider it wrong to destroy embryos for any reason, including research into life-saving therapies. Most, but not all, of these critics base their objections on religious beliefs, and also oppose abortion. Different countries have taken radically different approaches to the issue. The UK, China, Japan, India and Singapore are among the field's enthusiastic backers, both permitting ES cell research and supporting it with public money. Others, such as Germany and Italy, have banned or partially banned it.

The question has become most politically charged in the U.S., the world's strongest scientific superpower, but also a nation with an influential religious right. In 2001, President Bush announced that federal funds could be used only to study ES cell lines already in existence, a compromise that satisfied few. The embryo-rights lobby still sees all research as immoral. Scientists and patient groups regard the rules as unnecessarily restrictive, and point out that as existing lines were grown with mouse cells, they will be unsuitable for transplantation. Several states, notably California, set up their own funds for ES cell

Adult stem cells

Stem cells are not unique to embryos. Certain kinds of stem cells can also be found in the tissue of fetuses, children and adults, serving as a seed stock from which cells can be replenished and organs repaired. Bone marrow is especially rich in stem cells, as is blood from the umbilical cord.

As they do not require the destruction of embryos, research and therapy using adult stem cells are not controversial. They are already employed in treatments such as bone marrow transplants, and other applications have started clinical trials. Adult stem cells, however, are not as versatile as ES cells, as they have already taken steps towards differentiating into specialized tissue. Consequently, they may not be as useful for treating some conditions. Most scientists believe this promising branch of medical research should proceed in parallel with ES cell studies, not instead of them.

research, private companies continue to invest in it and, in March 2009, President Obama lifted the ban.

The road to therapy

No ES cells have yet been used to treat patients with any medical condition, though at the time of writing, Geron, a U.S. company, is poised to start clinical trials. The cells, however, have been differentiated into a wide range of tissue types in the laboratory, which have been used successfully to treat conditions like Parkinson's, muscular dystrophy and paralysis in animals. Genetic discoveries have also helped scientists to create a new kind of pluripotent stem cell by reprograming adult tissue, which could address some ethical objections to the technology.

As well as Nanog, several other genes that are expressed in a particular pattern in ES cells have been identified. These include genes called Oct-4, LIN28 and three gene "families" known as "Sox," "Myc" and "Klf." By genetically modifying adult tissue, so that these genes are switched on, it is now possible to turn back the clock on skin cells, so they acquire the pluripotency of embryonic cells. This was first achieved in mice in 2006, by a Japanese team led by Shinya Yamanaka, of the University of Kyoto, and in 2007, both Yamanaka and Thomson repeated the feat in humans. These induced pluripotent stem (IPS) cells have already been used to treat sickle-cell anemia in mice.

IPS cells could have several advantages over standard ES cells. They do not require human eggs or embryos, which are in short supply.

As they could be grown from the patient to be treated, they would be genetically identical and unlikely to be rejected by the immune system. They can also be made without destroying human embryos.

These advantages, however, do not make ES cell research obsolete. First, the techniques currently used to make IPS cells are too dangerous for therapy. The genetic modification is done with a virus that can promote cancer, as can one of the genes that is altered, called c-Myc. These cells also resolve ethical objections only in part. As Yamanaka and Thomson point out, they would not exist had scientists not been allowed to examine the genetics of ES cells in the first place.

The study of IPS cells is still in its infancy, and it is not yet known whether they behave in exactly the same way as ES cells. Stem cell scientists consider it critical to examine how both kinds of cells work, side by side. One kind may be better for some indications, and the other for others. It is too early to tell.

25 Cloning

The most celebrated sheep of all time, Dolly, was born in a Scottish laboratory on July 5, 1996. The ewe, created by Keith Campbell and Ian Wilmut, of the Roslin Institute in Edinburgh, was the first mammal to be cloned from an adult cell — a genetic copy of a living animal. As the cloned DNA was taken from a mammary gland, they named her after a famously ample-chested country and western singer: Dolly Parton.

Frogs and fish had been cloned decades before Dolly's birth, and in the 1980s Russian scientists had cloned a mouse called Masha, by moving the nucleus from an embryonic stem (ES) cell into an empty egg. Every effort to make a mammalian embryo bearing an adult's DNA, however, had failed. In mammals, certain genes that are essential to embryonic development are always switched off in adult somatic cells, by a process called methylation, and this seemed to make cloning impossible.

> " The potential of cloning to alleviate suffering . . . is so great in the medium term that I believe it would be immoral not to clone human embryos for this purpose. "
>
> Ian Wilmut

Campbell and Wilmut, however, proved otherwise. They took the nucleus from a somatic (adult) cell of a sheep and placed it into an egg from which the nucleus had been removed, then coaxed it into dividing with electrical stimulation. Somehow, and the precise reasons remain unclear, this method can reprogram the nucleus and undo methylation, allowing a cloned embryo to grow. Dolly shared all her nuclear DNA with her somatic cell donor. Only the DNA in her mitochondria came from the ewe that supplied the egg.

The technique, known as somatic cell nuclear transfer (SCNT), was not efficient — it took the Roslin scientists 277 attempts to make Dolly — but by showing that it can work, they opened exciting possibilities. Prize livestock might be cloned in agricultural breeding programmes. And if SCNT worked for human cells, there could be medical applications.

Therapeutic cloning

ES cells can grow into any tissue in the body, and could thus provide replacements for diseased or damaged cells. SCNT suggested that "therapeutic cloning" might further enhance their medical utility. If stem cells were grown from an embryo cloned from the patient in need of treatment, these would share that patient's genetic code. The cells could be transplanted without fear of rejection by the immune system.

The technique could also create models of disease. DNA from patients with conditions such as motor neuron disease could be used to make cloned ES cells, carrying genetic defects that influence the disorder. Such cells would be valuable for studying the condition and testing new drugs.

Such developments, however, first required human embryos to be cloned by SCNT — a task that faced two major hurdles, one ethical and one technical. Even some people who approve of ES cell research object to therapeutic cloning as it could advance efforts to clone a human baby. And more critically, though SCNT was soon used to clone mice, pigs, cattle and cats, it is much more difficult to accomplish in primates.

The Hwang affair

Countries that allow ES cell research have generally decided that the medical potential of therapeutic cloning outweighs the risks, and have permitted this application of SCNT while banning it for reproductive purposes. However, in February 2004, scientists from one of these nations, South Korea, claimed to have mastered the technical challenges, involved.

Cloning people

Though a UFO cult called the Raelians announced in 2002 that it had cloned a human being, it has never produced any evidence and its claim is ridiculed by scientists. Human reproductive cloning might one day become possible, however, even if it is for now considered too dangerous to be attempted.

What might a human clone be like? It's a misconception that it would be a carbon copy of its "clonee": while it would share its genome donor's DNA, it would not share any of its experiences. In fact, a clone would be less like its donor than identical twins are to one another: twins, at least, share a womb and a childhood environment.

In a paper published in the journal *Science*, a team led by Woo-Suk Hwang reported the creation of the world's first cloned human embryo and the extraction of ES cells. The following May, Hwang announced a still greater feat — the production of 11 cloned ES cell lines, each genetically matched to a different patient. Just as importantly, the group said it had refined SCNT so that fewer than 20 eggs were needed to generate a colony of cloned cells. With such a success rate, the technique could become medically viable.

It was all too good to be true. In November 2005, it emerged that Hwang had obtained eggs for his research unethically, and as his work fell under greater scientific scrutiny, it unraveled. His cloned stem cells had been faked: genetic tests proved that they had not been cloned at all. Of his purported achievements, only the creation of Snuppy, the first cloned dog, withstood independent genetic analysis. He had perpetrated a great scientific fraud.

Human embryos can be cloned, even if Hwang may not have managed it. In 2005, a team at Newcastle University succeeded, as have two American companies. None of these groups has yet made cloned stem cells, but such cells have been extracted from cloned monkey embryos. This ultimate goal is within reach.

Therapeutic cloning, however, has lost some of its luster since the Hwang affair. Human eggs, which are essential, are always likely to be in short supply as they cannot be donated without risk. This means that even if cells can be cloned from patients, they are likely to be prohibitively expensive. Scientists are thus looking elsewhere for alternatives. For therapy, many think it will be more practical to use induced pluripotent stem cells, made by reprograming adult tissue, or banks of ordinary ES cells from which suitable matches can be made to patients.

THERAPEUTIC CLONING

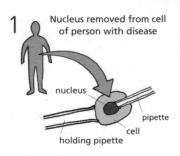

1 Nucleus removed from cell of person with disease

nucleus

pipette

holding pipette

cell

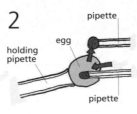

2

pipette

egg

holding pipette

pipette

Old nucleus removed from egg

New adult cell nucleus inserted

3

Resulting cloned embryo stimulated with electricity to start cell division

Jurassic Park

In the 1993 movie _Jurassic Park_, dinosaurs are brought back from extinction by cloning, using DNA from mosquitoes preserved in amber, that had fed on their blood. While this makes great science fiction, most scientists think it impossible in practice. DNA from creatures that lived tens of millions of years ago will almost certainly be too degraded to use for cloning. On top of that, these animals have no living relatives sufficiently close to provide eggs for injection with dinosaur DNA.

Cloning, however, might be used to revive creatures that have become extinct more recently. In Australia, a project is underway to clone the Tasmanian tiger, using DNA from the last known animal, which died in 1936. It might also be possible to re-create the mammoth: an exceptional specimen recovered from the Siberian permafrost is thought to contain DNA of sufficient quality to attempt cloning. Its modern relative, the elephant, could plausibly act as egg donor and surrogate mother.

In research, another approach is to use SCNT to put human nuclei into empty animal eggs, to make "cytoplasmic hybrids" carrying genetic material that is 99.9 percent human. While these would be unsuitable for therapy, they could generate cell models of diseases, which British and Chinese scientists are already trying to create. The SCNT technique that created Dolly may never be used to create cloned cells for transplanting into patients, but it could still be a medical tool of great value.

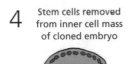

4 Stem cells removed from inner cell mass of cloned embryo

inner cell mass

5

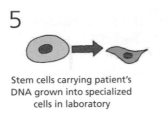

Stem cells carrying patient's DNA grown into specialized cells in laboratory

6

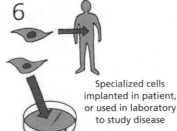

Specialized cells implanted in patient, or used in laboratory to study disease

26 Gene therapy

The development of gene therapy, in which genetically modified cells are administered to patients to counter inherited genetic deficiencies, has been widely recognized in recent years as a major advance in medical technology.

Ashanti DeSilva is an American college student in her early 20s. Yet when she was born in 1986 she was not expected to attend high school, let alone university. Ashanti suffered from a rare recessive disorder called severe combined immune deficiency (SCID). It meant she lacked a functioning immune system, leaving her dangerously exposed to each and every passing germ.

Children with SCID live forever on the edge of disaster. As they cannot fight off pathogens, even mild infections can be life-threatening. Many die in infancy, and those who survive are often shielded from the outside world in a sterile pouch — the condition is often known as "bubble baby" syndrome. They cannot attend school or mix with other children, and without a bone marrow transplant from a matching donor, few live to adulthood.

> **For patients without bone marrow donors, gene therapy now provides a remarkable, potentially curative, way forward.**
>
> Len Seymour,
> **British Society for Gene Therapy**

No suitable donor could be found for Ashanti, but in 1990, researchers at the National Institutes of Health came up with an alternative. A team led by French Anderson removed some of her ineffective white blood cells from her body, and infected them with a virus modified to carry a healthy copy of her faulty gene. When these treated cells were infused into her bloodstream, Ashanti's immune function improved by 40 percent. She became well enough to go to school, and even to receive vaccines, which cannot normally be given

Germline gene therapy

All the gene therapies attempted to date work on the somatic cells that make up the vast majority of the body's tissues and organs. They aim to correct genetic defects in an individual patient, but as they do not alter germ cells that produce eggs and sperm, those mutations can still be passed on to children.

Future technologies may go further, to create "germline gene therapies" that change genes of both patients and their offspring. This is more controversial, because people who are yet to be born have no say over genetic manipulations that might have unforeseen consequences. Proponents of germline gene therapy, however, do not see what the fuss is about — at least where disorders such as SCID or cystic fibrosis are involved. If it is possible to root a deleterious gene out of a family for good, they ask, why would it be wrong to do so?

to immune-compromised patients. She was the first patient to be successfully treated with gene therapy.

Our friend the virus

Gene therapy did not cure Ashanti: the genetically modified cells worked only for a few months at a time, and she has had to return for regular treatment. For that reason, the technique was at first used only when a bone marrow transplant was not an option. In 2000, a team at Great Ormond Street Hospital in London, and the Necker Hospital in Paris, advanced the procedure further, to correct the SCID mutation in the bone marrow of children, which should offer an indefinite cure. Its early success raised hope that the strategy might be effective against this and other inherited diseases.

The therapy works by harnessing the aggressive properties of one of humanity's microscopic enemies. When viruses infect us, they reproduce by introducing their genetic material into our cells, hijacking their replication machinery and forcing it to churn out more viruses. One class, the retroviruses, actually write themselves into our genome with specialized enzymes.

GENE THERAPY

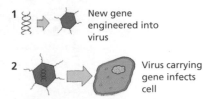

1 New gene engineered into virus

2 Virus carrying gene infects cell

3 Viral DNA, including new gene, inserted into cell's genome

4 New gene begins making protein, to treat disease

Medicine can exploit this viral talent and turn these pathogens into vectors for transporting fresh DNA into cells. Virulence genes are excised to make the virus harmless, then a normal copy of the defective human gene that needs to be replaced is spliced into its genetic code. When a patient's cells are infected with this modified virus, they take up the new gene, and should start making the normal protein. The principle is the same as downloading a "patch" to debug misfiring computer software.

With some viral vectors, such as adenoviruses that commonly cause tonsillitis, the new gene will be active only in those cells that were infected: when these die, their successors will not express the added trait. That is why Ashanti DeSilva needed repeated treatments. If a retrovirus is used, however, the new gene will be incorporated into the genome of infected cells, and passed on to its descendants. The genetic defect should be corrected for good.

Unintended consequences

Viral vectors are critical to existing methods of gene therapy, but they are also its greatest weakness. They can affect the human body in unpredictable ways, causing side-effects that have greatly limited the technique. The Anglo-French SCID trial, which used a retrovirus, might have corrected the disorder, but this success has come at considerable cost. Five of the 25 children treated so far have developed leukemia.

When a retrovirus incorporates itself into the genome of host cells, doctors cannot control the position at which it burrows its way in. Sometimes, it interrupts an oncogene, triggering uncontrolled cell

Gene doping

It is already hard enough to detect athletes who use performance-enhancing drugs such as human growth hormone. Gene therapy could make it even more difficult. Scientists have already used the technology to alter the genes of mice and monkeys, so that they produce increased amounts of proteins that boost strength or endurance, such as erythropoietin (EPO). Such "gene doping" by athletes could be virtually impossible to prove. Those found to have excess EPO in their systems could claim, correctly, that their genes were responsible. Sophisticated genetic tests, not yet available, would be needed to show that genetic enhancement had taken place.

division and cancer. As 80 percent of children with leukemia recover, while untreated SCID is invariably fatal, it can be argued that this risk is worth running. Only one of the affected children has died, while three are in remission, and one had been recently diagnosed at the time of writing. But it is by no means ideal for a therapeutic technique that has been claimed as a standard-bearer for genetic medicine.

Leukemia is not the only unwelcome consequence that viral vectors can have. In 1999, Jesse Gelsinger, an 18-year-old with a genetic liver disorder, took part in a University of Pennsylvania trial of a gene therapy designed to treat his condition. He suffered a massive immune reaction to the adenovirus vector, which killed him. As well as being a personal tragedy, it was a severe setback to the field.

Adenoviruses and retroviruses are now being replaced in gene therapy trials by a different vector, the adeno-associated viruses. Unlike retroviruses, these always incorporate themselves in the genome in the same, safe place, and unlike adenoviruses, they do not normally cause human diseases, making immune overreactions unlikely. A trial based on this approach has improved the sight of four patients with Leber's congenital amaurosis, a single-gene cause of blindness. Non-viral vectors, such as synthetic zinc-finger proteins, are another promising option.

But while these kinds of gene therapy are likely to prove safer, and perhaps more effective, scientists are not as excited about this technology as they once were. Although it is promising for a few single-gene disorders, it has failed in many more. It is one thing to modify enclosed tissues, such as bone marrow and retinal cells, but quite another to correct genetic defects that have more systemic effects, such as the cystic fibrosis mutation.

Most diseases, moreover, are not caused by single genes, but are influenced by multiple genetic variants that each slightly raise risk. Diabetes can be affected by two dozen genes, and it will be impractical to alter them all. While gene therapy will have a place in the clinic, it is no panacea for inherited disease.

27 Genetic testing

In the English city of Cambridge, a cycle path is decorated with more than 10,000 lines, each in one of four colors. The pattern follows the sequence of a gene on chromosome 13, which was identified in 1995. The gene is BRCA2, which takes its name from the disease that often results when it is defective: breast cancer.

One in nine women in developed countries will get breast cancer during their lifetimes. Among women who have mutations in the BRCA2 gene, however, up to four-fifths will develop the disease, and a similar risk applies to defects in another gene called BRCA1. Both are tumor suppressors, which normally stop cells from becoming cancerous. Women who are unlucky enough to inherit mutations lack a critical line of defence, which leaves them peculiarly vulnerable to cancers of the breasts and ovaries.

Thousands of women belong to families with a long history of breast cancer, who have often lost mothers, grandmothers, sisters and aunts. The isolation of the BRCA genes has meant that some of these people have been able to discover whether the family risk applies to them. If a BRCA mutation is known in a relative with breast cancer, women can be tested to determine whether they have inherited it too. For those who test negative, the procedure can offer peace of mind, while those with a positive result can take action to reduce

Tailor-made medicines

One of the most important benefits of genetic testing could be in choosing which drugs to give to particular patients. Genetic variation can affect the way medicines are metabolized in the body, making a drug effective in one person, worthless in another, and positively dangerous in a third. If these relationships can be teased out, it should be possible to select the right drug for individual patients using clues from their DNA. This is already happening in cancer: the breast cancer drug Herceptin is given only to patients whose tumors have a particular genetic profile that means it is likely to work.

their risk. Most have regular mammograms to ensure any incipient tumor is detected early, and some even opt for preventive mastectomies.

Testing dilemmas

BRCA1 and BRCA2 are just two of the genetic diseases for which it is now possible to test. Newborn babies, for instance, have their heels pricked with a sterile needle a week after birth, to collect blood that is then screened for inherited disorders such as phenylketonuria (PKU). In the UK, about 250 infants test positive each year, and can thus be protected against the neurological damage that PKU would otherwise inflict.

> **If, as a competent adult, you choose to look at your risk of developing Alzheimer's, that is your prerogative. But no one will force you to look at your Alzheimer's risk if you do not want to.**
> Kari Stefansson, of deCODEme

Other reliable tests are available for hundreds of disorders caused by defective single genes. Often, as with PKU or hemophilia, their results ensure that patients receive appropriate treatment. Even for incurable diseases, such as cystic fibrosis and Duchenne muscular dystrophy, genetic diagnosis can help doctors to manage symptoms, and parents to prepare themselves for the future.

Some genetic tests, however, are more problematic. Huntington's disease is a prime example. As it is caused by a dominant mutation, anybody with an affected parent has a 50 percent chance of having inherited it. Yet though there is a reliable test, many of those at risk refuse to have it — including Nancy Wexler, the scientist whose research led to the test's development (see Chapter 15). Huntington's is a late-onset condition that causes progressive cognitive decline, it is invariably fatal, and there is no cure. As a positive test is tantamount to a death sentence, many people would rather not know.

Another genetic dilemma is posed by amniocentesis, which can be used to test the developing fetus for abnormalities such as Down's syndrome. If the result is positive, nothing can be done. A couple must decide whether to progress with a pregnancy that will produce a disabled child, or to have an abortion.

Consumer tests

All the genetic tests described so far belong to clinical medicine: they are available only through doctors, and after appropriate counseling. They look for rare but major mutations that always result in disease, or which dramatically raise risk. Most genetic influences on health, however, do not work that way: they involve common variations that slightly raise or lower people's chances of developing diabetes or cardiovascular disease. Tests for these variants pose new challenges, not least because they are increasingly being marketed directly to consumers.

Two companies that offer such services, deCODEme and 23andMe, were launched in 2007. For $1,000, they will take DNA from a mouth swab and scan a million single nucleotide polymorphisms (SNPs) — points at which the genetic code varies between individuals. The results are used to assess customers' risk of more than 20 diseases, as well as other aspects of inherited physiology such as male pattern baldness.

In theory, such information should be of great value to people's health, allowing them to change diet or lifestyle to counter inherited risk, or to ensure they get regular screening. But these tests can create problems, too. The variants that are examined are not like the BRCA genes — they have only a small impact on disease risk, and environmental factors matter too. Only a handful of SNPs that influence these diseases are yet known, so the results are necessarily going to be incomplete.

66 This test could cause unnecessary worry about potential health risks, or give others a false sense of security. 99

Joanna Owens,
Cancer Research UK

This means that personal genotyping can very easily mislead. There is a danger of providing false reassurance that promotes a cavalier attitude to health: people with SNPs that suggest a lower risk of lung cancer might become less likely to quit smoking. Equally, apparently frightening results may provoke needless anxiety, especially when people are tested by online services that offer no counseling or medical

Personal genomics

When the human genome was first sequenced, the published results were averages, composed of data from several people. The falling cost of the technology has now allowed two individuals — Craig Venter and James Watson — to have their personal genomes decoded. Venter's genome, published in 2007, cost $10m, and Watson's, published a year later, cost just $1m. The price tag is coming down still further — in 2008, a company called Applied Biosystems mapped the genome of an anonymous Nigerian for $60,000.

The X Prize Foundation, which has already run a contest to launch the first private space flight, has now endowed a genomics award to stimulate further technological developments. The $10m prize will go to the first team to sequence 100 anonymous human genomes in 10 days, at a cost of no more than $10,000 apiece.

advice. If you have an allele such as ApoE e4, which raises the risk of Alzheimer's sixfold, would you really want to find out through a website? When James Watson's personal genome was sequenced (see box), he asked not to be told his results for this gene.

Personal genetic testing, though, is only going to become more common as the costs come down. It took $4 billion to sequence the entire human genome for the first time, but personal testing can now be done for as little as $100,000. Most scientists think the cost will fall to $1,000 or less within the next five years. That will open up exciting medical possibilities, but many of the clues revealed are going to be cryptic and devilishly difficult to interpret.

28 Designer babies

The ability to analyze the genetic quality of embryos has led in recent years to the possibility of guaranteeing the birth of children mercifully free of what would otherwise be likely, or even inevitable, genetic defects.

Debbie Edwards once thought she would never have children. Her nephew had inherited a genetic condition called adrenoleukodystrophy, and a test had revealed that she carried the mutation responsible on one of her X chromosomes. As she was a woman, with a second X chromosome bearing a working copy of the gene, Mrs. Edwards was perfectly healthy. Any son she might conceive, though, would have a 50 percent chance of developing progressive brain damage and dying young. She took the painful decision not to try to start a family.

Yet on July 15, 1990, Mrs. Edwards gave birth to twin girls, Natalie and Danielle, at Hammersmith Hospital in London. She had not had a change of heart about the danger of adrenoleukodystrophy: science had found a way to avert it. The development of an embryo-screening technique had allowed her to conceive, confident in the knowledge that she would not have sick children.

The Hammersmith team, led by Alan Handyside and Robert Winston, created embryos by in-vitro fertilization (IVF), and grew them in the laboratory until the eight-cell stage. One cell was then removed from each embryo, so the scientists could examine the sex

PRE-IMPLANTATION GENETIC DIAGNOSIS

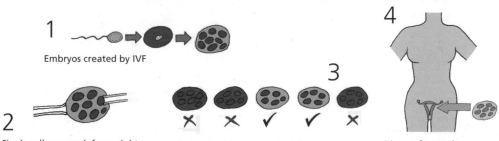

1 Embryos created by IVF

2 Single cell removed from eight-cell embryo for genetic analysis

3 Embryos without genetic defect selected for transfer to womb

4 Disease-free embryo transferred to womb

chromosomes to determine which were male and which were female. As adrenoleukodystrophy is X-linked, and affects only boys, only female embryos were implanted into Mrs. Edwards' womb.

The procedure is known as pre-implantation genetic diagnosis, or PGD, and Natalie and Danielle were the first examples of what the media called, a little misleadingly, "designer babies."

The PGD revolution

PGD babies have not, in truth, been designed. Their DNA has not been altered, but the term has stuck because the technology allows parents to do something that has never before been possible. They can choose between potential children on the basis of genetic qualities that they do or don't have — much as a shopper might choose a designer dress.

The advance gave couples who knew they might pass on a devastating inherited disease the opportunity to have healthy offspring. At first, PGD was sensitive enough only to prevent X-linked disorders such as hemophilia or Duchenne muscular dystrophy, by testing for sex, but it soon became possible to screen for autosomal conditions such as cystic fibrosis and Huntington's as well. More than 200 diseases can now be detected, and several thousand PGD babies have been born worldwide.

The technique, however, has stirred ethical controversy. Those who object to the destruction of embryos think PGD immoral because those embryos that carry genetic mutations are thrown away or donated for medical research. Particularly contentious is its application to genes

Pre-implantation genetic screening

The embryo biopsy technique also has applications in fertility treatment, to check the genetic quality of embryos with a view to improving the chances of a successful pregnancy. Most embryos with too many or too few chromosomes will miscarry, and the test can be used to count them so that only normal ones are selected for transfer.

In the UK, eight clinics are licensed to perform this pre-implantation genetic screening for women with a history of miscarriage or failed IVF, but there is controversy about whether it works. A 2007 Dutch study suggested it might actually reduce IVF success rates, probably because the biopsy can damage the embryo. The technique's advocates, however, point to methodological problems with that research, and argue that when properly performed, it has clear benefits for some women.

Artificial chromosomes

PGD is only capable of choosing between embryos with genetic profiles inherited from their parents, but true "designer babies" might one day become possible through advanced genetic engineering. If this does happen, one route may be to use synthetic chromosomes, deliberately engineered to carry beneficial genes and inserted into the cells of early-stage embryos.

This approach probably remains decades away from plausibility, but it would have two advantages. It would not interrupt the genetic sequence of existing chromosomes, reducing the risk of introducing an error that might cause a disease such as cancer. And biophysicist Gregory Stock, a proponent of genetic enhancements, has suggested it might be possible to activate artificial chromosomes at a later date. Children who receive them could thus choose whether to switch on their genetic modifications on reaching adulthood.

such as BRCA1. Mutations in this gene substantially raise the risk of breast cancer, but do not invariably cause it, and those who inherit them can protect themselves, albeit with mutilating prophylactic surgery. Critics see embryo screening as a form of eugenics, of rooting out the disease by eliminating the carrier.

PGD has also offered a lifeline to existing children who suffer from diseases such as leukemias and anemias, and need a cell transplant from a matching donor. If no such match exists, parents can try for another baby and screen their embryos to select those suitable to donate. In 2002, a young American girl named Molly Nash, who suffers from Fanconi anemia, became the first to be successfully treated with tissue from such a "savior sibling." She received stem cells from the umbilical cord of her newborn brother, Adam, whose tissue type had been screened as an embryo.

This extension of PGD has raised another concern. The biopsy process carries a very small risk to the embryo, and some people think it wrong to incur this when that embryo will not directly benefit. Britain's embryology watchdog, the Human Fertilization and Embryology Authority, originally ruled that savior siblings could be tissue-typed only if they were also being tested for a disease. It later reversed its decision, with the emergence of new evidence that the technique appears safe.

A slippery slope?

A further objection to PGD rests on a different argument: that while it is easy to sympathize with families who want to avoid having a baby with a serious disease, or to find a donor for a much-loved sick child, this sets an alarming precedent. Allowing even these applications, it is said, puts society on a slippery slope that will lead to embryos being screened for genes that affect intelligence, height or good looks. Children will come to be seen as commodities — at least, by those who can afford the technology.

> **We are all on the slippery slope. The question we should be asking is: skis or crampons?**
>
> John Harris, Professor of bioethics, University of Manchester

It is perfectly possible, however, for societies to decide that PGD is permissible for some purposes, but not for others. The UK, indeed, has banned its use to choose a baby's sex for social reasons, or for deliberate selection of disabled children, but not for preventing disease.

The technology's dystopian potential is also firmly limited by science. First, PGD always requires IVF, which holds little appeal for the naturally fertile. Then there is the issue of what to look for. The sort of traits that pushy parents might want — such as intellectual or athletic ability — are governed by dozens of genes that interact in complex ways, and by environmental factors as well. It is impossibly difficult to screen for them all, or to guarantee the desired outcome.

There is also the matter of the raw material. Embryologists using PGD can work only with what nature provides, which means genes already carried by each parent. It is all very well ordering a designer baby with the brains of Stephen Hawking and the body of Kate Moss. If Mum and Dad don't have those qualities, it isn't going to happen.

PGD is a great tool for preventing genetic diseases that cascade down the generations, blighting whole families with misery and suffering. But it is entirely unsuitable for the mass production of babies-to-order.

29 Junk DNA

The human genome contains three billion base pairs, the DNA letters in which the code of life is written. Yet only a tiny proportion of these letters — no more than 2 percent — are actually used to write our 21,500 or so genes. The remainder, which makes none of the proteins that drive the chemical reactions of life, has long been something of a mystery. Its apparent lack of function has led it to be dubbed "junk DNA."

The presence of large tracts of DNA with no purpose, however, would present an evolutionary puzzle. It takes energy to copy DNA, and if the vast quantities of junk found in all organisms were truly useless, it ought not to have survived the attentions of natural selection. Individuals that successfully eliminated inert genetic material should have an advantage over those that do not, producing thrifty genomes of more manageable size. The fact that they have not done so suggests that junk DNA is somehow important.

> If you think of the letters that make up the human genome as the alphabet, then you can think of genes as the verbs. We're identifying all of the other grammatical elements and the syntax of the language we need to read the genetic code completely.
> Manolis Dermitzakis, ENCODE consortium

A further clue to its significance was revealed when the Human Genome Project found many fewer protein-coding genes than the 100,000 that had once been predicted. This tally seemed far too low to explain all the differences between humans and other organisms, indicating that the genome must amount to more than the sum of its genes. Beyond the genes, what remained was the junk, which geneticists have now started to observe with fresh eyes.

What's in the junk?

Much of our junk DNA has origins that have been relatively simple to establish. A very large part of it belonged originally to viruses, which have incorporated their own genetic codes into our genome in order to reproduce. These human endogenous retroviruses are currently thought to make up about 8 percent of the total: they account for more of the book of humanity than human genes.

The legacy of our viral ancestors can also be seen in so-called retrotransposons. These repetitive chunks of DNA, which were originally deposited by viruses, have the ability to copy themselves into the human genome again and again, using an enzyme called reverse transcriptase. The commonest class are known as "LINEs" (for long interspersed nuclear elements) and on current estimates, they account for around 21 percent of all human DNA. Shorter retrotransposons, the commonest of which is the Alu family, make up even more of the genome, and still smaller elements include the short tandem repeats, which are used in DNA fingerprinting.

Other types of non-coding DNA include the introns that separate the protein-coding sections of genes, and the centromeres and telomeres that occur at the middle and ends of chromosomes, respectively. There are also pseudogenes — the rusting wrecks of genes that were important in our ancestors, but which have decayed through mutation. Hundreds of these genetic fossils can be found in the human genome (see box).

What does it do?

In some ways, the continued presence of this junk DNA is not surprising: DNA is "selfish," and will replicate itself regardless of utility to its host organism. But for it to withstand natural selection, some of it must surely be functional. Further evidence for a biological role comes from more than 500 regions of junk DNA that are highly conserved from species to species. These have probably been preserved because they perform a vital task, to which mutations are catastrophic.

One hypothesis for junk DNA's role is that it protects genes. If the genome contained nothing but protein-coding elements, many of these would be broken up and rendered meaningless by recombination errors. Non-coding DNA could provide a buffer, reducing the probability that a critical gene will be damaged. Another idea, also founded on recombination, is that it provides a reservoir from which new genes can evolve. When chromosomes cross over, some bits of junk might fall together in useful new combinations. This would make the junk

Fossil genes

Some of our junk DNA is made up of "pseudogenes" — sequences that were once working genes, but which have lost their ability to make proteins though lack of use. They are genetic fossils, which tell the story of evolution as faithfully as fossilized bones.

When important genes acquire mutations, these are usually weeded out by natural selection because they put individuals that carry them at a disadvantage. But when a gene codes for a protein that a species no longer needs, that disadvantage no longer applies. Animals that live underground, such as moles, will not suffer if a mutation knocks out an eyesight gene. As mutation occurs randomly, but at a consistent rate, such luxury genes will inevitably decay over time. Defunct versions of these genes, though, will continue to be preserved in their genome.

A good human example is the Vr1 family of genes, which are involved in detecting scent. Mice have more than 160 functional Vr1 genes, while people have just five. The dead Vr1 genes have not vanished from the human genome — they have fossilized, providing proof of our shared evolutionary heritage with mice.

analogy rather apposite — junk, after all, is not thrown away, but stored in case we can put it to good use in the future.

It is now clear, however, that much of our junk DNA is misnamed — it is not surplus to requirements at all, but carries out specialized and significant tasks. Large parts of it are thought to be involved in regulating gene activity, in generating messages that tell coding parts of the genome when and how to operate, and when to keep quiet.

The most telling evidence for this hidden biological function has emerged from the ENCODE (Encyclopedia of DNA Elements) consortium. This international effort to study the workings of the whole genome, not just the genes, is effectively compiling a "parts list" of the DNA that is biologically active in the body. Its pilot phase, which reported in 2007, has looked in detail at 30 million base pairs — 1 percent of the total.

What it found was remarkable. While only about 2 percent of the genome is made up of genes, at least 9 percent of it is transcribed into RNA — a sign that much of it is biologically active. Only a small proportion of this transcribed RNA is the messenger RNA that carries protein-making instructions. Junk DNA generates different kinds of RNA, the range of which will be explored in Chapter 32. These molecules, in turn, modify the expression of genes and proteins, to fine-tune human metabolism.

This fine-tuning has a profound effect on physiology. Single-letter DNA changes that influence the risk of disease have been found in non-coding parts of the genome, as well as genes. A rare mutation in a gene called MC4R, for instance, causes childhood obesity, but people with normal versions are also more susceptible to weight gain if they inherit a common variant in the surrounding junk. The variant seems to lie in a region that regulates MC4R, altering its activity.

> It's no longer the neat and tidy genome we thought we had. It would now take a very brave person to call non-coding DNA junk.
>
> John Greally, Albert Einstein College of Medicine

Variation in non-coding DNA could also explain differences between species. Some 99 percent of human and chimpanzee genes are identical, compared with only 96 percent of all DNA. As there is so much more diversity in the junk, this may well underlie uniquely human traits such as intelligence and language. The notion that protein-coding genes are the only contents of the genome that matter is manifestly mistaken.

30 Copy number variation

It has become a commonplace that humans are 99.9 percent identical at a genetic level. The mapping of the human genome revealed that while the human genome contains 3 billion DNA base pairs, only around 3 million of these, or 0.1 percent, typically vary in spelling. These one-letter changes are the single nucleotide polymorphisms or SNPs. A little genetic variation seems to go a very long way.

This estimate of genetic difference, however, has since been revealed to be wrong. SNPs, it turns out, are not the only way in which genomes vary. Whole genes and fragments of genes can also be duplicated, deleted, reversed and inserted into the genome. This new type of variation was found in 2006 to be extremely common, and it has as great a bearing on biology and health as the conventional kind.

> Each one of us has a unique pattern of gains and losses of complete sections of DNA. We now appreciate the immense contribution of this phenomenon to genetic differences between individuals.
> **Matthew Hurles**

This copy number variation, also known as "structural variation," suggests that the average genetic difference between any two people is not 0.1 percent, the figure derived from SNPs. It is actually at least three times greater, at 0.3 percent or more. This may partially explain why so few SNPs can produce such extensive human diversity: our knowledge of the genome's variability was incomplete. It is an insight

Investigating copy number variation

The first wave of genome-wide association studies, the powerful new tools for finding genes that affect disease that were introduced in Chapter 15, looked only at SNPs. The growing understanding that copy number variation is at least as important is changing the way this research is being designed. In April 2008, the Wellcome Trust announced a £30 million grant to fund a second phase of its successful Case Control Consortium, which will investigate two dozen new diseases. This time, it will cast the net beyond SNPs, using gene chips that can detect copy number variants as well.

that has forced a reappraisal of how DNA makes every one of us — and our species — unique.

Duplications and deletions

The standard model of genetics is that everybody inherits two copies of the genetic sequence, one from each parent. Yet a research team led by Matthew Hurles, of the Wellcome Trust Sanger Institute, and Charles Lee, of Harvard Medical School, has established that this is too simple. When they started to look in detail at the genomes of 270 people, originally recruited for the HapMap project we met in Chapter 15, they found that the double-copy paradigm was not universal at all.

Across about 12 percent of the genome, large portions of DNA, ranging in size from 10,000 base pairs to 5 million, were sometimes repeated or entirely absent. While most people have just two copies of these sequences, some have one or none, while others have several — as many as five to ten copies in some cases. Strings of DNA can also be inserted out of place, or inverted so they read back to front. The genome varies considerably in structure, as well as in spelling.

It has long been known that some stretches of human DNA are occasionally duplicated or deleted, as with the extra copy of chromosome 21 that causes Down's syndrome. Such changes, however, had been thought to be rare, and serious in their consequences. It is now known that variations of this sort are common — more common, in fact, than conventional SNPs.

Sometimes, this structural variation is trivial — as with SNPs, certain changes make no difference to genetic function. But it can be strongly linked to altered physiology or susceptibility to disease, and

New genetic territory

The Human Genome Project did not complete a comprehensive map of our genetic code. It produced an average sequence that provides a reference point against which scientists can compare the DNA of individuals and other species. Studies of copy number variation are now revealing whole segments of DNA that do not appear in this reference genome, but which are nevertheless reasonably common. A 2008 study that looked in detail at eight people's genomes found no fewer than 525 new sequences that are sometimes inserted into the code. Many more probably remain undiscovered.

may also account for differences between species. Once copy number variation is brought into play, we share just 96 to 97 percent of our DNA with chimps, not the 99 percent estimated from spelling alone.

Copy number and disease

The most exciting implications of copy number variation lie in its consequences for disease. Now that scientists have become aware that it is something worth looking at, all manner of associations are starting to emerge between individual health and DNA deletions, duplications, insertions and inversions.

A gene called CCL3L1, of which some Africans have multiple copies, is one of the most interesting early manifestations of this phenomenon. Those with a high copy number, it appears, are less susceptible to infection with HIV. While it is not yet known precisely how or why this happens, a working hypothesis is that extra copies boost production of a protein that is important to HIV resistance. This promises to open new approaches to treating the virus, and to preventing its spread.

Other copy number variations that have been reliably linked to disease include genes known as FCGR3B, in which a low number predisposes to the auto-immune disorder lupus, and EGFR, which is often repeated in patients with non-small cell lung cancer. People of southeast Asian origin often tend to have multiple copies of another gene, which seems to offer a degree of protection against malaria. An examination of structural variation in genes

> **"Copy number variations are the most common cause of autism we can identify today, by far."**
> **Arthur Beaudet,**
> **Baylor College of Medicine**

expressed in the brain has found possible associations with 17 conditions of the nervous system, including Parkinson's and Alzheimer's.

Copy number variation is also offering insights into the genetic origins of two of the most confusing conditions in which inheritance is involved: schizophrenia and autism. Twin and family studies have proved that both disorders are highly heritable, yet the search for genetic variants and mutations that are responsible has drawn little success. Recent research, much of it led by Jonathan Sebat at Cold Spring Harbor Laboratory, has suggested that copy number variation is often involved — particularly in sporadic cases that occur among people with no family history of these conditions.

Deletions or duplications in certain "hot spots" of the genome are much more common among children with autistic spectrum disorders than in the general population. Many have variations that are absent in their non-autistic parents. For schizophrenia, Sebat's group found that rare copy number variations are present in 15 percent of people who developed the mental illness as adults, and among 20 percent of teenage patients, compared with just 5 percent of healthy controls. Many of the changes in copy number that affect both disorders may be unique to the individuals who carry them — which could explain why their genetic roots have been so hard to pin down.

These discoveries are changing the way in which scientists think about genetic diversity. As Hurles puts it: "The variation that researchers had seen before was simply the tip of the iceberg, while the bulk lay submerged, undetected." This vast repository of difference is only beginning to give up its secrets, but science now knows that it is there.

31 Epigenetics

It was previously thought that a person's genetic makeup is arrived at independently of any environmental influences, but it now appears that sometimes acquired characteristics may, in fact, be passed on in the genes to succeeding generations.

In the autumn of 1944, railway workers in the Netherlands, then under German occupation, went on strike to assist the advancing Allies. When the initial British and American assault failed, the Nazis retaliated by imposing a devastating food embargo. At least 20,000 Dutch citizens starved or died of malnutrition in the ensuing famine.

The effects of the *Hongerwinter* or "Hunger Winter" were to last long beyond the country's liberation in 1945. Mothers who were pregnant during the famine had children with an elevated risk of a wide range of health problems, including diabetes, obesity and cardiovascular disease. In some cases, even their grandchildren were more likely to be born underweight. While damage to the first generation's health might be explained by malnutrition during pregnancy, the Netherlands was a rich nation by the time the second generation was born. Yet an inherited effect remained.

The story of the Dutch famine is not unique. The village of Överkalix, in northern Sweden, boasts meticulous historical records of

Suicide

Epigenetic effects may explain how dreadful experiences leave their mark on human behavior, making adults more likely to suffer depression and even to commit suicide. A team led by Moshe Szyf, of McGill University, examined DNA from the brains of 13 men who committed suicide, and found that while their genetic sequences were normal, their epigenetic programing was different from that seen in men who had died of other causes. All the men in the study had been abused as children, which could have triggered this epigenetic change. "It's possible the changes in epigenetic markers were caused by the exposure to childhood abuse," Professor Szyf says.

harvests, births and deaths, which have allowed Marcus Pembrey, of the Institute of Child Health in London, to conduct an exhaustive study of food availability and life expectancy. He found that when boys grew up during periods of plenty, their grandsons were more likely to die at a young age. Further analysis revealed that this reflected a predisposition to diabetes and heart disease, and confirmed that the effect was passed only through the male line.

Both examples suggest that people's health can be affected by the diets followed by their grandparents. Yet according to orthodox evolutionary theory, such an effect should be impossible. Acquired characteristics are not supposed to be inherited — that was the Lamarckian heresy, which has not been fashionable since Darwin.

> **We are changing the view of what inheritance is. You can't, in life, in ordinary development and living, separate out the gene from the environmental effect. They're so intertwined.**
> Marcus Pembrey

Genetic memory

The Dutch and Swedish experiences can be explained by a phenomenon known as epigenetics, by which the genome appears to "remember" certain environmental influences to which it has been exposed. Normally, these epigenetic effects act only on the somatic cells of the adult body, switching genes off or otherwise adjusting their activity. Some, however, can also alter sperm and eggs, to be inherited by future generations. Acquired characteristics, it turns out, can sometimes be passed on after all.

Epigenetics owes its prefix to the ancient Greek for "over" or "on," and it generally relies on two broad mechanisms. One is methylation, the process we met in Chapter 20, which silences genes by adding part of a molecule called a methyl group to the DNA base cytosine, or C. The other is modification of chromatin, the combination of DNA and histones (kinds of protein), of which chromosomes are made up. Changes to chromatin structure can affect which genes are made available for transcription into messenger RNA and protein, and which are hidden away out of reach. In neither case is the actual sequence of

DNA altered at all, but changes in its organization can still be passed on from one cell to its offspring.

These epigenetic processes are central to normal growth, development and metabolism. Every cell contains the full set of genetic instructions that are needed by every type of tissue, and epigenetics determines which of these are actually issued and executed. It ensures that genes that are required for rapid cell division in the embryo are switched off in the adult body, so they cannot cause cancer, and controls the patterns of gene expression that tell a cell it belongs to a kidney or brain.

Epigenetic effects also allow nurture to guide nature, by changing the way genes act in the body in response to environmental cues. Experiments with mice have demonstrated this lucidly: changes in diet while females are pregnant affect the coat color of their offspring, by modifying the way genes are methylated. This effect, indeed, could account for the surprising observation that many cloned animals differ from their parents in color and markings. While their genomes are identical, their "epigenomes" are not.

Normally, such epigenetic changes are stripped away from the genome during embryonic development, so that they are not passed on to offspring. But they can sometimes be retained, causing environmental effects on health and behavior that cascade down the generations. That could account for what happened in Holland

Stem cells

Though embryonic stem cells can grow into any type of tissue, their genetic code is no different from that of the specialized adult cells to which they give rise. Their uniquely malleable properties seem to be derived from their epigenetic character. Adult cells from the skin or bone carry all the genetic instructions that are needed to make any other cell type, but most of these instructions are switched off by epigenetics. Only in embryonic stem cells are all the genes required for pluripotency unmethylated and active.

It has recently become possible to reprogram adult cells into a pluripotent state (see Chapter 24), but only by replacing silenced genes with active copies — a technique which carries a risk of causing cancer. This hazard might plausibly be reduced by reprograming these cells' epigenomes instead.

and Sweden. Parental diets seem to have changed the epigenetic programing of their children and grandchildren, to modify metabolism to cope with the prevailing nutritional environment. This, in turn, has influenced health risks such as diabetes.

The importance of the epigenome

As with copy number variation and junk DNA, science is starting to understand that epigenetic effects are just as significant to biology as conventional genetic mutations. Epigenetics, for example, plays an important role in cancer. Several chemicals are known to be carcinogenic, even though they are not mutagens that directly damage DNA. They induce epigenetic effects, silencing important tumor suppressor genes, or altering chromatin structure so that oncogenes become more active.

Epigenetic markings also ensure that when cancer cells divide, their daughter cells are cancerous too. An understanding of how these processes work could open a new approach to medicine. The first cancer drug that works by removing methylation, Vidaza, was approved by the U.S. Food and Drug Administration in 2004.

The Human Epigenome Project, recently started by a European consortium, should help to bring more epigenetic therapies into medicine. This ambitious initiative aims to map all the possible methylation patterns of every gene in every type of human tissue. A pilot project has already achieved this for the major histo-compatibility complex — a cluster of genes on chromosome 6 that affect immune response.

Once these methylation sites have been identified, it should be possible to link variations to diseases, in much the same way as can be done for SNPs. Doctors, indeed, may well find their patients' epigenomes more useful than their genomes. The genetic code, as the early history of gene therapy has proved, is painfully difficult to correct in living organisms; methylation should be comparatively simple to undo. Many of the medicines of the future could be designed to exploit this natural method of genetic control to treat and prevent disease.

32 The RNA revolution

The role of DNA in genetics has long overshadowed that of RNA, but it may well be that in reality it is the latter that plays the more significant part in the creation of life.

Since its discovery by Friedrich Miescher, and especially since the discovery of its structure by Francis Crick and James Watson, DNA has been considered to be the queen of the nucleic acids. It is the stuff of which genes are made, the encoded language in which the instruction manual of life is written.

Ribonucleic acid suffers by comparison to this giant among molecules: RNA has often been seen as DNA's servant. It is the signaling chemical that runs cellular errands for its master, and the fetcher and carrier that collects amino acids so that DNA's muse can find expression in protein.

> **RNA interference provides a simple tool for manipulating gene expression in the laboratory, and has great promise for altering gene expression to treat diseases such as viral infections and cancer.**
>
> Chris Higgins,
> UK Medical Research Council

RNA, however, now looks rather more interesting than the first generation of molecular biologists had assumed. So interesting, in fact, that some scientists think it necessary to reassess the priority of the two nucleic acids that between them drive every form of life on the planet. DNA might contain the genome's core information, but it is through its chemical sister that it shapes organisms and their life cycles. RNA

The origin of life

The question of how life on Earth began, about 4 billion years ago, is one to which there remains no answer. One leading hypothesis is that some of the first self-replicating life forms, if not the first, were based on RNA. It is simpler than DNA, usually occurring in one strand rather than two, and it can both replicate itself and catalyze chemical reactions from surrounding molecules. This led figures like American microbiologist Carl Woese and Francis Crick to suggest that primitive "ribo-organisms" might have used chemicals from their environment to create new copies of themselves. Only later did life move beyond this "RNA world," and start using the more robust DNA molecule to encode its genetic information.

is anything but passive: it is a dynamic and versatile molecule that exists in dozens of forms, the vital functions of which science is only beginning to understand. It may even be the root of life itself.

The many faces of RNA

We have already met the most basic type of RNA — the single-stranded messenger RNA (mRNA) molecules into which DNA is transcribed, and which carry instructions for making proteins. Only about 2 percent of our RNA, however, is mRNA.

Many other varieties are involved in protein manufacture alone. The important bits of mRNA — the exons — are interspersed with stretches of nonsense called introns. An RNA-based structure called the spliceosome snips away the introns, and stitches the exons into a meaningful message. This then travels to a cell's ribosomes, or protein factories, which are themselves made chiefly of ribosomal RNA, another specialized type. Transfer RNA, a cross-shaped variety, then identifies and collects amino acids for threading onto protein chains.

RNA is not just a protein-making tool. It also comes in small molecules such as micro RNAs (miRNAs), which are tiny single strands of between 21 and 23 bases in length. These are transcribed from DNA, but from the junk DNA that does not code for proteins, and their function seems to be to regulate how genes work in the body. They switch them on and off, or tweak their activity so that levels of protein manufacture go up or down. These are now thought to explain much of the complexity of human life.

There are thousands of different types of human miRNA — the total number may well exceed the tally of around 21,500 genes. Each can modify not only the activity of single genes, but that of groups of genes and other RNA molecules as well. This means that when thrown together in combination, miRNAs can manipulate gene expression in subtle ways that are almost limitless. They allow a relatively small range of genes, many shared with other animals, plants and even microbes, to produce structures as complicated as the human brain. There is good evidence, indeed, that the number of miRNAs increases with an organism's degree of complexity. While people have only a couple of thousand more genes than do nematode worms, we have many times the number of miRNAs. These molecules seem to be responsible for building more sophisticated forms of life.

RNA interference

Growing recognition of the importance of RNA is also shedding light on disease, and on how it might be treated — particularly through a process called RNA interference (RNAi). This natural phenomenon, first discovered in petunia plants in the early 1990s, is thought to have evolved as a defense against viruses. It has become one of the most exciting frontiers in medicine, in such short order that two of its pioneers, Andrew Fire and Craig Mello, won the 2006 Nobel Prize for Medicine, just eight years after their key work was published.

RNAi relies on double-stranded RNA molecules called short interfering RNAs (siRNAs), each about 21 units in length. Working with nematode worms, Fire and Mello established that when siRNAs

An RNAi contraceptive?

One exciting future application of RNAi could be in a new type of contraceptive pill that does not rely on hormones. Zev Williams, of Brigham and Women's Hospital in Boston, has shown that the technique can be used to silence a gene called ZP3, which is active in eggs before ovulation. When ZP3 is switched off, the egg forms without its outer membrane, to which sperm must bind for fertilization to occur.

As ZP3 is expressed only in growing eggs, the technique should be reversible: undeveloped eggs would be untouched, and could be ovulated normally once a woman has stopped taking the drug. It should also be free from side-effects, as ZP3 is not active in any other types of tissue.

with a particular sequence are injected into a cell, they interfere with the activity of genes that generate the same sequence in messenger RNA, so that lower quantities of protein are produced.

What happens is that once inside a cell, siRNAs are unzipped into single strands, which then bind to pieces of mRNA that match their sequence. When mRNAs are tagged in this way, they are torn up by cellular enzymes. The protein-making instructions that they carry are destroyed, and protein manufacture is impaired.

The technique's medical potential lies in its ability to target particular genes and their protein products with great precision. The 21-letter code of siRNAs can be written to match a specific set of mRNA instructions, so that production of one protein is inhibited without affecting the synthesis of others. RNAi can therefore be used to switch off rogue genes, of the sort that drive cancer or other disorders, without messing up the chemistry of healthy cells. It also allows scientists to manipulate the activity of genes in the laboratory, to establish how they work.

No RNAi drug is yet available, but several are in the advanced stages of development. Clinical trials are assessing treatments for age-related macular degeneration, a common form of blindness, which work by targeting a growth factor that is over-expressed in the eyes. Another study has shown that siRNAs can make breast tumor cells 10,000 times more sensitive to chemotherapy, by silencing genes that confer resistance to the drug Taxol. Scientists also hope to exploit the technique against HIV, to knock out a gene the virus needs to reproduce.

As science reveals more about how genes and the proteins they produce affect the course of disease, RNAi is likely to become ever more important to medicine. It is promising to provide something that clinical geneticists have long craved — a precision instrument with which to switch off genes that cause disease.

33 **Artificial life**

Mycoplasma genitalium is a bacterium that makes its home in the human urethra, where it sometimes causes a mild sexually transmitted infection. Until recently, its only distinction was to have the smallest genome of any free-living bacterium — but no longer. It has become the template for the first attempt to create artificial life.

The prospect of breathing life into inanimate matter has long intrigued humanity, as the enduring popularity of Mary Shelley's *Frankenstein* shows. A project led by Craig Venter, the maverick who led the private effort to sequence the human genome, is now promising to move it beyond science fiction and into reality.

Since 1999, he has been studying *M. genitalium* with a view to identifying the qualities of what he calls the "minimal genome" — the smallest set of genes that is sufficient to sustain life. And now that he has reached an answer — the bacterium can survive with just 381 of

Bioerror

Though mousepox is a relative of the smallpox virus, it does not normally make the rodents that contract it especially ill. That changed, however, when scientists at the Australian National University introduced a small genetic change to the virus in 2001. Though they had no intention of making the pathogen more virulent — they were investigating a contraceptive vaccine — the genetic modification had catastrophic effects. All the animals that were infected in the experiment died — victims not of bioterror, but of bioerror.

Critics of synthetic biology argue that if such an accident can happen when just one gene is altered in a microorganism, the potential for unintended disaster could be vast when whole genomes are being designed from scratch. Supporters say that such organisms would not be released from the laboratory until they had been proven safe, and that any that escaped accidentally would not survive in the wild.

the 485 genes it possesses in nature — he is seeking to synthesize such an organism in the laboratory, with a man-made genetic code. If it works, life will have been made out of chemicals from a bottle. As one of Venter's critics has put it, "God has competition."

Creating Synthia

Venter calls the organism he plans to make *Mycoplasma labatorium*, but the ETC Group, an anti-biotech organization, has come up with a catchier name: Synthia. She would not be quite the first synthetic organism: Eckard Wimmer, of Stony Brook University, has assembled the genome of the polio virus, and Venter's team has recreated a different virus, Phi-X174, from scratch. Viruses, however, are relatively easy pickings for synthetic biology. Their genomes are tiny, and as they need to hijack host cells to reproduce, they are not normally considered to be properly alive.

> " I want to take us far from shore into unknown waters, to a new phase of evolution, to the day when one DNA-based species can sit down at a computer to design another. I plan to show that we understand the software of life by creating new artificial life. "
> Craig Venter

Though Synthia will have a genetic code 18 times longer than any virus, her genome will also be animated, in part, by another form of life. While her DNA will be stitched together in the laboratory, scientists cannot yet reproduce the complicated cellular machinery that exists outside the nucleus, so the artificial genome will have to be transplanted into the shell of a similar bacterium. In 2007, Venter showed that such a transplant was possible, by moving the genome from one *Mycoplasma* bacterium into a close relative. This silenced the host's genome, essentially transforming one species into another. Use the same procedure to transplant a synthetic genome, and an artificial organism would enter the world.

The next stage — the construction of such a synthetic genome — has also been accomplished. Venter has rebuilt the single circular chromosome of *M. genitalium*, which contains almost 583,000 base

pairs, from bottled DNA. The bacterium's code was first split into
101 chunks or "cassettes" of 5,000 to 7,000 nucleotides, then these
components were ordered from companies that manufacture short
DNA sequences and assembled. The result was identical to the wild
bacterium's genome, except in one important respect. A single gene,
which allows wild-type *M. genitalium* to infect mammalian cells, was
knocked out as insurance against accidents.

All that remains, at the time of writing, is for this synthetic
chromosome to be successfully transplanted into a bacterial casing.
The resulting organism will have natural hardware, but the genetic
software it "boots up" will be laboratory-made.

Use and abuse

Venter's experiments with synthetic biology have two main goals.
The first is intellectual — to understand a little more about the
mystery of what separates living from non-living things. The second is
practical — he sees it as a tool with which to manufacture organisms
that benefit humankind.

Hydrogen, which some bacteria make naturally, is widely
considered to be one of the energy sources of the future, as when burned
it emits only water as waste. Venter plans to use synthetic biology to
design microorganisms that make this clean fuel efficiently. His work is
partially supported by the U.S. Department of Energy. Other prospects
include building organisms that consume and thus clean up toxic waste
that is not normally biodegradable by natural bacteria, or which absorb
carbon dioxide from the atmosphere to counter climate change.

Genetic engineering of existing bacteria could help to address this technological challenge, but is limited by the natural properties of the microorganisms that can be engineered. If it works, synthetic biology would open a much more powerful approach, allowing genomes to be designed from scratch with a particular purpose in mind.

Any technology, however, can be employed for both good and evil ends. Quite apart from raising moral objections from those who think it wrong to meddle with life, synthetic biology has stirred great concern about how it might be abused. As Hamilton Smith, Venter's collaborator, admitted when the team rebuilt the Phi-X174 virus: "We could make the smallpox genome." A deadly pathogen that has been eradicated in the wild could potentially be revived by bioterrorists or rogue states.

Just as troubling is the prospect of "bioerror" — the accidental creation of a germ of new virulence or infectivity, against which our unhabituated bodies would have no defense. Some biologists have argued for a temporary moratorium on this research until implications and safety protocols can be worked out, as the Asilomar conference did for recombinant DNA in the 1970s (see Chapter 8).

Some of these fears are misplaced, at least for now. Venter interrupted his work for 18 months while it was reviewed by an independent ethics panel, and the microorganisms that Venter is designing will be so weak that they could not survive outside the laboratory. Genetic engineering of bacteria has also been happening for more than three decades, without a single notable accident. But as science moves forward, this technology is certain to create challenges and threats, as well as opportunities. There is a good case for proceeding with caution.

PHYSICS

Physics is a practical science. Although rooted in the logical language of mathematics, physicists ask basic questions of the world. Why do things fall down and not up? Why when we put a cup on a table does it not fall through it? Why does ice melt? Why is the sky blue?

Careful measurement is the first step in gaining understanding. The ancient Greeks measured the shape and extent of the Earth's surface using their knowledge of geometry. Early astronomers assembled navigational tools by which the globe was explored and populated. The design of more intricate instruments let us pinpoint the stars more accurately, and telescopes let us see farther and ponder our place beneath the heavens. As our horizons have expanded over the centuries, our view of humankind has been forever changed by physics: our planet is a rocky ball floating in dark, empty space, one of many planets in our solar system, the Milky Way and the universe. Yet it is not austere, but the home of life.

Brave theories followed, often challenging conventional thinking. Copernicus' proposition that the Earth orbited the Sun was unpopular with the church of the 16th and 17th centuries, which wanted to keep man central in the universe. Kepler and Galileo made it worse: they tracked the paths of the planets on the sky and

420 B.C. Democritus postulates the existence of atoms

335 B.C. Aristotle states that objects move due to the action of forces

A.D. 150 Ptolemy suggests planets move in epicycles

1543 Copernicus proposes planets orbit the sun

1609 Kepler discovers that planets move in elliptical orbits

watched other moons revolve around Jupiter. If Earth wasn't central, neither was the Sun the focus of all motions of celestial bodies. Kepler worked out the physics of orbits, inspiring Newton to think that the same force that caused an apple to fall from a tree spun planets around the Sun. The concept of gravity was thus born in a grand leap of imagination.

Physics around us: harnessing forces

The physics of forces plays out all around us. Newton translated projectiles' parabolic paths into mathematical laws, defining concepts of mass, velocity and acceleration. Forces, and weight, have been harnessed by engineers to build bridges, skyscrapers, cars and airplanes and even thrilling rollercoaster rides.

Experimenting in his rooms with light beams and prisms led Newton to explain the character of white light as a blend of the colors of the rainbow. Light was later shown to be related to electricity and magnetism, an electromagnetic wave. In the 18th and 19th centuries, scientists like Franklin and Faraday isolated electrical sparks in the laboratory. Franklin famously flew kites into thunderstorms to conduct current to the ground. Magnetism, once a mysterious force inhabiting some iron-rich rocks, turned out to be closely tied to electricity. Light turned out to have a fixed speed of 300 million meters per second; a rate which nothing can exceed.

Experiments can also be imagined. In the 20th century Einstein wondered what would happen if the train he sat in traveled at the speed of light. What would he see out of the window? What would he see if another train traveling at light speed passed in the

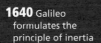

1640 Galileo formulates the principle of inertia

1672 Newton explains the rainbow

1678 Huygens' treatise on the wave theory of light is published

1687 Newton publishes his law of gravitation

1752 Franklin conducts his lightning experiment

opposite direction? Neither could travel relative to one another at twice light speed — something was wrong. Einstein, a visual thinker, worked out a novel concept of space and time to accommodate the conundrum. His general relativity distorts space and time to make sure that nothing goes faster than light. The entire landscape twists and time slows down so that the ultimate speed is not breached. It sounds far-fetched but has been proven by experiments, including clocks flown in space and the bending of light from distant stars around the Sun's limb.

Physics within us: hidden worlds

Modern physics contains many concepts which seem implausible; yet they have survived experimental tests. The atom is one area where many surprises lurk. Originally a Greek idea, that nature was made out of tiny building blocks that could be configured into any form, modern physics has dissected atoms to reveal a zoo of fundamental particles, from quarks to bosons to gluons. As the layers of the atom were unwrapped, it became clear that most of what we think of as solid matter is in fact empty space.

Like a Russian doll, the atom has gradually revealed finer and finer detail. Late in the 19th century, the lightweight electron was spotted in its outer reaches, giving rise to negative electric charge and tying chemical bonds between atoms in molecules. To balance, a hard nucleus of positive charge was envisaged, and the proton was discovered a decade later. But the weights of different elements hinted at more — the charge-free neutron made up the mass of an atom. Modern particle physics has shown that protons and neutrons

1820 Orsted links electricity and magnetism

1824 Carnot lays foundations of thermodynamics

1839 Becquerel observes the photoelectric effect

1850 Clausius defines entropy and the second law of thermodynamics

1873 Maxwell's equations show light is an electromagnetic wave

can be further broken down into quarks, and forces are mediated by a weird menagerie of subatomic particles.

The chemical behavior of an element is dictated by how many protons and neutrons it hosts. Running like a grid on a snakes and ladders board, from light to heavy atomic weights, the periodic table links similar families of elements and graphically suggests how they can be combined to build molecules and chemical compounds.

In the 20th century, the wishes of earlier alchemists to transform one element into another was realized when nuclear fission was achieved. In fission a large cumbersome nucleus splits into two smaller more stable ones. Although used catastrophically in bombs, since the 1950s fission has also provided us with a reliable energy source. The opposite reaction of fusion, to push lightweight nuclei together, is what makes the Sun and stars shine. Most chemical elements around us on Earth were made in the stars.

Physics beyond us: uncertainty and fate

Quantum physics was perhaps the most revolutionary — and bizarre — idea of 20th-century physics. It grew from trying to explain simultaneously conflicting evidence, such as the idea that particles and waves were similar. At the atomic scale matter and energy can behave as either. Even more oddly, it seems to choose which mode only when an experimenter decides in which form to measure it. Yet there are clear rules. Quantum mechanics has conjured up a weird subatomic world, of fuzziness and uncertainty. Nothing is set in stone until it is observed.

1895 Röntgen discovers x-rays

1897 J.J. Thomson discovers the electron

1901 Planck publishes his law of black-body radiation and energy quanta

1905 Einstein publishes his special theory of relativity

1915 Einstein publishes his general theory of relativity

Investigating scales that are far removed from our everyday experience, quantum physics has become like philosophy. If the act of measurement locks in a system's final state, what happens to the other options that weren't measured? Is that information lost, or buried? Does it get spun off into a parallel universe? The idea of parallel universes often pops up in science fiction and film: we take a decision, but what would have happened if we'd taken a different path? Teleportation, though, is a quantum reality. Because quantum information follows tight rules, quantum properties can be transmitted across space instantaneously.

Twentieth-century physics's greatest breakthroughs have concerned the very small and very large, astronomy and particle physics. Both fields continue to provide a wealth of new information and adjustments to our vision of what it means to be human in this universe. We are a tiny speck in a vast cosmos; we may be unique or we may be part of a wider community of life-forms existing in space; but we have managed to leave our own planet to explore other worlds. Particle physics uses vast machines such as the Large Hadron Collider to smash atoms to find their smallest bits. And nuclear physics has given us the power to destroy life on our planet, should we decide. It has also given us a new energy source. And one day we may get fusion power, like the Sun, a clean, limitless and powerful source of energy.

Physics for us: benefiting society

Some think physics, perhaps because of the atom bomb, is a hard and unemotional subject. But the physics of the 21st century is humane

1918 Rutherford isolates the proton

1927 Copenhagen Interpretation of quantum mechanics formulated

1928 Dirac derives the existence of antimatter

1929 Hubble establishes expansion of universe

1932 Chadwick discovers the neutron

and blends with other areas of science. Climate change mitigation hinges on understanding the physics of the atmosphere, including complexity, fluid mechanics, forces and chemistry. Geophysicists seek not only to understand earthquakes, tsunami and the interior structure of the Earth, but also to prevent deaths from natural disasters. Medical physicists use atomic physics to scan our bodies, with CAT, ultrasound and nuclear magnetic resonance scanners. New energy sources rely on basic concepts of physics. Even astronomers are looking for life on other worlds like never before — discovering distant planets and visiting exotic worlds like Titan and Europa in our own solar system.

Although some might say that physics has distanced us from ourselves, opening up the vastness of the universe and the empty space of the subatomic world, our views of the Earth have shown us how important we are and the laws of physics are being used like never before to solve major problems of society.

1933 Zwicky measures dark matter

1938 Atomic fission is observed

1956 Neutrinos are detected

1965 Cosmic microwave background discovered by Penzias and Wilson

1998 Supernova data suggest dark energy

34 Mach's principle

A child whirling on a merry-go-round is tugged outwards by the distant stars. This is Mach's principle that "mass there influences inertia here." Through gravity, objects far away affect how things move, and spin, nearby. But why is this and how can you tell if something is moving or not?

If you have ever sat in a train at a station and seen through the window a neighboring carriage pull away from yours, you will know that sometimes it is hard to tell whether it is your own train leaving the station or the other arriving. Is there a way that you could measure for sure which one is in motion?

Ernst Mach, an Austrian philosopher and physicist, grappled with this question in the 19th century. He was treading in the footsteps of the great Sir Isaac Newton who had believed, unlike Mach, that space was an absolute backdrop. Like graph paper, Newton's space contained an engraved set of coordinates and he mapped all motions as movements with respect to that grid. Mach, however, disagreed, arguing instead that motion was only meaningful if measured with respect to another object, not the grid. What does it mean to be moving if not relative to something else? In this sense Mach, who was influenced by the earlier ideas of Newton's competitor Gottfried Leibniz, was a forerunner to Albert Einstein in preferring to think that only relative motions made sense. Mach argued that because a ball rolls in the same way whether it is in France or Australia, the grid of space is irrelevant. The only thing that can conceivably affect how the ball rolls is gravity. On the Moon the ball might well roll differently because the gravitational force pulling on the ball's mass is weaker there. Because every object in the universe exerts a gravitational pull on every other, each object will feel each other's presence through their mutual attractions. So motion must ultimately depend on the distribution of matter, or its mass, not on the properties of space itself.

Mass

What exactly is mass? It is a measure of how much matter an object contains. The mass of a lump of metal would be equal to the sum of the masses of all the atoms in it. Mass is subtly different from weight. Weight is a measure of the force of gravity pulling a mass down — an astronaut weighs less on the Moon than on Earth because the gravitational force exerted by the smaller Moon is less. But the astronaut's mass is the same — the number of atoms he contains has not changed. According to Albert Einstein, who showed that energy and mass are interchangeable, mass can be converted into pure energy. So mass is, ultimately, energy.

Inertia

Inertia, named after the Latin word for "laziness," is very similar to mass but tells us how hard it is to move something by applying a force. An object with large inertia resists movement. Even in outer space a massive object takes a large force to move it. A giant rocky asteroid on a collision course with the Earth may need a huge shove to deflect it, whether it is created by a nuclear explosion or a smaller force applied for a longer time. A smaller spacecraft, with less inertia than the asteroid, might be maneuvered easily with tiny jet engines.

> **Absolute space, of its own nature without reference to anything external, always remains homogenous and immovable.**
> Sir Isaac Newton, 1687

The Italian astronomer Galileo Galilei proposed the principle of inertia in the 17th century: if an object is left alone, and no forces are applied to it, then its state of motion is unchanged. If it is moving, it continues to move at the same speed and in the same direction. If it is standing still it continues to do so. Newton refined this idea to form his first law of motion.

Newton's bucket

Newton also codified gravity. He saw that masses attract one another. An apple falls from a tree to the ground because it is attracted by the Earth's mass. Equally, the Earth is attracted by the apple's mass, but we

Ernst Mach 1838–1916

As well as for Mach's principle, Austrian physicist Ernst Mach is remembered for his work in optics and acoustics, the physiology of sensory perception, the philosophy of science and particularly his research on supersonic speed. He published an influential paper in 1877 that described how a projectile moving faster than the speed of sound produces a shock wave, similar to a wake. It is this shockwave in air that causes the sonic boom of supersonic aircraft. The ratio of the speed of the projectile, or jet plane, to the speed of sound is now called the Mach number, such that Mach 2 is twice the speed of sound.

would be hard pressed to measure the microscopic shift of the whole Earth towards the apple.

Newton proved that the strength of gravity falls off quickly with distance, so the Earth's gravitational force is much weaker if we are floating high above it rather than on its surface. But nevertheless we would still feel the reduced pull of the Earth. The farther away we go the weaker it would get, but it could still tweak our motion. In fact, all objects in the universe may exert a tiny gravitational pull that might subtly affect our movement.

Newton tried to understand the relationships between objects and movement by thinking about a spinning bucket of water. At first when the bucket is turned, the water stays still even though the bucket moves. Then the water starts to spin as well. Its surface dips as the liquid tries to escape by creeping up the sides but it is kept in place by the bucket's confining force. Newton argued that the water's rotation could only be understood if seen in the fixed reference frame of absolute space, against its grid. We could tell if the bucket was spinning just by looking at it because we would see the forces at play on it producing the concave surface of the water.

Centuries later Mach revisited the argument. What if the water-filled bucket were the only thing in the universe? How could you know it was the bucket that was rotating? Couldn't you equally well say the water was rotating relative to the bucket? The only way to make sense of it would be to place another object into the bucket's universe, say the wall of a room, or even a distant star. Then the bucket would clearly be spinning relative to that. But without the frame of a stationary room, and the fixed stars, who could say whether it was the bucket or the water that rotates? We experience the same thing when we watch the Sun and stars arc across the sky. Is it the stars or the Earth that is rotating? How can we know?

According to Mach, and Leibniz, motion requires external reference objects for us to make sense of it, and therefore inertia as a concept is meaningless in a universe with just one object in it. So if the universe were devoid of any stars, we'd never know that the Earth was spinning. The stars tell us we're rotating relative to them.

The ideas of relative versus absolute motion expressed in Mach's principle have inspired many physicists since, notably Einstein (who actually coined the name "Mach's principle"). Einstein took the idea that all motion is relative to build his theories of special and general relativity. He also solved one of the outstanding problems with Mach's ideas: rotation and acceleration must create extra forces, but where were they? Einstein showed that if everything in the universe were rotating relative to the Earth, we should indeed experience a small force that would cause the planet to wobble in a certain way.

The nature of space has puzzled scientists for millennia. Modern particle physicists think it is a seething cauldron of subatomic particles being continually created and destroyed. Mass, inertia, forces and motion may all in the end be manifestations of a bubbling quantum soup.

35 Newton's laws of motion

Sir Isaac Newton was one of the most prominent, contentious and influential scientists of all time. He helped to invent Calculus, explained gravity and identified the constituent colors of white light. His three laws of motion describe why a golf ball follows a curving path, why we are pressed against the side of a cornering car and why we feel the force through a baseball bat as it strikes the ball.

Although motorcycles had yet to be invented in Newton's time, his three laws of motion explain how a stunt rider can mount the vertical wall of death, and how Olympic cyclists race on inclined tracks.

Newton, who lived in the 17th century, is considered one of the foremost intellects of science. It took his highly inquisitive character to understand some of the most seemingly simple yet profound aspects of our world, such as how a thrown ball curves through the air, why things fall down rather than up and how the planets move around the Sun.

An average student at Cambridge in the 1660s, Newton began by reading the great works of mathematics. Through them he was drawn away from civic law into the laws of physics. Then, on sabbatical at home when the university was closed due to an outbreak of plague, Newton took the first steps to developing his three laws of motion.

Forces

Borrowing Galileo's principle of inertia, Newton formulated his first law. It states that bodies do not move or change their speed unless a force acts. Bodies that are not moving will remain stationary unless a force is applied; bodies that are moving with some constant speed keep moving at that same speed unless acted upon by a force. A force (for instance a push) supplies an acceleration that changes the velocity of the object. Acceleration is a change in speed over some time.

Newton's laws of motion

First law Bodies move in a straight line with a uniform speed, or remain stationary, unless a force acts to change their speed or direction.

Second law Forces produce accelerations that are in proportion to the mass of a body ($F = ma$).

Third law Every action of a force produces an equal and opposite reaction.

This is hard to appreciate in our own experience. If we throw a hockey puck it skims along the ice but eventually slows due to friction with the ice. Friction causes a force that decelerates the puck. But Newton's first law may be seen in a special case where there is no friction. The nearest we might get to this is in space, but even here there are forces like gravity at work. Nevertheless, this first law provides a basic touchstone from which to understand forces and motion.

Acceleration

Newton's second law of motion relates the size of the force to the acceleration it produces. The force needed to accelerate an object is proportional to the object's mass. Heavy objects — or rather ones with large inertia — need more force to accelerate them than lighter objects. So to accelerate a car from standing still to 100 kilometers an hour in one minute would take a force equal to the car's mass times its increase in speed per unit time.

Newton's second law is expressed algebraically as "$F = ma$," force (F) equals mass (m) times acceleration (a). Turning this definition around, the second law expressed in another way says that acceleration is equal to force per unit mass. For a constant acceleration, force per unit mass is also unchanged. So the same amount of force is needed to move a kilogram mass whether it is part of a small or large body. This explains Galileo's imaginary experiment that asks which would hit the ground first if dropped together: a cannonball or a feather? Visualizing it, we may think that the cannonball would arrive ahead of the drifting feather. But this is simply due to the air resistance that wafts the feather. If there were no air, then both would fall at the same rate, hitting the ground together. They experience the same acceleration, gravity, so they fall side by side. *Apollo 15* astronauts showed in 1971 that on the Moon, where there is no atmosphere to slow it down, the feather falls at the same rate as a geologist's heavy hammer.

Action equals reaction

Newton's third law states that any force applied to a body produces an equal and opposite reaction force in that body. In other words, for every action there is a reaction. The opposing force is felt as recoil. If one roller-skater pushes another, then she will also roll backwards as she pushes against her partner's body. A marksman feels the kick of the rifle against his shoulder as he shoots. The recoil force is equal in size to that originally expressed in the shove or the bullet. In crime films the victim of a shooting often gets propelled backwards by the force of the bullet. This is misleading. If the force was really so great then the shooter should also be hurled back by the recoil of his gun. Even if we jump up off the ground, we exert a small downward force on the Earth,

Sir Isaac Newton 1643–1727

Sir Isaac Newton was the first scientist to be honored with a knighthood in Britain. Despite being "idle" and "inattentive" at school, and an unremarkable student at Cambridge University, Newton flourished suddenly when plague closed the university in the summer of 1665. Returning to his home in Lincolnshire, Newton devoted himself to mathematics, physics and astronomy, and even laid the foundations for calculus. There he produced early versions of his three laws of motion and deduced the inverse square law of gravity. After this remarkable outburst of ideas, Newton was elected to the Lucasian Chair of Mathematics in 1669 at just 27 years old. Turning his attention to optics, Newton discovered with a prism that white light was made up of rainbow colors, quarreling famously with Robert Hooke and Christiaan Huygens over the matter. Newton wrote two major works, *Philosophiae naturalis Principia mathematica*, or *Principia*, and *Opticks*. Late in his career, Newton became politically active. He defended academic freedom when King James II tried to interfere in university appointments and entered Parliament in 1689. A contrary character, on the one hand desiring attention and on the other withdrawn and trying to avoid criticism, Newton used his powerful position to fight bitterly against his scientific enemies and remained a contentious figure until his death.

but because the Earth is so much more massive than we are, it barely shows.

With these three laws, plus gravity, Newton could explain the motion of practically all objects, from falling acorns to balls fired from a cannon. Armed with these three equations he could confidently have climbed aboard a fast motorbike and sped up onto the wall of death, had such a thing existed in his day. How much trust would you place in Newton's laws? The first law says that the cycle and its rider want to keep traveling in one direction at a certain speed. But to keep the cycle moving in a circle, according to the second law, a confining force needs to be provided to continually change its direction, in this case applied by the track through the wheels. The force needed is equal to the mass of the cycle and rider multiplied by their acceleration. The third law then explains the pressure exerted by the cycle on the track, as a reactionary force is set up. It is this pressure that glues the stunt rider to the inclined wall, and if the bike goes fast enough it can even ride on a vertical wall.

Even today knowledge of Newton's laws is pretty much all you need to describe the forces involved in driving a car fast around a bend or, heaven forbid, crashing it. Where Newton's laws do not hold is for things moving close to the speed of light or with very small masses. It is in these extremes that Einstein's relativity and the science of quantum mechanics take over.

36 Kepler's laws

Johannes Kepler looked for patterns in everything. Peering at astronomical tables describing the looped motions of Mars projected on the sky, he discovered three laws that govern the orbits of the planets. Kepler described how planets follow elliptical orbits and how more distant planets orbit more slowly around the Sun. As well as transforming astronomy, Kepler's laws laid the foundations for Newton's law of gravity.

As the planets move around the Sun, the closest ones move more quickly around it than those farther away. Mercury circles the Sun in just 80 Earth days. If Jupiter traveled at the same speed it would take about 3.5 Earth years to complete an orbit when, in fact, it takes 12. As all the planets sweep past each other, when viewed from the Earth some appear to backtrack as the Earth moves forwards past them. In Kepler's time these "retrograde" motions were a major puzzle. It was solving this puzzle that gave Kepler the insight to develop his three laws of planetary motion.

> ❝ It suddenly struck me that that tiny pea, pretty and blue, was the Earth. I put up my thumb and shut one eye, and my thumb blotted out the planet Earth. I didn't feel like a giant. I felt very, very small. ❞
> **Neil Armstrong, b.1930**

Patterns of polygons

The German mathematician Johannes Kepler sought patterns in nature. He lived in the late 16th and early 17th centuries, a time when astrology was taken very seriously and astronomy as a physical science was still in its infancy. Religious and spiritual ideas were just as important in revealing nature's laws as observation. A mystic who believed that the underlying structure of the universe was built from perfect geometric forms, Kepler devoted his life to trying to tease out the patterns of imagined perfect polygons hidden in nature's works.

Kepler's work came a century after Polish astronomer Nicolaus Copernicus proposed that the Sun lies at the center of the universe and the Earth orbits the Sun, rather than the other way around. Before then, going back to the ancient Greek philosopher Ptolemy, it was believed

Johannes Kepler 1571–1630

Johannes Kepler liked astronomy from an early age, recording in his diary a comet and a lunar eclipse before he was ten. While teaching at Graz, Kepler developed a theory of cosmology that was published in the *Mysterium Cosmographicum* (*The Sacred Mystery of the Cosmos*). He later assisted astronomer Tycho Brahe at his observatory outside Prague, inheriting his position as Imperial Mathematician in 1601. There Kepler prepared horoscopes for the emperor and analyzed Tycho's astronomical tables, publishing his theories of noncircular orbits, and the first and second laws of planetary motion, in *Astronomia Nova* (*New Astronomy*). In 1620, Kepler's mother, a herbal healer, was imprisoned as a witch and only released through Kepler's legal efforts. However, he managed to continue his work and the third law of planetary motion was published in *Harmonices Mundi* (*Harmony of the Worlds*).

that the Sun and stars orbited the Earth, carried on solid crystal spheres. Copernicus dared not publish his radical idea during his lifetime, leaving it to his colleague to do so just before he died, for fear that it would clash with the doctrine of the church. Nevertheless Copernicus caused a stir by suggesting that the Earth was not the center of the universe, implying that humans were not the most important beings in it, as favored by an anthropocentric god.

Kepler adopted Copernicus' heliocentric idea, but nevertheless still believed that the planets orbited the Sun in circular orbits. He envisaged a system in which the planets' orbits lay within a series of nested spheres spaced according to mathematical ratios that were derived from the sizes of three-dimensional shapes that would fit within them. So he imagined a series of polygons with increasing numbers of sides that fit within the spheres. The idea that nature's laws followed basic geometric ratios had originated with the Ancient Greeks.

The word planet comes from the Greek for "wanderer." Because the other planets in our solar system lie much closer to the Earth than the distant stars, they appear to wander across the sky. Night after night they pick out a path through the stars. Every now and again, however, their path reverses and they make a little backwards loop. These retrograde motions were thought to be bad omens. In the Ptolemaic model of planetary motion this behavior was impossible to understand, and so astronomers added "epicycles" or extra loops to the orbit of a planet that mimicked this motion. But the epicycles did not

Kepler's laws

First law Planetary orbits are elliptical with the Sun at one focus.

Second law A planet sweeps out equal areas in equal times as it orbits the Sun.

Third law The orbital periods scale with ellipse size, such that the period squared is proportional to the semi-major axis length cubed.

work very well. Copernicus' Sun-centered universe needed fewer epicycles than the older Earth-centered one, but still could not explain the finer details.

Trying to model the orbits of the planets to support his geometric ideas, Kepler used the most accurate data available, intricate tables of the planets' motions on the sky, painstakingly prepared by Tycho Brahe. In these columns of numbers Kepler saw patterns that suggested his three laws.

> ❝We are just an advanced breed of monkeys on a minor planet of a very average star. But we can understand the universe. That makes us something very special.❞
> Stephen Hawking, 1989

Kepler got his breakthrough by disentangling the retrograde motions of Mars. He recognized that the backward loops would fit if the planets' orbits were elliptical around the Sun and not circular as had been thought. Ironically this meant that nature did not follow perfect shapes. Kepler must have been overjoyed at his success in fitting the orbits, but also shocked that his entire philosophy of pure geometry had been proved wrong.

Orbits

In Kepler's first law, he noted that the planets move in elliptical orbits with the Sun at one of the two foci of the ellipse.

Kepler's second law describes how quickly a planet moves around its orbit. As the planet moves along its path, it sweeps out an equal area segment in an equal time. The segment is measured using the angle drawn between the Sun and the planet's two positions (*AB* or *CD*), like a slice of pie. Because the orbits are elliptical, when the planet is close to the Sun it needs to cover a larger distance to sweep out the same area than when it is farther away. So the planet moves faster near the Sun than when it is distant. Kepler's second law ties its speed with its distance from the Sun. Although Kepler didn't realize it at the time, this behavior is ultimately due to gravity accelerating the planet faster when it is near the Sun's mass.

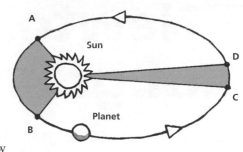

Kepler's third law goes one step further again and tells us how the orbital periods scale up for different sized ellipses at a range of distances from the Sun. It states that the squares of the orbital periods are inversely proportional to the cube power of the longest axis of the elliptical orbit. The larger the elliptical orbit, the slower the period, or time taken to complete an orbit. A planet in orbit twice as far away from the Sun as the Earth would take eight times longer to go around. So planets farther from the Sun orbit more slowly than nearby planets. Mars takes nearly two Earth years to go around the Sun, Saturn 29 years and Neptune 165 years.

> **" I measured the skies, now the shadows I measure, Sky-bound was the mind, earth-bound the body rests. "**
> Kepler's epitaph, 1630

In these three laws, Kepler managed to describe all the planets' orbits in our solar system. His laws apply equally to any body in orbit around another, from comets, asteroids and moons in our solar system to planets around other stars and even artificial satellites whizzing around the Earth. Kepler succeeded in unifying the principles into geometric laws but he did not know why these laws held. He believed that they arose from the underlying geometric patterns of nature. It took Newton to unify these laws into a universal theory of gravity.

37 **Newton's law of gravitation**

Sir Isaac Newton made a giant leap when he connected the motions of cannonballs and fruit falling from trees to the movements of the planets, thus linking heaven and earth. His law of gravitation remains one of the most powerful ideas of physics, explaining much of the physical behavior of our world. Newton argued that all bodies attract each other through the force of gravity and the strength of that force drops off with distance squared.

The idea of gravity supposedly came to Sir Isaac Newton when he saw an apple fall from a tree. We don't know if this is true or not, but Newton stretched his imagination from earthly to heavenly motions to work out his law of gravitation.

Newton perceived that objects were attracted to the ground by some accelerating force (see Chapter 35). If apples fall from trees, what if the tree were even higher? What if it reached to the Moon? Why doesn't the Moon fall to the Earth like an apple?

All fall down

Newton's answer lay first in his laws of motion linking forces, mass and acceleration. A ball blasted from a cannon travels a certain distance before falling to the ground. What if it were fired more quickly? Then it would travel farther. If it was fired so fast that it traveled far enough in a straight line that the Earth curved away beneath it, where would it fall? Newton realized that it would be pulled towards Earth but would then follow a circular orbit. Just like a satellite constantly being pulled but never reaching the ground.

> **Gravity is a habit that is hard to shake off.**
> Terry Pratchett, 1992

When Olympic hammer-throwers spin on their heels, it is the pull on the string that keeps the hammer rotating. Without this pull the hammer would fly off in a straight line, just as it does on its release. It's just the same with Newton's cannonball — without the centrally directed force tying the projectile to Earth, it would fly off into space. Thinking further, Newton reasoned that the Moon also hangs in the sky because it is held by the invisible tie of gravity. Without gravity it too would fly off into space.

Inverse square law

Newton then tried to quantify his predictions. After exchanging letters with Robert Hooke, Newton showed that gravity follows an inverse square law — the strength of gravity decreases by the square of the distance from a body. So if you travel twice some distance from a body its gravity is four times less; the gravity exerted by the Sun would be four times less for a planet in an orbit twice as far from it as the Earth, or a planet three times distant would experience gravity nine times less.

Newton's inverse square law of gravity explained in one equation the orbits of all the planets as described in the three laws of Johannes Kepler (see Chapter 36). Newton's law predicted that they traveled quicker near the Sun as they followed their elliptical paths. A planet feels a stronger gravitational force from the Sun when it travels close to it, which makes it speed up. As its speed increases the planet is thrown away from the Sun again, gradually slowing back down. Thus, Newton pulled together all the earlier work into one profound theory.

> " Every object in the universe attracts every other object along a line of the centers of the objects, proportional to each object's mass, and inversely proportional to the square of the distance between the objects. "
>
> Sir Isaac Newton, 1687

Universal law

Generalizing boldly, Newton then proposed that his theory of gravity applied to everything in the universe. Any body exerts a gravitational force in proportion to its mass, and that force falls off as the inverse square of distance from it. So any two bodies attract each other. But because gravity is a weak force we only really observe this for very massive bodies, such as the Sun, Earth and planets.

If we look closer, though, it is possible to see tiny variations in the local strength of gravity on the surface of the Earth. Because massive mountains and rocks of differing density can raise or reduce the strength of gravity near them, it is possible to use a gravity meter to map out geographic terrains and to learn about the structure of the Earth's crust. Archeologists also sometimes use tiny gravity changes to spot buried settlements. Recently, scientists have used gravity-measuring space satellites to record the (decreasing) amount of ice covering the Earth's poles and also to detect changes in the Earth's crust following large earthquakes.

> **It has been said that arguing against globalization is like arguing against the laws of gravity.**
> Kofi Annan, b.1938

Back in the 17th century, Newton poured all his ideas on gravitation into one book, *Philosophiae naturalis principia mathematica*, known as the *Principia*. Published in 1687, the *Principia* is still revered as a scientific milestone. Newton's universal law of gravity theory explained the motions not only of planets and moons but also of

The discovery of Neptune

The planet Neptune was discovered thanks to Newton's law of gravitation. In the early 19th century, astronomers noticed that Uranus did not follow a simple orbit but acted as though another body was disturbing it. Various predictions were made based on Newton's law and in 1846 the new planet, named Neptune after the sea god, was discovered close to the expected position. British and French astronomers disagreed over who had made the discovery, which is credited to both John Couch Adams and Urbain Le Verrier. Neptune has a mass 17 times that of the Earth and is a "gas giant" with a thick dense atmosphere of hydrogen, helium, ammonia and methane smothering a solid core. The blue color of Neptune's clouds is due to methane. Its winds are the strongest in the solar system, reaching as much as 2,500 kilometers per hour.

Tides

Newton described the formation of ocean tides on the Earth in his *Principia*. Tides occur because the Moon pulls differently on oceans on the near and far sides of the Earth, compared with the solid Earth itself. The different gravitational pull on opposite sides of the Earth causes the surface water to bulge both towards and away from the Moon, leading to tides that rise and fall every 12 hours. Although the more massive Sun exerts a stronger gravitational force on the Earth than the smaller Moon, the Moon has a stronger tidal effect because it is closer to the Earth. The inverse square law means that the gravitational gradient (the difference felt by the near and far sides of the Earth) is much greater for the closer Moon than the distant Sun. During a full or new Moon, the Earth, Sun and Moon are all aligned and especially high tides result, called "spring" tides. When these bodies are misaligned, at 90 degrees to one another, weak tides result called "neap" tides.

projectiles, pendulums and apples. He explained the orbits of comets, the formation of tides and the wobbling of the Earth's axis. This work cemented Newton's reputation as one of the great scientists of all time.

Newton's universal law of gravitation has stood for hundreds of years and still today gives a basic description of the motion of bodies. However, science does not stand still, and 20th-century scientists built upon its foundations, notably Einstein with his theory of general relativity. Newtonian gravity still works well for most objects we see and for the behavior of planets, comets and asteroids in the solar system that are spread over large distances from the Sun where gravity is relatively weak. Although Newton's law of gravitation was powerful enough to predict the position of the planet Neptune, discovered in 1846 at the expected location beyond Uranus, it was the orbit of another planet, Mercury, that required physics beyond that of Newton. Thus, general relativity is needed to explain situations where gravity is very strong, such as close to the Sun, stars and black holes.

38 Conservation of energy

Energy is an animating force that makes things move or change. It comes in many guises and may manifest itself as a change in height or speed, traveling electromagnetic waves or the vibrations of atoms that cause heat. Although energy can metamorphose between these types, the overall amount of energy is always conserved. More cannot be created and it can never be destroyed.

We are all familiar with energy as a basic drive. If we are tired, we lack it; if we are leaping around with joy, we possess it. But what is energy? The energy that fires up our bodies comes from combusting chemicals, changing molecules from one type into another with energy being released in the process. But what types of energy cause a skier to speed down a slope or a light bulb to shine? Are they really the same thing?

Coming in so many different guises, energy is difficult to define. Even now, physicists do not know intrinsically what it is, even though they are expert at describing what it does and how to handle it. Energy is a property of matter and space, a sort of fuel or encapsulated drive with the potential to create, to move or to change. Philosophers of nature going back to the ancient Greeks had a vague notion of energy as a force or essence that gives life to objects, and this idea has stuck with us through the ages.

Energy exchange

It was Galileo who first spotted that energy might be transformed from one type to another. Watching a pendulum swinging back and forth, he saw that the bob exchanges height for forward motion, and vice versa as the speed then brings the pendulum back up again before it falls and repeats the cycle. The pendulum bob has no sideways velocity when it is at either peak of its swing, and moves most quickly as it passes through the lowest point.

Galileo reasoned that there are two forms of energy being swapped by the swinging bob. One is gravitational potential energy, which may raise a body above the Earth in opposition to gravity. Gravitational energy needs to be added to lift a mass higher, and is released when it falls. If you have ever cycled up a steep hill you will know it takes a lot of energy to combat gravity. The other type of energy in the bob is kinetic energy — the energy of motion that accompanies speed. So the pendulum converts gravitational potential energy into kinetic energy and vice versa. A canny cyclist uses exactly the same mechanism. Riding down a steep hill, she could pick up speed and race to the bottom even without pedaling, and may use that speed to climb some of the way up the next hill (see box).

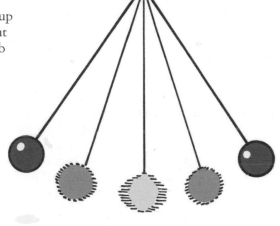

Likewise, the simple conversion of potential into kinetic energy can be harnessed to power our homes. Hydroelectric schemes and tidal barrages release water from a height, using its speed to drive turbines and generate electricity.

Many faces of energy

Energy manifests itself as many different types that can be held temporarily in different ways. A compressed spring can store within it elastic energy that can be released on demand. Heat energy increases the vibrations of atoms and molecules in the hot material. So a metal pan on a cooker heats up because the atoms within it are being made to

Energy formulae

Gravitational potential energy (PE) is written algebraically as $PE = mgh$, or mass (m) times gravitational acceleration (g) times height (h). This is equivalent to force ($F = ma$ from Newton's second law) times distance. So a force is imparting energy.

Kinetic energy (KE) is given by $KE = \frac{1}{2}mv^2$ so the amount of energy scales with the square of velocity (v). This also comes from working out the average force times the distance moved.

wobble faster by the input of energy. Energy can also be transmitted as electric and magnetic waves, such as light or radio waves, and stored chemical energy may be released by chemical reactions, as happens in our own digestive systems.

Einstein revealed that mass itself has an associated energy that can be released if the matter is destroyed. So, mass and energy are equivalent. This is his famous $E = mc^2$ equation — the energy (E) released by the destruction of a mass (m) is m times the speed of light (c) squared. This energy is released in a nuclear explosion or in the fusion reactions that power the Sun (see Chapters 54 and 55). Because it is scaled by the speed of light squared, which is very large (light travels at 300 million meters per second in a vacuum), the amount of energy released by destroying even a few atoms is enormous.

We consume energy in our homes and use it to power industry. We talk about energy being generated, but in reality it is being transformed from one type to another. We take chemical energy from coal or natural gas and convert it into heat that spins turbines and creates electricity. Ultimately even the chemical energy in coal and gas comes from the Sun, so solar energy is the root of everything that operates on Earth. Even though we worry that energy supplies on Earth are limited, the amount of energy that can be derived from the Sun is more than enough to power our needs, if we can only harness it.

Energy conservation

Energy conservation as a rule of physics is much more than reducing our use of household energy; it states that the total amount of energy is unchanged even though it may switch between different types. The concept appeared relatively recently only after many types of energies were studied individually. At the start of the 19th century, Thomas Young introduced the word energy; before then this life force was called *vis viva* by Gottfried Leibniz who originally worked out the mathematics of the pendulum.

It was quickly noticed that kinetic energy alone was not conserved. Balls or flywheels slowed down and did not move forever. But fast motions did often cause machines to heat up by friction, such as when boring metal cannon tubes, so experimenters deduced that heat was one destination for released energy. Gradually, on accounting for all the different types of energy in built machines, the scientists began to show that energy is transferred from one type to another and is not destroyed or created.

Momentum

The idea of conservation in physics is not limited to energy. Two other concepts are closely related — the conservation of linear momentum and the conservation of angular momentum. Linear momentum is defined as the product of mass and velocity, and describes the difficulty of slowing a moving body. A heavy object moving quickly has high momentum and is difficult to deflect or stop. So a truck moving at 60 kilometers an hour has more momentum than a car moving at the same speed, and would do even more damage if it hit you. Momentum has not just a size but, because of the velocity, it also acts in a specific direction. Objects that collide exchange momentum such that overall it is conserved, both in amount and direction. If you have ever played billiards or pool you have used this law. As two balls collide, they transfer motion from one to the other so as to conserve momentum. So if you hit a still ball with a moving one, the final paths of both balls will be a combination of the velocity and direction of the initial moving ball. The speed and direction of both can be worked out assuming that momentum is conserved in all directions.

Angular momentum conservation is similar. Angular momentum, for an object spinning about a point, is defined as the product of the object's linear momentum and the distance it is away from the rotation point. Conservation of angular momentum is used to effect in performances by spinning ice skaters. When their arms and legs are stretched out they whirl slowly, but just by pulling their limbs in to their body they can spin faster. This is because the smaller dimensions require an increased rotation speed to compensate. Try doing this in an office chair; it works too.

Conservation of energy and momentum are still basic tenets of modern physics. They are concepts that have found a home even in contemporary fields such as general relativity and quantum mechanics.

39 Second law of thermodynamics

The second law of thermodynamics is a pillar of modern physics. It says that heat travels from hot to cold bodies, and not the other way around. Because heat measures disorder, or entropy, another way of expressing the concept is that entropy always increases for an isolated system. The second law is tied to the progression of time, the unfolding of events and the ultimate fate of the universe.

When you add hot coffee to a glass of ice, the ice heats up and melts and the coffee is cooled. Have you ever asked why the temperature doesn't become more extreme? The coffee could extract heat from the ice, making itself hotter and the ice even cooler. Our experience tells us this doesn't happen, but why is this so?

The tendency of hot and cold bodies to exchange heat and move towards an even temperature is captured in the second law of thermodynamics. It says that, overall, heat cannot flow from a cold to a hot object.

So how do refrigerators work? How can we chill a glass of orange juice if we cannot transfer its warmth to something else? The second law allows us to do this in special circumstances only. As a by-product of cooling things down, refrigerators also generate a lot of heat, as you can tell if you put your hand behind the back of one. Because they liberate heat, they do not in fact violate the second law if you look at the total energy of the refrigerator and its surroundings.

Entropy

Heat is really a measure of disorder and, in physics, disorder is often quantified as "entropy," which measures the ways in which a number of items can arrange themselves. A packet of uncooked spaghetti, a bundle of aligned pasta sticks, has low entropy because it shows high order. When the spaghetti is thrown into a pan of boiling water and becomes tangled, it is more disordered and so has higher entropy. Similarly, neat rows of toy soldiers have low entropy, but their distribution has higher entropy if they are scattered across the floor.

What has this got to do with refrigerators? Another way of stating the second law of thermodynamics is that, for a bounded system, entropy increases; it never decreases. Temperature is directly related to entropy and cold bodies have low entropy. Their atoms are less disordered than those in hot bodies, which jiggle around more. So any changes in the entropy of a system, considering all its parts, must produce a net effect that is an increase.

> " Just as the constant increase of entropy is the basic law of the universe, so it is the basic law of life to be ever more highly structured and to struggle against entropy. "
>
> Václav Havel, 1977

In the case of the refrigerator, cooling the orange juice decreases its entropy, but this is compensated for by the hot air that the appliance produces. In fact the entropy increase of the hot air actually exceeds any drop due to chilling. If you consider the whole system, refrigerator and surroundings, then the second law of thermodynamics still holds true. Another way of stating the second law is that entropy always increases.

The second law is true for an isolated system, a sealed one where there is no influx into or outflow of energy from it. Energy is conserved within it. The universe itself is an isolated system, in that nothing exists outside it, by definition. So for the universe as a whole, energy is conserved and entropy must always increase. Small regions might experience a slight decrease in entropy, such as by cooling, but this has to be compensated for, just like the refrigerator, by other regions heating up and creating more entropy so that the sum increases overall.

What does an increase in entropy look like? If you pour chocolate syrup into a glass of milk, it starts off with low entropy; the milk and syrup are distinct swathes of white and brown. If you increase the disorder by stirring the drink, then the molecules become mixed up together. The end point of maximum disorder is when the syrup is completely mixed into the milk and it turns a pale caramel color.

The (un)fashionable universe?

Astronomers recently tried to calculate the average color of the universe, by adding up all the starlight in it, and found it is not sunshine yellow or pink or pale blue, but a rather depressing beige. In billions of years, when entropy finally wins out over gravity, the universe will become a uniform sea of beige.

Thinking again of the whole universe, the second law likewise implies that atoms gradually become more disordered over time. Any clumps of matter will slowly disperse until the universe is awash with their atoms. So the eventual fate of the universe, which starts out as a multicolor tapestry of stars and galaxies, is a gray sea of mixed atoms. When the universe has expanded so much that galaxies are torn apart and its matter is diluted, all that will remain is a blended soup of particles. This end state, presuming the universe continues to expand, is known as "heat death."

Perpetual motion

Because heat is a form of energy, it can be put to work. A steam engine converts heat into mechanical movement of a piston or turbine, which may produce electricity. Much of the science of thermodynamics was developed in the 19th century from the practical engineering of steam engines, rather than first being deduced by physicists on paper. Another implication of the second law is that steam engines, and other engines that run off heat energy, are not perfect. In any process that changes heat into another form of energy a little energy is lost, so that the entropy of the system as a whole increases.

The idea of a perpetual motion machine, an engine that never loses energy and so can run forever, has been tantalizing scientists since medieval times. The second law of thermodynamics scotched their hopes, but before this was known many of them put forward sketches of possible machines. Robert Boyle imagined a cup that drained and refilled itself, and the Indian mathematician Bhaskara proposed a wheel that propelled its own rotation by dropping weights along spokes as it rolled. In fact, on closer inspection, both machines lose energy. Ideas like these were so widespread that even in the 18th century perpetual motion machines garnered a bad name. Both the French Royal Academy of Sciences and the American Patent Office banned consideration of perpetual motion machines. Today they remain the realm of eccentric backyard inventors.

Maxwell's demon

One of the most controversial attempts to violate the second law was proposed as a thought experiment by the Scottish physicist James Clerk Maxwell, in the 1860s. Imagine two boxes of gas, side by side, both at the same temperature. A small hole is placed between the boxes, so that particles of gas can pass from one box to the other. If one side was warmer than the other, particles would pass through and gradually even out the temperature. Maxwell imagined that there was a tiny demon, a microscopic devil, who could grab only fast molecules from one box and push them through into the other. In this way the average speed of molecules in that box would increase, at the expense of the other. So, Maxwell postulated, heat could be moved from the colder to the hotter box. Wouldn't this process violate the second law of thermodynamics? Could heat be transferred into the hotter body by selecting the right molecules?

An explanation of why Maxwell's demon could not work has puzzled physicists ever since. Many have argued that the process of measuring the particles' velocities and opening and closing any trap door would require work and therefore energy, so this would mean that the total entropy of the system would not decrease. The nearest anyone has come to a "demon machine" is the nanoscale work of Edinburgh physicist David Leigh. His creation has indeed separated fast- and slow-moving particles, but requires an external power source to do so. Because there is no mechanism that could move particles without using extra energy, even today's physicists have not found a way to violate the second law. Thus far, at least, it is holding fast.

Another view of the laws of thermodynamics

First law You can't win (see Chapter 38).
Second law You can only lose (see page 172).
Third law You can't get out of the game (see Chapter 40).

40 Newton's theory of color

We've all wondered at the beauty of a rainbow — Isaac Newton explained how they form. Passing white light through a glass prism, he found it split into rainbow hues and showed that the colors were embedded in the white light rather than imprinted by the prism. Newton's color theory was contentious at the time but has influenced generations of artists and scientists ever since.

Shine a beam of white light through a prism and the emerging ray spreads out into a rainbow of colors. Rainbows in the sky appear in the same way as sunlight is split by water droplets into the familiar spectrum of hues: red, orange, yellow, green, blue, indigo and violet.

All in the mix

Experimenting with light and prisms in his rooms in the 1660s, Sir Isaac Newton demonstrated that light's many colors could be mixed together to form white light. Colors were the base units rather than being made by later mixing or by the prism glass itself, as had been

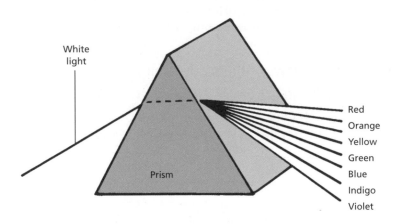

White light

Prism

Red
Orange
Yellow
Green
Blue
Indigo
Violet

thought. Newton separated beams of red and blue light and showed that these single colors were not split further if they were passed through more prisms.

Although so familiar today, Newton's theory of color proved contentious at the time. His peers argued vociferously against it, preferring to believe instead that colors arose from combinations of white light and darkness, as a type of shadow. Newton's fiercest battles were with his equally famous contemporary, Robert Hooke. The pair fought publicly over color theory throughout their lifetimes. Hooke believed instead that colored light was an imprint, just as if you look through stained glass. He cited many examples of unusual colored light effects in real life to back up his claim and criticized Newton for not performing more experiments.

> ❝ Nature and nature's laws lay hid in night; God said 'Let Newton be' and all was light. ❞
> Alexander Pope, 1727 (Newton's epitaph)

Newton also realized that objects in a lit room appear colored because they scatter or reflect light of that color, rather than color being somehow a quality of the object. A red sofa reflects primarily red light, and a green table reflects green light. A turquoise cushion reflects blue and a little yellow light. Other colors arise from mixtures of these basic types of light.

Light waves

For Newton, understanding color was a means of interrogating the physics of light itself. Experimenting further, he concluded that light behaves in many ways like water waves. Light bends around obstacles in a similar way to sea waves around a harbor wall. Light beams can also be added together to reinforce or cancel out their brightness, as overlapping water waves do. In the same way that water waves are large-scale motions of invisible water molecules, Newton believed that light waves were ultimately ripples of miniscule light particles, or "corpuscles," which were even smaller than atoms. What Newton did not know, because it was not discovered until centuries later, was that light waves are in fact electromagnetic waves — waves of coupled electric and magnetic fields — and not the reverberation of solid particles. When the electromagnetic wave behavior of light was discovered, Newton's corpuscle idea was put on ice. It was resurrected, however, in a new form when Einstein showed that light may also behave sometimes like a stream of particles that can carry energy but have no mass.

Wave motions appear in many guises. There are two basic wave types: longitudinal and transverse waves. Longitudinal, or compression, waves result when the pulses that produce the wave act along the same direction in which the wave travels, causing a series of high and low pressure crests. Sound waves, caused for example by a drum skin vibrating in air, are longitudinal, as are the ripples of a millipede's legs as they crunch up close and then spread apart as the creature shuffles forwards. Light and water waves, on the other hand, are transverse waves where the original disturbance acts at a right angle to the direction of travel of the wave. If you sweep one end of a slinky spring from side to side a transverse wave will travel along the length of the spring even though your hand's motion is perpendicular to it. Similarly, a snake makes a transverse wave as it slithers, using side to side motion to propel it forwards. Water waves are also transverse because individual water molecules float up and down whereas the wave itself travels towards the horizon. Unlike water waves, the transverse motion of light waves is due to changes in the strength of electric and magnetic fields that are aligned perpendicular to the direction of wave propagation.

Across the spectrum

The different colors of light reflect the different wavelengths of these electromagnetic waves. Wavelength is the measured distance between

Color wheel

Newton arranged the colors of the rainbow in order from red to blue and painted them onto a circular color wheel, so he could show the ways in which colors combined. Primary colors — red, yellow and blue — were spaced around it, and when combined in different proportions could make all the other colors in between. Complementary colors, such as blue and orange, were placed opposite one another. Many artists became interested in Newton's color theory and especially in his color wheel that helped them depict contrasting hues and illumination effects. Complementary colors achieved maximum contrast, or were useful for painting shadows.

consecutive crests of a wave. As it passes through a prism, the white light separates into many hues because each hue is associated with a different wavelength and so they are deflected to varying degrees by the glass. The prism bends the light waves by an angle that depends on the wavelength of light, where red light is bent least and blue most, to produce the rainbow color sequence. The spectrum of visible light appears in order of wavelength, from red with the longest through green to blue with the shortest.

What lies at either end of the rainbow? Visible light is just one part of the electromagnetic spectrum. It is so important to us because our eyes have developed to use this sensitive part of the spectrum. As the wavelengths of visible light are on roughly the same scale as atoms and molecules (hundreds of billionths of a meter), the interactions between light and atoms in a material are large. Our eyes have evolved to use visible light because it is very sensitive to atomic structure. Newton was fascinated by how the eye worked; he even stuck a darning needle round the back of his own eye to see how pressure affected his perception of color.

Beyond red light comes infrared, with wavelengths of millionths of a meter. Infrared rays carry the Sun's warmth and are also collected by night-vision goggles to "see" the heat from bodies. Longer still are microwaves, with millimeter-to-centimeter wavelengths, and radio waves, with wavelengths of meters and longer. Microwave ovens use microwave electromagnetic rays to spin water molecules within food, heating them up. At the other end of the spectrum, beyond blue, comes ultraviolet light. This is emitted by the Sun and can damage our skin, although much of it is stopped by the Earth's ozone layer. At even shorter wavelengths are x-rays — used in hospitals because they travel through human tissue — and at the smallest wavelengths are gamma rays.

Developments

While Newton elucidated the physics of light, philosophers and artists remained interested in our perception of colors. In the 19th century, German polymath Johann Wolfgang von Goethe, investigated how the human eye and mind interprets colors placed next to one another. Goethe introduced magenta to Newton's color wheel (see box) and noticed that shadows often take on the opposite color of the illuminated object, so that a blue shadow appears to fall behind a red object. Goethe's updated color wheel remains the choice for artists and designers today.

41 Huygens' principle

If you drop a stone into a pond, it produces a circular expanding ripple. Why does it expand? And how might you predict its behavior if it then flows round an obstacle, such as a tree stump, or reflects back from the edge of the pond? Huygens' principle is a tool for working out how waves flow by imagining that every point on a wavefront is a new ripple source.

Dutch physicist Christiaan Huygens devised a practical way for predicting the progression of waves. Let's say you have cast a pebble into a lake, and rings of ripples result. If you imagine freezing a circular ripple at a moment in time, then each point on the circular wave can be thought of as a new source of circular waves whose properties match those of the frozen ripple. It is as if a ring of stones was dropped simultaneously into the water following the outline of the first wave. This next set of disturbances widens the ripple further, and the new locus marks the starting points for another set of sources of spreading wave energy. By repeating the principle many times the evolution of the wave can be tracked.

Step by step

The idea that every point on a wavefront acts like a new source of wave energy with matching frequency and phase is called Huygens' principle. The frequency of a wave is the number of wave cycles that occur in some time period and the phase of a wave identifies where you are in the cycle. For example, all wave crests have the same phase, and all troughs are half a cycle away from them. If you imagine an ocean wave, the distance between two wave peaks, known as its wavelength, is maybe 100 meters. Its frequency, or the number of wavelengths that pass some point in one second, might be one wavelength of 100 meters in 60 seconds, or 1 cycle per minute. The fastest ocean waves are tsunami that can reach 800 kilometers per hour, the speed of a jet aircraft, slowing down to tens of kilometers per hour and rising up as they reach and swamp the coast.

To map the progress of a wave, Huygens' principle can be applied again and again as it encounters obstacles and crosses the paths of other waves. If you draw the position of a wavefront on a piece of paper, then the subsequent position can be described by using pairs of compasses to draw circles at many points along the wavefront, and drawing a smooth line through their outer edges to plot the next wave position.

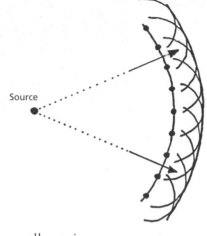

Source

The simple approach of Huygens describes waves in many circumstances. A linear wave remains straight as it propagates because the circular wavelets it produces along its length add together to form a new linear wavefront ahead of the first. If you watch sets of parallel linear ocean waves as they pass through a small opening in a harbor wall, however, they distort into arcs once they pass through the gap. Only a very short length of straight wave passes through, and the arcs are formed at the edges of this unaffected remnant where, according to Huygens' principle, new circular ripples are born. If the gap is small compared with the distance between the waves then the rounded edges dominate the pattern and the transmitted wave may look almost semi-circular. This spreading out of the wave energy either side of the gap is called diffraction.

Christiaan Huygens 1629–95

Son of a Dutch diplomat, Christiaan Huygens was an aristocratic physicist who collaborated widely with scientists and philosophers across Europe in the 17th century, including such famous names as Newton, Hooke and Descartes. Huygens' first publications were on mathematical problems, but he also studied Saturn. He was a practical scientist who patented the first pendulum clock and tried to devise a nautical clock that could be taken to sea to calculate longitude. Huygens traveled throughout Europe, especially Paris and London, meeting and working with prominent scientists on the pendulum, circular motion, mechanics and optics. Although he worked on centrifugal force alongside Newton, Huygens thought Newton's theory of gravity, with its concept of action at a distance, "absurd." In 1678 Huygens published his treatise on the wave theory of light.

In 2004, a catastrophic tsunami created by a huge earthquake off Sumatra sped across the entire Indian Ocean. Its force in some places was diminished because the wave energy was spread out by diffraction as it traveled past and between strings of islands.

Believe your ears?

Huygens' principle also explains why if you shout to someone in another room, they hear your voice as if you are standing in the doorway rather than elsewhere in the adjacent room. According to Huygens, when the waves arrive at the doorway, just like the harbor opening, a new set of point-like sources of wave energy is created there. So all the listening person knows is that these waves were generated at the doorway, the past history of the waves in the other room is lost.

Likewise, if you watch a circular ripple as it reaches the edge of a pond, its reflection produces inverted circles. The first wave point to reach the edge acts as a new source, so the backward propagation of a new circular ripple begins. Thus wave reflections can also be described using Huygens' principle.

If ocean waves move into shallower water, such as near a beach, their speed changes and the wavefronts bend inwards towards the shallows. Huygens described this "refraction" by altering the radius of the wavelets so that slower waves produced smaller wavelets. The slow wavelets do not travel as far as faster ones, so the new wavefront is at an angle to the original.

One unrealistic prediction of Huygens' principle is that if all these new wavelets are sources of wave energy then they should generate a reverse wave as well as a forward wave. So why does a wave propagate

Huygens on Titan

The Huygens space probe landed on the surface of Titan on January 14, 2005, after a seven-year journey. Contained inside a protective outer shell a few meters across, the Huygens probe carried a suite of experiments that measured the winds, atmospheric pressure, temperature and surface composition as it descended through the atmosphere to land on an icy plain. Titan is a weird world whose atmosphere and surface is damp with liquid methane. It is, some think, a place that could harbor primitive life forms such as methane-eating bacteria. Huygens was the first space probe to land on a body in the outer solar system.

only forwards? Huygens did not have an answer and simply assumed that wave energy propagates outwards and the backwards motion is ignored. Therefore, Huygens' principle is really only a useful tool for predicting the evolution of waves rather than a fully explanatory law.

> " Each time a man stands up for an ideal . . . he sends forth a tiny ripple of hope, and crossing each other from a million different centers of energy and daring, those ripples build a current that can sweep down the mightiest walls of oppression and resistance. "
> Robert Kennedy, 1966

Saturn's rings

As well as wondering about ripples, Huygens also discovered Saturn's rings. He was the first to demonstrate that the planet was girdled by a flattened disk rather than flanked by extra moons or a changing equatorial bulge. He deduced that the same physics that explained the orbits of moons, Newton's gravity, would apply to many smaller bodies that would orbit in a ring. In 1655, Huygens also discovered Saturn's largest moon, Titan. Exactly 350 years later a spaceship called Cassini reached Saturn, carrying with it a small capsule, named after Huygens, which descended through the clouds of Titan's atmosphere to land on its surface of frozen methane. Titan has continents, sand dunes, lakes, and perhaps rivers, made of solid and liquid methane and ethane, rather than water. Huygens would have been amazed to think that a craft bearing his name would one day travel to that distant world, but the principle named after him can still be used to model the alien waves found there.

42 Bragg's law

The double helix structure of DNA was discovered using Bragg's law. It explains how waves traveling through an ordered solid reinforce one another to produce a pattern of bright spots whose spacing depends on the regular distances between the atoms or molecules in the solid. By measuring the emergent spot pattern the architecture of the crystalline material can be deduced.

If you are sitting in a lit room, put your hand close to the wall and you will see behind it a sharp silhouette. Move your hand farther away from the wall and the shadow's outline becomes fuzzy. This is due to light diffracting around your hand. The rays of light spread inwards around your fingers as they pass by, smudging their outline. All waves behave like this. Water waves diffract around the edges of harbor walls and sound waves curve out beyond the edges of concert stages.

> The important thing in science is not so much to obtain new facts as to discover new ways of thinking about them.
>
> **Sir William Bragg, 1968**

Diffraction can be described using Huygens' principle, which predicts the passage of a wave by considering each point on a wavefront to be a point source of further wave energy. Each point produces a circular wave, and these waves add together to describe how the wave progresses forward. If the wavefront is restricted, then the circular waves at the end points spread unimpeded. This happens when a series of parallel waves pass around an obstacle, such as your hand, or through an aperture, such as a harbor entrance or doorway.

X-ray crystallography

Australian physicist Sir William Lawrence Bragg discovered that diffraction even happens for waves traveling through crystals. A crystal is made up of many atoms stacked in a neat lattice structure, with regular rows and columns. When Bragg shone x-rays through a crystal and out onto a screen, the rays scattered off the rows of atoms. The outgoing rays piled up more in certain directions than others, gradually building up a pattern of spots. Different spot patterns appeared depending on the type of crystal used.

Sir William Lawrence Bragg 1890–1971

William Lawrence Bragg was born in Adelaide, where his father William Henry was a professor of mathematics and physics. Bragg junior became the first Australian to have a medical x-ray when he fell off his bicycle and broke his arm. He studied physical sciences and after graduating he followed his father to England. At Cambridge Bragg discovered his law on the diffraction of x-rays by crystals. He discussed his ideas with his father, but was upset that many people thought his father had made the discovery rather than him. During the First and Second World Wars, Bragg joined the army and worked on sonar. Afterwards, he returned to Cambridge where he set up several small research groups. In his later career, Bragg became a popular science communicator, setting up lectures for school children in London's Royal Institution and appearing regularly on television.

X-rays, discovered by German physicist Wilhelm Röntgen in 1895, were needed to see this effect because their wavelength is tiny, a 1,000 times less than the wavelength of visible light, and smaller than the spacing of atoms in the crystal. So x-ray wavelengths are small enough to travel through, and be strongly diffracted by, the crystal layers.

The brightest x-ray spots are generated when rays traverse paths through the crystal that result in their signals being "in phase" with one another. In-phase waves, where the peaks and troughs are aligned, can add together to reinforce their brightness and produce spots. When "out of phase," with the peaks and troughs misaligned, they cancel out

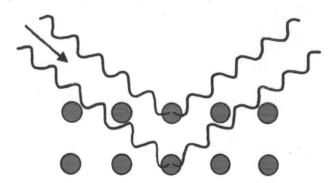

and no light emerges. So you see a pattern of bright dots whose spacing tells you the distances between rows of atoms in the crystal. This effect of reinforcement and canceling of waves is called "interference."

Bragg wrote this down mathematically by considering two waves, one reflecting from the crystal surface and the other having penetrated just one layer of atoms inside the crystal. For the second wave to be in phase and reinforce the first wave it must travel an extra distance that is a whole number of wavelengths longer than the first wave. This extra distance depends on the angle at which the rays strike and the separation between the layers of atoms. Bragg's law states how the observed interference and crystal spacings are related for a certain wavelength.

Deep structure

X-ray crystallography is widely used to determine the structure of new materials and by chemists and biologists investigating molecular architectures. In 1953 it was used to identify the double helix structure of DNA; Francis Crick and James Watson famously got their idea from looking at Rosalind Franklin's x-ray interference patterns from DNA and realizing that the molecules that produced them must be arranged as a double helix (See Chapter 6).

For the first time, the discovery of x-rays and crystallography techniques gave physicists tools to look at the deep structure of matter

The DNA double helix

In the 1950s researchers were puzzling over the structure of DNA, one of the building blocks of life. British physicists James Watson and Francis Crick published its double helix structure in 1953, which was a major breakthrough. They acknowledged inspiration from researchers at King's College London, Maurice Wilkins and Rosalind Franklin, who had made x-ray crystallography photographs of DNA using Bragg's law. Franklin made exquisitely clear photographs showing the interference array of bright spots that ultimately gave away DNA's structure. Crick, Watson and Wilkins received the Nobel Prize for their work, but Franklin missed out because she died young. Some also think her role in the discovery was played down, perhaps due to sexist attitudes at the time. Franklin's results may also have been leaked to Watson and Crick without her awareness. Her contribution has since been acknowledged.

Wilhelm Röntgen 1845–1923

Wilhelm Röntgen was born in Germany's Lower Rhine, moving as a young child to the Netherlands. He studied physics at Utrecht and Zurich, and worked in many universities before his major professorships at the Universities of Würzburg and then Munich. Röntgen's work centered on heat and electromagnetism, but he is most famous for his discovery of x-rays in 1895. While passing electricity through a low-pressure gas he saw that a chemical-coated screen fluoresced even when the experiment was carried out in complete darkness. These new rays passed through many materials, including the flesh of his wife's hand placed in front of a photographic plate. He called the rays "x-rays" because their origin was unknown. Later, it was shown that they are electromagnetic waves like light except that they have much higher frequency.

and even inside the body. Many techniques used today for medical imaging rely on similar physics. Computed tomography reassembles many x-ray slices of the body into a realistic internal picture; ultrasound maps high-frequency echoes from organs in the body; magnetic resonance imaging (MRI) scans water throughout the body's tissues, identifying molecular vibrations set up using powerful magnets; and positron emission tomography (PET) follows radioactive traces as they flow through the body. So doctors and patients alike are grateful to physicists such as Bragg for developing these tools.

Bragg's law is written mathematically as:

$$2d \sin \theta = n\lambda$$

where d is the distance between the atomic layers, θ is the angle of incidence of the light, n is a whole number and λ is the wavelength of light.

43 Doppler effect

We've all heard the drop in pitch of an ambulance siren's wail as it speeds past. Waves coming from a source that is moving towards you arrive squashed together and so seem to have a higher frequency. Similarly, waves become spread out and so take longer to reach you from a source that is receding, resulting in a frequency drop. This is the Doppler effect. It has been used to measure speeding cars, blood flow and the motions of stars and galaxies in the universe.

When an ambulance races past you on the street its siren wail changes in pitch, from high when it approaches to low as it recedes. This change in tone is the Doppler effect, proposed by Austrian mathematician and astronomer Christian Doppler in 1842. It arises because of the motion of the emitting vehicle relative to you, the observer. As the vehicle approaches, its sound waves pile up, the distance between each wavefront is squashed and the sound gets higher. As it speeds away, the wavefronts consistently take a little longer to reach you, the intervals get longer and the pitch drops. Sound waves are pulses of compressed air.

Christian Doppler 1803–53

Christian Doppler was born into a family of stonemasons in Salzburg, Austria. He was too frail to continue the family business and went to university in Vienna instead to study mathematics, philosophy and astronomy. Before finding a university job in Prague, Doppler had to work as a book-keeper and even considered emigrating to America. Although promoted to professor, Doppler struggled with his teaching load, and his health suffered. One of his friends wrote: "It is hard to believe how fruitful a genius Austria has in this man. I have written to . . . many people who can save Doppler for science and not let him die under the yoke. Unfortunately I fear the worst." Doppler eventually left Prague and moved back to Vienna. In 1842, he presented a paper describing the color shift in the light of stars, that we now call the Doppler effect.

"It is almost to be accepted with certainty that this will in the not too distant future offer astronomers a welcome means to determine the movements and distances of such stars which, because of their unmeasurable distances from us and the consequent smallness of the parallactic angles, until this moment hardly presented the hope of such measurements and determinations."

Although regarded as imaginative, he received a mixed reception from other prominent scientists. Doppler's detractors questioned his mathematical ability, whereas his friends thought very highly of his scientific creativity and intuition.

To and fro

Imagine if someone on a moving platform, or train, was throwing balls to you continually at a frequency of one ball every three seconds, prompted by their wristwatch timer. If they are motoring towards you it will always take a little less than three seconds for the balls to arrive because they are launched a little closer to you each time. So the rate will seem quicker to the catcher. Similarly, as the platform moves away, the balls take a little more time to arrive, traveling a little extra distance each throw, so their arrival frequency is lower. If you could measure that shift in timing with your own watch then you could work out the speed of the thrower's train. The Doppler effect applies to any objects moving relative to one another. It would be the same if it was you moving on the train and the ball thrower was standing still on a stationary platform. As a way of measuring speed, the Doppler effect has many applications. It is used in medicine to measure blood flow and also in roadside radars that catch speeding drivers.

Motion in space

Doppler effects also appear frequently in astronomy, showing up wherever there is moving matter. For example, light coming from a planet orbiting a distant star would show Doppler shifts. As the planet moves towards us the frequency rises, and as it spins away its light frequency drops. Light from the approaching planet is said to be "blue-shifted"; as it moves away it has a "red-shift." Hundreds of planets have been spotted around distant stars since the 1990s by finding this pattern imprinted in the glow of the central star.

Redshifts can arise not only due to planets' orbital motions, but also from the expansion of the universe itself, when it is called cosmological redshift. If the intervening space between us and a distant galaxy swells steadily as the universe expands, it is equivalent to the galaxy moving away from us with some speed. Similarly, two dots on a balloon being inflated look as if they are moving apart.

Extrasolar planets

More than 200 planets orbiting around stars other than our Sun have been discovered. Most are gas giants similar to Jupiter but orbiting much closer to their central stars. But a few possible rocky planets, similar to the Earth in size, have been spotted. About one in ten stars have planets, and this has fueled speculation that some may even harbor forms of life. The great majority of planets have been found by observing the gravitational tug of the planet on its host star. Planets are tiny compared to the stars they orbit, so it is hard to see them against their star's glare. But the mass of a planet swings the star around a little, and this wobble can be seen as a Doppler shift in the frequency of a characteristic feature in the spectrum of the star.

The first extrasolar planets were detected around a pulsar in 1992 and around a normal star in 1995. Their detection is now routine but astronomers are still seeking Earth-like solar systems and trying to figure out how different planetary configurations occur. New space observatories, namely the 2006 European telescope COROT and Nasa's Kepler (in 2008), are expected to identify many Earth-like planets in the near future.

Consequently the galaxy's light is shifted to lower frequencies because the waves must travel farther and farther to reach us. So very distant galaxies look redder than ones nearby. Strictly speaking cosmological redshift is not a true Doppler effect because the receding galaxy is not actually moving relative to any other objects near it. The galaxy is fixed in its surroundings and it is the intervening space that is actually stretching.

> **Perhaps when distant people on other planets pick up some wavelength of ours all they hear is a continuous scream.**
> **Iris Murdoch, 1919–99**

To his credit, Doppler himself saw that his discovery could be useful to astronomers but even he could not have foreseen how much would flow from it. He claimed to have seen it recorded in the colors of light from paired stars, but this was disputed in his day.

Doppler was an imaginative and creative scientist but sometimes his enthusiasm outstripped his experimental skill. Decades later, however, redshifts were measured for galaxies by astronomer Vesto Slipher, setting the stage for the development of the big bang model of the universe. And now, the Doppler effect may help identify worlds around distant stars that could even turn out to harbor life.

44 Ohm's law

Why are you safe when flying in a thunderstorm? How do lightning conductors save buildings? Why do the light bulbs in your house not dim every time you switch another one on? Ohm's law has the answers.

Electricity arises from the movement of electric charge. Electric charge is a basic property of subatomic particles that dictates how they interact with electromagnetic fields. These fields create forces that move electrically charged particles. Charge, like energy, is conserved overall; it cannot be created or destroyed but may be moved around.

Charge can be a positive or negative property. Particles of opposite charge attract one another; those with like charges repel. Electrons have a negative charge (measured by Robert Millikan in 1909) and protons a positive charge. Not all subatomic particles are charged, however. Neutrons, as the name suggests, have no charge and so are "neutral."

Static electricity

Electricity may remain static, as a fixed distribution of charges, or flow, as an electric current. Static electricity builds up when charged particles move, so that opposite charges accumulate in different places. If you rub a plastic comb on your sleeve, for example, it becomes charged and can pick up other small items that carry opposing charge, such as small scraps of paper.

Lightning is formed in a similar way, as friction between molecules in turbulent storm clouds builds up electricity that is discharged suddenly as a lightning bolt. Lightning sparks can reach several miles in length and tens of thousands of degrees Celsius in temperature.

On the move

Electric current, as used in the home, is a flow of charge. Metal wires conduct electricity because the electrons in metals are not tied to particular atomic nuclei and can easily be set in motion. Metals are said to be conductors of electricity. The electrons move through a metal wire like water through a pipe. In other materials, it may be positive charges that move. When chemicals are dissolved in water both electrons and

Benjamin Franklin 1706–90

Benjamin Franklin was born in Boston, Massachusetts, the 15th and youngest son of a tallow chandler. Although pushed to become a clergyman, Benjamin ended up working as a printer. Even after Franklin had achieved fame, he signed his letters modestly "B. Franklin, Printer." Franklin issued *Poor Richard's Almanac*, which made him famous through memorable quotes like "Fish and visitors stink in three days." Franklin was a prodigious inventor — developing the lightning rod, glass harmonica, bifocal glasses and many more — but was fascinated most of all by electricity. In 1752 he conducted his most famous experiment, extracting sparks from a thundercloud by flying a kite in a storm. Franklin contributed to public life in his later years, introducing public libraries, hospitals and volunteer fire fighters to America and working to abolish slavery. He became a politician, conducting diplomacy between the United States, Great Britain and France during and after the American Revolution. He was a member of the Committee of Five that drafted the Declaration of Independence in 1776.

positively charged nuclei (ions) float freely. Conducting materials, such as metals, allow charges to move easily through them. Materials that do not allow electricity to pass through, such as ceramics or plastics, are called insulators. Those that conduct electricity in certain circumstances only are termed semiconductors.

Like gravity, an electrical current can be created by a gradient, in this case in an electrical field or an electrical potential. So just as a change in height (gravitational potential) causes a river to run downhill, a change in electrical potential between two ends of a conducting material causes a current of charge to flow through it. This "potential difference," or voltage, drives the current flow and also gives energy to the charges.

Resistance

When lightning strikes, the electric discharge flows very quickly through the ionized air to the ground. In doing so it is canceling out the potential difference that drove it, so a lightning strike carries a huge current. It is the huge current, not the voltage, which can kill you as it surges through your body. In practice, charges cannot move at such large speeds through most materials because they encounter resistance. Resistance limits the size of current by dissipating the electrical energy as heat. To avoid being killed by lightning you could stand on an insulator, perhaps a rubber mat, which has very high resistance. Or you could hide inside a metal cage, as the lightning can flow more easily through the metal bars than your body, which, being mostly water, is not a good conductor. This construction is known as the Faraday cage, after Michael Faraday who built one in 1836. The electrical field pattern set up with a Faraday cage — a hollow conductor — means that all the charge is carried on the outside of the cage, and inside the cage it is completely neutral. Faraday cages were useful safety devices for 19th-century scientists performing with artificial lightning displays. Today they still protect electronic equipment and explain why, when you are flying through an electrical storm in a metal plane, you are safe — even if the plane scores a direct lightning hit. You are equally safe in a metal car, as long as you don't park near a tree.

Benjamin Franklin's lightning conductor operates in a similar way, providing a low-resistance pathway for lightning's current to follow, rather than releasing its energy into the high-resistance building that it hits. Sharp pointed rods work best because they compress the electric field onto their tip, making it more likely that the electricity will be funneled via this route to the ground. Tall trees also concentrate the electric field, so it's not a good idea to shelter beneath one in a storm.

In Philadelphia in 1752, Benjamin Franklin successfully "extracted" electricity from a storm cloud with a kite.

Circuits

Electrical flows follow loops called circuits. The movement of current and energy through circuits can be described in the same way that water flows through a series of pipes. Current is similar to flow speed, voltage to the pressure of water, and resistance to the pipe width or aperture restrictions placed within it.

Georg Ohm published one of the most useful laws for interpreting circuits in 1826. Ohm's law is written algebraically as $V = IR$, which states that the voltage drop (V) is equal to the product of the current (I) and the resistance (R). According to Ohm's law, voltage is proportional to current and resistance. Double the voltage across a circuit and you double the current flowing through it if the resistance is unchanged; to maintain the same current you need a resistance twice as large. Current and resistance are inversely related, so increasing the resistance slows the current. Ohm's law applies to even quite complex circuits with many loops. The simplest circuit can be imagined as a single light bulb

Lightning

It may not strike the same place twice but, on average, lightning strikes the Earth's surface a hundred times every second, or 8.6 million times a day. In the U.S. alone, as many as 20 million lightning bolts hit the ground per year from 100,000 thunderstorms.

connected by wire to a battery. The battery supplies the potential difference needed to drive the current through the wire, and the bulb's tungsten filament provides some resistance as it converts electrical energy into light and heat. What would happen if you inserted a second light bulb into the circuit? According to Ohm's law, if the two light bulbs were placed next to one another you would have doubled the resistance and so the voltage across each of them, and thus the energy available to each, must be split in two making both bulbs glow more faintly. This wouldn't be much use if you were lighting a house — every time you plugged in another light bulb in a room they would all dim.

However, by connecting the second bulb in a linked loop directly around the first, each light bulb can be made to experience the full potential drop. The current diverts at the junction and passes through both bulbs separately before coming back together again, so the second bulb shines as brightly as the first. This sort of circuit is called a "parallel" circuit. The former, where resistors are linked side by side, is a "series" circuit. Ohm's law can be used throughout any circuit to calculate the voltages and currents at any point.

45 Maxwell's equations

Maxwell's four equations are a cornerstone of modern physics and the most important advance since the universal theory of gravitation. They describe how electric and magnetic fields are two sides of the same coin. Both types of field are manifestations of the same phenomenon — the electromagnetic wave.

Early 19th-century experimenters saw that electricity and magnetism could be changed from one form into the other. But James Clerk Maxwell completed one of the major achievements of modern physics when he managed to describe the whole field of electromagnetism in just four equations.

Electromagnetic waves

Electric and magnetic forces act on charged particles and magnets. Changing electric fields generate magnetic fields, and vice versa. Maxwell explained how both arise from the same phenomenon, an electromagnetic wave, which exhibits both electric and magnetic characteristics. Electromagnetic waves contain a varying electric field, accompanied by a magnetic field that varies similarly but lies at right angles to the other.

Maxwell measured the speed of electromagnetic waves traveling through a vacuum, showing it to be essentially the same as the speed of light. Combined with the work of Hans Christian Ørsted and Faraday, this confirmed that light was also a propagating electromagnetic disturbance. Maxwell showed that light waves, and all electromagnetic waves, travel at a constant speed in a vacuum of 300 million meters per second. This speed is fixed by the absolute electric and magnetic properties of free space.

Electromagnetic waves can have a range of wavelengths and cover a whole spectrum beyond visible light. Radio waves have the longest wavelengths (meters or even kilometers), visible light has wavelengths that are similar to the spacing between atoms, while at the highest

> ❝We can scarcely avoid the conclusion that light consists in the transverse undulations of the same medium which is the cause of electric and magnetic phenomena.❞
>
> James Clerk Maxwell, c.1862

frequencies are x-rays and gamma rays. Electromagnetic waves are used mainly for communications, via the transmission of radio waves, television and mobile phone signals. They can provide heat energy, such as in microwave ovens, and are often used as probes (e.g. medical x-rays and in electronic microscopes).

The electromagnetic force exerted by electromagnetic fields is one of the four fundamental forces, along with gravity and the strong and weak nuclear forces, that hold atoms and nuclei together. Electromagnetic forces are crucial in chemistry where they bind charged ions together to form chemical compounds and molecules.

Fields

Maxwell started out by trying to understand Faraday's work describing electric and magnetic fields experimentally. In physics, fields are the way in which forces are transmitted across distances. Gravity operates across even the vast distances of space, where it is said to produce a gravitational field. Similarly, electric and magnetic fields can affect charged particles quite far away. If you have played with iron filings sprinkled over a sheet of paper with a magnet below it, you will have seen that the magnetic force moves the iron dust into looped contours stretching from the north to the south pole of the magnet. The magnet's strength also falls off as you move farther away from it. Faraday had mapped these "field lines" and worked out simple rules. He also mapped similar field lines for electrically charged shapes but was not a trained mathematician. So it fell to Maxwell to try to unite these various ideas into a mathematical theory.

Four equations

To every scientist's surprise, Maxwell succeeded in describing all the various electromagnetic phenomena in just four fundamental equations. These equations are now so famous that they feature on some T-shirts followed by the comment "and so god created light." Although we now think of electromagnetism as one and the same thing, at the time this idea was radical, and as important as if we united quantum physics and gravity today.

$$\nabla \cdot D = \rho$$
$$\nabla \times H = J + (\delta D/\delta t)$$
$$\nabla \cdot B = 0$$
$$\nabla \times E = -(\delta B/\delta t)$$

Maxwell's equations

The first of Maxwell's equations is Gauss' law, named after 19th-century physicist Carl Friedrich Gauss, which describes the shape and strength of the electric field generated by a charged object. Gauss' law is an inverse square law, mathematically similar to Newton's law of gravity. Like gravity, an electric field drops off away from the surface of a charged object in proportion to the square of the distance. So the field is four times weaker if you move twice as far away from it.

Although there is no scientific evidence that mobile phone signals are bad for your health, the inverse square law explains why it might be safer to have a mobile phone mast close to your home rather than far away. The field from the transmitter mast drops off rapidly with distance, so it is very weak by the time it reaches you. In comparison, the field from the mobile phone is strong because it is held so close to your head. So, the closer the mast the less power the potentially more dangerous phone uses when you talk on it. Nevertheless, people are often irrational and fear masts more.

James Clerk Maxwell 1831–79

James Clerk Maxwell was born in Edinburgh, Scotland. He grew up in the countryside where he became curious about the natural world. After his mother died, he was sent to school in Edinburgh where he was given the nickname "dafty," because he was so absorbed in his studies. As a student at Edinburgh University and later at Cambridge, Maxwell was thought clever, if disorganized. After graduation, he extended Michael Faraday's work on electricity and magnetism and condensed it into equations. Maxwell moved back to Scotland when his father became ill and tried to get a job again at Edinburgh. Losing out to his old mentor, he went to King's College London where he carried out his most famous work. Around 1862 he calculated that the speed of electromagnetic waves and light were the same and 11 years later he published his four equations of electromagnetism.

The second of Maxwell's equations describes the shape and strength of the magnetic field, or the pattern of the magnetic field lines, around a magnet. It states that the field lines are always closed loops, from the north to the south pole. In other words, all magnets must have both a north and south pole — there are no magnetic monopoles and a magnetic field always has a beginning and an end. This follows from atomic theory, where even atoms can possess magnetic fields and grand-scale magnetism results if these fields are all aligned. If you chop a bar magnet in half, you always reproduce the north and south poles on each half. No matter how much you divide the magnet, the smaller shards retain both poles.

> " Any intelligent fool can make things bigger and more complex . . . It takes a touch of genius — and a lot of courage to move in the opposite direction. "
> attributed to Albert Einstein, 1879–1955

The third and fourth equations are similar to one another and describe electromagnetic induction. The third equation tells how changing currents produce magnetic fields, and the fourth how changing magnetic fields produce electric currents. The latter is otherwise familiar as Faraday's law of induction.

Describing so many phenomena in such a few simple equations was a major feat that led to Einstein rating Maxwell's achievement on a par with that of Newton. Einstein took Maxwell's ideas and incorporated them further into his relativity theories. In Einstein's equations, magnetism and electricity were manifestations of the same thing seen by viewers in different frames of reference; an electric field in one moving frame would be seen as a magnetic field in another. Perhaps it was Einstein then who ultimately contrived that electric and magnetic fields are truly one and the same thing.

46 Planck's law

Why do we say a fire is red hot? And why does steel glow first red, then yellow, then white, when it is heated? Max Planck described these color changes by knitting together the physics of heat and light. Describing light statistically rather than as a continuous wave, Planck's revolutionary idea seeded the birth of quantum physics.

In a famous 1963 speech, British Prime Minister Harold Wilson marveled at "the white heat of this [technological] revolution." But where does this phrase "white heat" come from?

Heat's color

We all know that many things glow when they are heated up. Barbecue coals and electric stove rings turn red, reaching hundreds of degrees Celsius. Volcanic lava, approaching 1,000 degrees Celsius (similar to the temperature of molten steel), can glow more fiercely — sometimes orange, yellow or even white hot. A tungsten lightbulb filament reaches over 3,000 degrees Celsius, similar to the surface of a star. In fact, with increasing temperature, hot bodies glow first red, then yellow and eventually white. The light looks white because more blue light has been added to the existing red and yellow. This spread of colors is described as a black-body curve.

Stars also follow this sequence: the hotter they are, the bluer they look. The Sun, at 6,000 kelvins, is yellow, while the surface of the red giant Betelgeuse (found in Orion) has a temperature of only half that. Hotter stars like Sirius, the brightest star in the sky, whose scorching surface reaches 30,000 kelvins, look blue–white. As the temperatures increase, more and more high-frequency blue light is given off. In fact, the strongest light from hot stars is so blue that most of it radiates in the ultraviolet part of the spectrum.

Hot

Blue   Cool   Red

Max Planck 1858–1947

Max Planck was schooled in Munich, Germany. Hoping for a career in music, he sought advice on what to study from a musician, but was told if he needed to ask the question he should study something else. His physics professor was no more encouraging, telling him physics as a science was complete and nothing more could be learned. Luckily Planck ignored him and continued his research, instigating the concept of quanta. Later in Planck's life he suffered the deaths of his wife and several children, including two sons killed in the world wars. Nevertheless, Planck remained in Germany and tried to rebuild physics research there following the wars. Today, many prestigious research institutes are named after Max Planck.

Black-body radiation

Nineteenth-century physicists were surprised to find that the light emitted when objects were heated followed the same pattern, irrespective of the substance they tested. Most of the light was given off at one particular frequency. When the temperature was raised, the peak frequency shifted to bluer (shorter) wavelengths, moving from red through yellow to blue–white.

> ❝ [the black-body theory was] an act of despair because a theoretical interpretation had to be found at any price, no matter how high that might be. ❞
> Max Planck, 1901

We use the term black-body radiation for a good reason. Dark materials are best able to radiate or absorb heat. If you've worn a black T-shirt on a hot day you'll know it heats up in the sun more than a white one. White reflects sunlight better, which is why houses in hot climates are often painted white. Snow reflects sunlight too. Climate scientists worry that the Earth will heat up more rapidly should the polar ice caps melt and reflect less sunlight back out into space. Black objects not only absorb but also release heat more quickly than white ones. This is why the surfaces of stoves or hearths are painted black — not just to hide the soot!

Planck's legacy in space

The most perfect black-body spectrum hails from a cosmic source. The sky is bathed in the faint glow of microwaves that are the afterglow of the fireball of the big bang itself, redshifted by the expansion of the universe to lower frequencies. This glow is called the cosmic microwave background radiation. In the 1990s, NASA's COBE satellite (COsmic Background Explorer) measured the temperature of this light — it has a black-body spectrum of 2.73 K, and is so uniform that this is still the purest black-body curve measured. No material on Earth has such a precise temperature. The European Space Agency recently honored Planck by naming their new satellite after him. It will map the cosmic microwave background in great detail.

A revolution

Although physicists had measured the black-body graphs, they could not fathom them or explain why the frequency peaked at a single color. Leading thinkers Wilhelm Wien, Lord Rayleigh and James Jeans worked out partial solutions. Wien described the dimming at bluer frequencies mathematically, while Rayleigh and Jeans explained the rising red spectrum, but both formulae failed at the opposite ends. Rayleigh and Jeans' solution, in particular, raised problems because it predicted that an infinite amount of energy would be released at ultraviolet wavelengths and above, due to the ever rising spectrum. This obvious problem was dubbed the "ultraviolet catastrophe."

In trying to understand black-body radiation, German physicist Max Planck joined the physics of heat and light together. Planck was a physics purist who liked returning to basics to derive physical principles. He was fascinated by the concept of entropy and the second law of thermodynamics. He considered this and Maxwell's equations to be fundamental laws of nature and set about proving how they were linked. Planck had complete faith in mathematics — if his equations told him something was true, it didn't matter if everyone else thought differently.

Planck reluctantly applied a clever fix to make his equations work. His insight was to treat electromagnetic radiation in the same way as thermodynamics experts treated heat. Just as temperature is the sharing of heat energy amongst many particles, Planck described light by allocating electromagnetic energy among a set of electromagnetic oscillators, or tiny subatomic units of electromagnetic field.

To fix the mathematics, Planck scaled the energy of each electromagnetic unit with frequency, such that $E = h\nu$, where E is energy, ν is light frequency, and h is a constant scaling factor now known as Planck's constant. These units were called "quanta," from the Latin for "how much."

In the new picture of energy quanta, the high-frequency electromagnetic oscillators each took on high energy. So, you couldn't have many of them in any system without blowing the energy limit. Likewise, if you received your monthly salary in 100 bank notes of mixed denominations, you'd receive mostly medium denominations plus a few of higher and lower value. By working out the most probable way of sharing electromagnetic energy between the many oscillators, Planck's model put most of the energy in the middle frequencies — it fitted the peaked black-body spectrum. In 1901, Planck published this law, linking light waves with probability, to great acclaim. And it was soon seen that his new idea solved the "ultraviolet catastrophe" problem.

Planck's quanta were just a construction for working out the mathematics of his law; he didn't for a moment imagine his oscillators were real. But, at a time when atomic physics was developing fast, Planck's novel formulation had surprising implications. Planck had planted a seed that would grow to become one of the most important areas of modern physics: quantum theory.

47 Photoelectric effect

When ultraviolet light shines on a copper plate, electricity is produced. This "photoelectric" effect remained a mystery until Albert Einstein, inspired by Max Planck's use of energy quanta, concocted the idea of the light particle, or photon. Einstein showed how light could behave as a stream of photon pellets as well as a continuous wave.

The dawn of the 20th century opened a new window onto physics. In the 19th century it was well known that ultraviolet light mobilized electrons to produce currents in a metal; understanding this phenomenon led physicists to invent a whole new language.

Blue batters

The photoelectric effect generates electric currents in metals when they are illuminated by blue or ultraviolet light, but not red light. Even a bright beam of red light fails to trigger a current. Charge flows only when the light's frequency exceeds some threshold, which varies for different metals. The threshold indicated that a certain amount of energy needed to be built up before the charges could be dislodged. The energy to free them must come from the light but, at the end of the 19th century, the mechanism by which this happened was unknown. Electromagnetic waves and moving charges seemed to be very different physical phenomena, and uniting them was a major puzzle.

> There are two sides to every question.
>
> Protagoras, 485–421 B.C.

Photons

In 1905, Albert Einstein came up with a radical idea to explain the photoelectric effect. It was this work, rather than relativity, that won him the Nobel Prize in 1921. Inspired by Max Planck's earlier use of quanta to budget the energy of hot atoms, Einstein imagined that light too could exist in little energy packets. Einstein borrowed wholesale Planck's mathematical definition of quanta, the proportionality of energy and frequency linked by Planck's constant, but applied it to light rather than atoms. Einstein's light quanta were later named photons. Photons have no mass and travel at the speed of light.

Rather than bathing the metal with continuous light waves, Einstein suggested that individual photon bullets hit electrons in the metal into motion to produce the photoelectric effect. Because each photon carries a certain energy, scaling with its own frequency, the bumped electron's energy also scales with the light's frequency. A photon of red light (with a low frequency) cannot carry enough energy to dislodge an electron, but a blue photon (light with a higher frequency) has more energy and can set it rolling. An ultraviolet photon has more energy still, so it can slam into an electron and donate even more speed. Turning up the brightness of light changes nothing, it doesn't matter that you have more red photons if each is incapable of shifting the electrons. It's like firing ping pong balls at a weighty sports utility vehicle. Einstein's idea of light quanta was unpopular at first, because it opposed the wave description of light summarized in Maxwell's equations that most physicists revered. However, the climate altered when experiments showed Einstein's wacky idea to be true. They confirmed the energies of the liberated electrons scaled proportionally with the frequency of light.

Wave–particle duality

Einstein's proposal was not only controversial but it raised the uncomfortable idea that light was both a wave and a particle, called wave–particle duality. Light's behavior up until Maxwell wrote down his equations had always followed that of a wave, bending round obstacles, diffracting, reflecting and interfering. Here, Einstein really rocked the boat by showing that light was also a stream of photon torpedoes.

Physicists are still struggling with this tension. Today, we even know that light seems to know whether to behave as one or the other under different circumstances. If you set up an experiment to measure its wave properties, such as passing it through a diffraction grating, it behaves as a wave. If instead you try to measure its particle properties it is similarly obliging.

Physicists have tried to devise clever experiments to catch light out, and perhaps reveal its true nature, but so far they have all failed. Many are variants of Young's double-slit experiment but with

Albert Einstein 1879–1955

1905 was an *annus mirabilis* for a part-time German-born physicist working as a clerk in the Swiss Patent Office. Albert Einstein published three physics papers in the German journal, *Annalen der Physik*. They explained Brownian motion, the photoelectric effect and special relativity, and each one was groundbreaking work. Einstein's reputation grew until, in 1915, he produced his theory of general relativity, confirming him as one of the greatest scientists of all time. Four years later, observations made during a solar eclipse verified his general relativity theory and he became world famous. In 1921 Einstein was awarded the Nobel Prize for his work on the photoelectric effect, which influenced the development of quantum mechanics.

> **"**The body's surface layer is penetrated by energy quanta whose energy is converted at least partially into kinetic energy of the electrons. The simplest conception is that a light quantum transfers its entire energy to a single electron.**"**
> **Albert Einstein, 1905**

components that can be switched in and out. Imagine a light source whose rays pass through two narrow slits onto a screen. With both slits open you see the familiar dark and light stripes of interference fringes. So light, as we know, is a wave. However, by dimming the light enough, at some point the level becomes so low that individual photons pass through the apparatus one by one, and a detector can catch the flashes as they arrive at the screen. Even if you do this, the photons continue to pile up into the striped interference pattern.

So, how does a single photon know whether to go through one or the other slit to build up the interference pattern? If you're quick, you could close one of the slits after the photon has left the light source, or even after it has traveled through the slits but before it hits the screen. In every case physicists have been able to test, the photons know whether there were one or two slits present at the time they went through. And even though only single photons are flying across, it appears as if each photon goes through both slits simultaneously.

Put a detector in one of the slits (so you know whether the photon went through that one or the other) and strangely the interference pattern disappears — you're left with a simple pile up of photons on the screen and no interference stripes. So no matter how you try to catch them out, photons know how to act. And they act as both waves and particles, not one or the other.

Matter waves

In 1924, Louis-Victor de Broglie suggested the converse idea that particles of matter could also behave as waves. He proposed that all bodies have an associated wavelength, implying that particle–wave duality was universal. Three years later the matter-wave idea was confirmed when electrons were seen to diffract and interfere just like light. Physicists have now also seen larger particles behaving like waves, such as neutrons, protons and recently even molecules, including microscopic carbon footballs or "bucky balls." Bigger objects, like ball bearings, have minuscule wavelengths, too small to see, so we cannot spot them behaving like waves. A tennis ball flying across a court has a wavelength of 10^{-34} meters, much smaller than a proton's width of 10^{-15} m.

As we have seen that light is also a particle and electrons are sometimes waves, the photoelectric effect has come full circle.

48 Schrödinger's wave equation

How can we say where a particle is if it is also spread out as a wave? Erwin Schrödinger wrote down a landmark equation that describes the probability of a particle being in some location while behaving as a wave. His equation went on to illustrate the energy levels of electrons in atoms, launching modern chemistry as well as quantum mechanics.

According to Einstein and Louis-Victor de Broglie, particles and waves are closely entwined. Electromagnetic waves, including light, take on both characteristics and even molecules and subatomic particles of matter can diffract and interfere as waves.

But waves are continuous, and particles are not. So how can you say where a particle is if it is spread out in the form of a wave? Schrödinger's equation, devised by Austrian physicist Erwin Schrödinger in 1926, describes the likelihood that a particle that is behaving as a wave is in a certain place, using the physics of waves and probability. It is one of the cornerstones of quantum mechanics, the physics of the atomic world.

Schrödinger's equation was first used to describe the positions of electrons in an atom. Schrödinger tried to describe electrons' wave-like behavior and also incorporated the concept of energy quanta introduced by Max Planck, the idea that wave energy comes in basic building blocks whose energy scales with wave frequency. Quanta are the smallest blocks, giving a fundamental graininess to any wave.

Bohr's atom

It was Danish physicist Niels Bohr who applied the idea of quantized energy to electrons in an atom. Because electrons are easily liberated from atoms, and negatively charged, Bohr thought that, like planets in orbit around the Sun, electrons were held in orbit about a positively charged nucleus. However, electrons could exist only with certain energies, corresponding to multiples of basic energy quanta. For electrons held within an atom, these energy states should restrict the electrons to distinct layers (or "shells") according to energy. It is as if the planets could only inhabit certain orbits, defined by energy rules.

Bohr's model was very successful, especially in explaining the simple hydrogen atom. Hydrogen contains just one electron orbiting around a single proton, a positively charged particle that acts as the nucleus. Bohr's hierarchy of quantized energies explained conceptually the characteristic wavelengths of light that were emitted and absorbed by hydrogen.

Just like climbing a ladder, if the electron in a hydrogen atom is given an energy boost, it can jump up to a higher rung, or shell. To hop up to the higher rung the electron must absorb energy from a photon with exactly the right energy to do it. So a particular frequency of light is needed to raise the electron's energy level. Any other frequency will not work. Alternatively, once boosted, the electron could jump back down to the rung below, emitting a photon of light of that frequency as it does so.

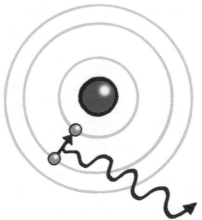

Spectral fingerprints

Moving electrons up the energy ladder, hydrogen gas can absorb a series of photons of characteristic frequencies corresponding to the energy gaps between rungs. If white light is shone through the gas, these frequencies appear blacked out because all the light at each gap frequency is absorbed. Bright lines result instead if the hydrogen is hot and its electrons started out high on the ladder. These characteristic energies for hydrogen can be measured, and they agree with the predictions of Bohr. All atoms produce similar lines, at different characteristic energies. So they are like fingerprints that can identify individual chemical species.

Wave functions

Bohr's energy levels worked well for hydrogen, but less well for other atoms with more than one electron and with heavier nuclei. Moreover, there was still de Broglie's conundrum that electrons should also be thought of as waves. So each electron orbit could equally be considered a wavefront. But, thinking of it as a wave meant it was impossible to say where the electron was at any time.

Schrödinger, inspired by de Broglie, wrote down an equation that could describe the position of a particle when it was behaving as a wave. He was only able to do this statistically by incorporating probability. Schrödinger's important equation is a fundamental part of quantum mechanics.

Schrödinger introduced the idea of a wave function to express the probability of the particle being in a given place at some time, and to include all the knowable information about that particle. Wave

Boxed in

A lone particle floating in free space has a wave function that looks like a sine wave. If it is trapped inside a box, then its wave function must drop to zero at the box walls, and outside, because it cannot be there. The wave function inside the box can be determined by considering the allowed energy levels, or energy quanta, of the particle, which must always be greater than zero. Because only specific energy levels are allowed by quantum theory, the particle will be more likely to be in some places than others and there are places within the box where the particle would never be found, where the wave function is zero. More complicated systems have wave functions that are a combination of many sine waves and other mathematical functions, like a musical tone made up of many harmonics. In conventional physics, we would use Newton's laws to describe the motion of a particle in a box (such as a miniature ball bearing). At any instant, we would know exactly where it is and the direction in which it is moving. In quantum mechanics, however, we can only talk about the probability of the particle being in some place at some time and, because energy quantization seeps in on atomic scales, there are favored places where the particle will be found. But we cannot say exactly where it is, because it is also a wave.

functions are notoriously difficult to comprehend as we do not witness them in our own experience and find it very hard to visualize and even interpret them philosophically.

The breakthrough that Schrödinger's equation heralded also led to models of electron orbitals in atoms. These are probability contours, outlining regions where electrons are 80–90 percent likely to be located (raising the issue that with some small probability they could be somewhere else entirely). These contours turned out not to be spherical shells, as envisaged by Bohr, but rather more stretched shapes, such as dumb-bells or donuts. Chemists now use this knowledge to engineer molecules.

> God runs electromagnetics by wave theory on Monday, Wednesday, and Friday, and the Devil runs them by quantum theory on Tuesday, Thursday, and Saturday.
> Sir William Bragg, 1862–1942

Schrödinger's equation revolutionized physics by bringing the wave–particle duality idea not only to atoms but to all matter. Together with Werner Heisenberg and others, Schrödinger truly is one of the founding fathers of quantum mechanics.

49 Heisenberg's uncertainty principle

Heisenberg's uncertainty principle states that the speed (or momentum) and position of a particle at an instant cannot both be known exactly — the more precisely you measure one, the less you can find out about the other. Werner Heisenberg argued that the very act of observing a particle changes it, making precise knowledge impossible. So neither the past nor the future behavior of any subatomic particle can be predicted with certainty. Determinism is dead.

In 1927, Heisenberg realized that quantum theory contained some strange predictions. It implied that experiments could never be done in complete isolation because the very act of measurement affected the outcome. He expressed this connection in his "uncertainty principle" — you cannot simultaneously measure both the position and momentum of a subatomic particle (or equivalently its energy at an accurate time). If you know one then the other is always uncertain. You can measure both within certain bounds, but the more tightly these bounds are specified for one, the looser they become for the other. This uncertainty, he argued, was a deep consequence of quantum mechanics — it had nothing to do with a lack of skill or accuracy in measuring.

Uncertainty

In any measurement, there is an element of uncertainty in the answer. If you measure the length of a table with a tape measure, you can say it is one meter long but the tape can only say so to within one millimeter because that is the size of the smallest tick mark on it. So the table could really be 99.9 centimeters or 100.1 centimeters long and you wouldn't know.

It is easy to think of uncertainties as being due to the limitations of your measuring device, such as the tape, but Heisenberg's statement is profoundly different. It states that you can never know both quantities, momentum and position, exactly at the same time no matter how accurate an instrument you use. It is as if when you measure a swimmer's position you cannot know her speed at the same instant. You can know both roughly, but as soon as you tie one down the other becomes more uncertain.

Measurement

How does this problem arise? Heisenberg imagined an experiment that measured the motion of a subatomic particle such as a neutron. A radar could be used to track the particle, by bouncing electromagnetic waves off it. For maximum accuracy you would choose gamma rays, which have very small wavelengths. However, because of wave–particle duality the gamma ray beam hitting the neutron would act like a series of photon bullets. Gamma rays have very high frequencies and so each photon would carry a great deal of energy. As a hefty photon hit the neutron, it would give it a big kick that would alter its speed. So, even if you knew the position of the neutron at that instant, its speed changes unpredictably because of the very process of observation.

If you used softer photons with lower energies, to minimize the velocity change, then their wavelengths would be longer and so the accuracy with which you could measure the position would now be degraded. No matter how you optimize the experiment, you cannot learn both the particle's position and speed simultaneously. There is a fundamental limit expressed in Heisenberg's uncertainty principle.

Heisenberg's uncertainty principle

$$\Delta x \Delta p > \frac{\hbar}{2} \qquad \Delta E \Delta t > \frac{\hbar}{2}$$

In reality, what is going on is more difficult to comprehend, because of the coupled wave–particle behavior of both subatomic particles and electromagnetic waves. The definitions of particle position, momentum, energy and time are all probabilistic. Schrödinger's equation describes the probability of a particle being in a certain place or having a certain energy according to quantum theory, as embodied in the wave function of the particle that describes all its properties.

Heisenberg was working on quantum theory at about the same time as Schrödinger. Schrödinger preferred to work on the wave-like aspects of subatomic systems, whereas Heisenberg investigated the stepped nature of the energies. Both physicists developed ways of describing quantum systems mathematically according to their own biases; Schrödinger using the mathematics of waves and Heisenberg using matrices, or two-dimensional tables of numbers, as a way of writing down the sets of properties.

> " The more precisely the position is determined, the less precisely the momentum is known in this instant, and vice versa. "
> **Werner Heisenberg, 1927**

The matrix and wave interpretations both had their followers, and both camps thought the other group was wrong. Eventually they pooled their resources and came up with a joint description of quantum theory that became known as quantum mechanics. It was while trying to formulate these equations that Heisenberg spotted uncertainties that could not go away. He brought these to the attention of a colleague, Wolfgang Pauli, in a letter in 1927.

Indeterminism

The profound implications of the uncertainty principle were not lost on Heisenberg and he pointed out how it challenged conventional physics. First, it implied that the past behavior of a subatomic particle was not constrained until a measurement of it was made. According to Heisenberg "the path comes into existence only when we observe it." We have no way of knowing where something is until we measure it. He also noted that the future path of a particle cannot be predicted either. Because of these deep uncertainties about its position and speed, the future outcome was also unpredictable.

Both of these statements caused a major rift with the Newtonian physics of the time, which assumed that the external world existed independently and it was just down to the observer of an experiment to see the underlying truth. Quantum mechanics showed that at an atomic

Werner Heisenberg 1901–76

Werner Heisenberg lived in Germany through two world wars. An adolescent during the First World War, Heisenberg joined the militarized German youth movement that encouraged structured outdoor and physical pursuits. Heisenberg worked on farms in the summer, using the time to study mathematics. He studied theoretical physics at Munich University, finding it hard to shuttle between his love of the countryside and the abstract world of science. After his doctorate, Heisenberg took up academic posts, and met Einstein on a visit to Copenhagen. In 1925, Heisenberg invented the first form of quantum mechanics, known as matrix mechanics, receiving the Nobel Prize for this work in 1932. Nowadays, he is best known for the uncertainty principle, formulated in 1927.

During the Second World War Heisenberg headed the unsuccessful German nuclear weapons project, and worked on a nuclear fission reactor. It is debatable whether the German inability to build a nuclear weapon was deliberate or simply due to lack of resources. After the war he was arrested by the Allies and interned with other German scientists in England before returning to research in Germany afterwards.

level, such a deterministic view was meaningless and one could only talk about probabilities of outcomes instead. We could no longer talk about cause and effect but only chance. Einstein, and many others, found this hard to accept, but had to agree that this is what the equations showed. For the first time, physics moved well beyond the laboratory of experience and firmly into the realm of abstract mathematics.

50 Copenhagen interpretation

The equations of quantum mechanics gave scientists the right answers, but what did they mean? Danish physicist Niels Bohr developed the Copenhagen interpretation of quantum mechanics, blending the wave equation of Schrödinger and the uncertainty principle of Heisenberg. Bohr argued that there is no such thing as an isolated experiment — that the observer's interventions fix the outcomes of quantum experiments. In doing so, he challenged the very objectivity of science.

In 1927, competing views of quantum mechanics were rife. Erwin Schrödinger argued that the physics of waves underlay quantum behavior, which could all be described using wave equations. Werner Heisenberg, on the other hand, believed that the particle nature of electromagnetic waves and matter, described in his tabular matrix representation, was of the foremost importance in comprehending nature. Heisenberg had also shown that our understanding was fundamentally limited by his uncertainty principle. He believed that both the past and future were unknowable until fixed by observation because of the intrinsic uncertainty of all the parameters describing a subatomic particle's movement.

> We are in a jungle and find our way by trial and error, building our road behind us as we proceed.
> **Max Born, 1882–1970**

Another man tried to pull together all the experiments and theories to form a new picture that could explain the whole. This was Niels Bohr, the head of Heisenberg's department at the University of Copenhagen, and the scientist who had explained the quantum energy states of electrons in the hydrogen atom. Bohr, together with Heisenberg, Max Born and others, developed a holistic view of quantum mechanics that became known as the Copenhagen interpretation. It is still the favorite interpretation of most physicists, although other variations have been suggested.

Niels Bohr 1885–1962

Niels Bohr lived through two world wars, and worked with some of the best physicists around. Young Niels pursued physics at Copenhagen University, performing award-winning physics experiments in his father's physiology laboratory. He moved to England after his doctorate but clashed with J.J. Thomson. After working with Ernest Rutherford in Manchester, he returned to Copenhagen, completing his work on the "Bohr atom" (still how most people picture an atom today). He won the Nobel Prize in 1922, just before quantum mechanics fully appeared. To escape Hitler's Germany in the 1930s, scientists flocked to Bohr's Institute of Theoretical Physics in Copenhagen, where they were entertained in a mansion donated by Carlsberg, the Danish brewers. When the Nazis occupied Denmark in 1940, Bohr fled via a fishing boat to Sweden and then England.

Two sides

Niels Bohr brought a philosophical approach to bear on the new science. In particular, he highlighted the impact that the observer themselves has on the outcomes of quantum experiments. First he accepted the idea of "complementarity," that the wave and particle sides of matter and light were two faces of the same underlying phenomenon and not two separate families of events. Just as pictures in a psychological test can switch appearance depending on how you look at them — two wiggly mirrored lines appearing either as a vase outline or two faces looking at one another — wave and particle properties were complementary ways of seeing the same phenomenon. It was not light that changed its character, but rather how we decided to view it.

To bridge the gap between quantum and normal systems, including our own experiences on human scales, Bohr also introduced the "correspondence principle," that quantum behavior must disappear for larger systems that we are familiar with, when Newtonian physics is adequate.

Unknowability

Bohr realized the central importance of the uncertainty principle, which states that one cannot measure both the position and momentum (or speed) of any subatomic particle at the same time. If one quantity is measured accurately, then the other is inherently uncertain. Heisenberg thought that the uncertainty came about because of the mechanics of the measurement act itself. To measure something, even to look at it, we must bounce photons of light off it. Because this always involves the transfer of some momentum or energy, then this act of observation disturbed the original particle's motion.

Bohr, on the other hand, thought Heisenberg's explanation flawed. He argued that we can never completely separate the observer from the system he or she is measuring. It was the act of observation itself that set the system's final behavior, through the probabilistic wave–particle behavior of quantum physics and not due to simple energy transfer. Bohr thought that an entire system's behavior needed to be considered as one; you could not separate the particle, the radar and even the observer themselves. Even if we look at an apple, we need to consider the quantum properties of the whole system, including the visual system in our own brain that processes the photons from the apple.

Bohr also argued that the very word "observer" is wrong because it conjures up a picture of an external viewer separated from the world that is being watched. A photographer such as Ansel Adams may capture the pristine natural beauty of the Yosemite wilderness, but is it really untouched by man? How can it be if the photographer himself is there too? The real picture is of a man standing within nature, not separate from it. To Bohr, the observer was very much part of the experiment.

> **Anyone who is not shocked by quantum theory has not understood it.**
> Niels Bohr, 1885–1962

This concept of observer participation was shocking to physicists, because it challenged the very way that their science had always been done and the fundamental concept of scientific objectivity. Philosophers also balked. Nature was no longer mechanical and predictable but, deep down, was inherently unknowable. What did this mean for concepts of basic truth, let alone simple ideas such as past and future? Einstein, Schrödinger and others had difficulty dropping their firm beliefs in an external, deterministic and verifiable universe. Einstein believed that, because it could only be described with statistics, the theory of quantum mechanics must be at the very least incomplete.

Collapsing wave functions

Given that we observe subatomic particles and waves as either one or the other entity, what decides how they manifest themselves? Why does light passing through two slits interfere like waves on Monday, but switch to particle-like behavior on Tuesday if we try to catch the photon as it passes through one slit? According to Bohr and supporters of the Copenhagen interpretation, the light exists in both states simultaneously, both as a wave and a particle. It only dresses itself as one or the other when it is measured. So we choose in advance how it turns out by deciding how we would like to measure it.

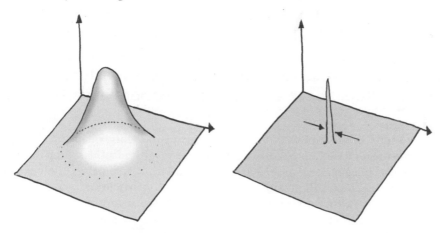

At this point of decision-making, when the particle- or wave-like character is fixed, we say that the wave function has collapsed. All the probabilities for outcomes that are contained in Schrödinger's wave function description crush down so that everything apart from the eventual outcome is lost. So, according to Bohr, the original wave function for a beam of light contains all the possibilities within it, whether the light appears in its wave or particle guise. When we measure it, it appears in one form, not because it changes from one type of substance to another but because it is truly both at the same time. Quantum apples and oranges are neither, but instead a hybrid.

Physicists still have trouble in understanding intuitively what quantum mechanics means and others since Bohr have offered new ways of interpreting it. Bohr argued that we needed to go back to the drawing board to understand the quantum world, and could not use concepts that were familiar in everyday life. The quantum world is something else strange and unfamiliar, and we must accept that.

51 Schrödinger's cat

Schrödinger's cat is both alive and dead at the same time. In this hypothetical experiment, a cat sitting inside a box may or may not have been killed by a poison capsule, depending on some random trigger. Erwin Schrödinger used this metaphor to show how ridiculous he found the Copenhagen interpretation of quantum theory, which predicted that, until the outcome was actually observed, the cat should be in a state of limbo, both alive and dead.

In the Copenhagen interpretation of quantum theory, quantum systems exist as a cloud of probability until an observer flicks the switch and selects one outcome for his or her experiment. Before being observed, the system takes on all possibilities. Light is both particle and wave until we decide which form we want to measure — then it adopts that form.

While a probability cloud may sound like a plausible concept for an abstract quantity like a photon or light wave, what might it mean for something larger that we might be able to be aware of? What really is the nature of this quantum fuzziness?

In 1935, Erwin Schrödinger published an article containing a hypothetical experiment which tried to illustrate this behavior with a more colorful and familiar example than subatomic particles. Schrödinger was highly critical of the Copenhagen view that the act of observation influenced its behavior. He wanted to show how daft the Copenhagen interpretation was.

Quantum limbo

Schrödinger considered the following situation, which was entirely imaginary. No animals were harmed.

"A cat is penned up in a steel chamber, along with the following diabolical device (which must be secured against direct interference by the cat): in a Geiger counter there is a tiny bit of radioactive substance, so small that perhaps in the

course of one hour one of the atoms decays, but also, with equal
probability, perhaps none; if it happens, the counter tube discharges and
through a relay releases a hammer which shatters a small flask of
hydrocyanic acid. If one has left this entire system to itself
for an hour, one would say that the cat still lives if
meanwhile no atom has decayed. The first atomic decay
would have poisoned it."

So there is a 50:50 probability of the cat being either alive (hopefully) or dead when the box is opened after that time. Schrödinger argued that, following the logic of the Copenhagen interpretation, we would have to think of the cat as existing in a fuzzy blend of states, being both alive and dead at the same time, while the box was closed. Just as the wave or particle view of an electron is only fixed on the point of detection, the cat's future is only determined when we choose to open the box and view it. On opening the box we make the observation and the outcome is set.

Surely, Schrödinger grumbled, this was ridiculous and especially so for a real animal such as a cat. From our everyday experience we know that the cat must be either alive or dead, not a mixture of both, and it is madness to imagine that it was in some limbo state just because we were not looking at it. If the cat lived, all it would remember was sitting in the box being very much alive, not being a probability cloud or wave function.

Amongst others, Einstein agreed with Schrödinger that the Copenhagen picture was absurd. Together they posed further questions. As an animal, was the cat able to observe itself, and so collapse its own wave function? What does it take to be an observer? Need the observer be a conscious being like a human or would any animal do? How about a bacterium?

Going even further, we might question whether anything in the world exists independently of our observation of it. If we ignore the cat in the box and just think of the decaying radioactive particle, will it have decayed or not if we keep the box closed? Or is it in quantum limbo until we open the box's flap, as the Copenhagen interpretation requires? Perhaps the entire world is in a mixed fuzzy state and that nothing resolves itself until we observe it, causing the wave function to collapse when we do. Does your workplace disintegrate when you are away from it at weekends, or is it protected by the gazes of passers by?

Erwin Schrödinger 1887–1961

Austrian physicist Erwin Schrödinger pursued quantum mechanics and tried (and failed), with Einstein, to unify gravity and quantum mechanics into a single theory. He favored wave interpretations and disliked wave–particle duality, leading him into conflict with other physicists.

As a boy Schrödinger loved German poetry but nevertheless decided to pursue theoretical physics at university. Serving on the Italian front during the First World War, Schrödinger continued his work remotely and even published papers, returning afterwards to academia. Schrödinger proposed his wave equation in 1926, for which he was awarded the Nobel Prize with Paul Dirac in 1933. Schrödinger then moved to Berlin to head Max Planck's old department, but with Hitler's coming to power in 1933 he decided to leave Germany. He found it hard to settle, and worked for periods in Oxford, Princeton and Graz. With Austria's annexation in 1938, he fled again, moving finally to a bespoke position created for him at the new Institute for Advanced Studies in Dublin, Ireland, where he remained until retiring to Vienna. Schrödinger's personal life was as complicated as his professional life; he fathered children with several women, one of whom lived with him and his wife for a time in Oxford.

If no one is watching, does your cabin in the woods cease to exist in reality? Or does it wait in a blend of probability states, as a superposition of having being burned down, flooded, invaded by ants or bears, or sitting there just fine, until you return to it? Do the birds and squirrels count as observers? As odd as it is, this is how Bohr's Copenhagen interpretation explains the world on the atomic scale.

Many worlds

The philosophical problem of how observations resolve outcomes has led to another variation on the interpretation of quantum theory — the many worlds hypothesis. Suggested in 1957 by Hugh Everett, the alternative view avoids the indeterminacy of unobserved wave functions by saying instead that there are an infinite number of parallel universes. Every time an observation is made, and a specific outcome noted, a new universe splits off. Each universe is exactly the same as the other, apart from the one thing that has been seen to change. So the probabilities are all the same, but the occurrence of events moves us on through a series of branching universes.

In a many worlds interpretation of Schrödinger's cat experiment, when the box is opened the cat is no longer in a superposition of all possible states. Instead it is either alive in one universe or dead in another parallel one. In one universe the poison is released, in the other it is not.

Whether this is an improvement on being in wave function limbo is arguable. We may well avoid the need for an observer to pull us out of being just a probability cloud sometimes, but the cost is to invoke a whole range of alternative universes where things are very slightly different. In one universe I am a rock star, in another just a busker. Or in one I am wearing black socks, in another grey. This seems a waste of a lot of good universes (and hints at universes where people have garish wardrobes). Other alternative universes might be more significant — in one Elvis still lives, in another John F. Kennedy wasn't shot, in another Al Gore was President of the U.S. This idea has been borrowed widely as a plot device in movies, like *Sliding Doors* where Gwyneth Paltrow lives two parallel lives in London, one successful, one not.

Some physicists today argue that Schrödinger's thinking on his metaphorical cat experiment was invalid. Just as with his exclusively wave-based theory, he was trying to apply familiar physics ideas to the weird quantum world, when we just have to accept that it is strange down there.

52 Rutherford's atom

Atoms are not the smallest building blocks of matter as once thought. Early in the 20th century, physicists like Ernest Rutherford broke into them, revealing first layers of electrons and then a hard core, or nucleus, of protons and neutrons. To bind the nucleus together a new fundamental force — the strong nuclear force — was invented. The atomic age had begun.

The idea that matter is made up of swarms of tiny atoms has been around since the ancient Greeks. But whereas the ancient Greeks thought the atom was the smallest indivisible component of matter, 20th-century physicists realized this was not so and began to probe the inner structure of the atom itself.

Plum pudding model

The first layer to be tackled was that of the electron. Electrons were liberated from atoms in 1887 by Joseph John (J.J.) Thomson who fired an electric current through gas contained in a glass tube. In 1904, Thomson proposed the "plum pudding model" of the atom, where negatively charged electrons were sprinkled like prunes or raisins through a sponge dough of positive charge. Today it might have been called the blueberry muffin model. Thomson's atom was essentially a cloud of positive charge containing electrons, which could be set free relatively easily. Both the electrons and positive charges could mix throughout the "pudding."

The nucleus

Not long after, in 1909, Ernest Rutherford puzzled over the outcome of an experiment he had performed in which heavy alpha particles were fired through very thin gold foil, thin enough that most of the particles passed straight through. To Rutherford's astonishment a small fraction of particles ricocheted straight back off the foil, heading towards him.

Ernest Rutherford 1871–1937

New Zealander Rutherford was a modern-day alchemist, transmuting one element, nitrogen, into another, oxygen, through radioactivity. An inspiring leader of the Cavendish Laboratory in Cambridge, England, he mentored numerous future Nobel Prize winners. He was nicknamed "the crocodile," and this animal is the symbol of the laboratory even today. In 1910, his investigations into the scattering of alpha rays and the nature of the inner structure of the atom led him to identify the nucleus.

> ❝It was almost as incredible as if you fired a 15-inch shell at a piece of tissue paper and it came back to hit you.❞
> **Ernest Rutherford, 1936**

They reversed direction by 180 degrees, as if they had hit a brick wall. He realized that within the gold atoms that made up the foil sheet lay something hard and massive enough to repel the heavy alpha particles.

Rutherford understood that Thomson's plum pudding model could not explain this. If an atom was just a paste of mixed positively and negatively charged particles then none would be heavy enough to knock back the bigger alpha particle. So, he reasoned, the gold atoms must have a dense core, called the "nucleus" after the Latin word for the "kernel" of a nut. Here began the field of nuclear physics, the physics of the atomic nucleus.

Isotopes

Physicists knew how to work out the masses of different elements of the periodic table, so they knew the relative weights of atoms. But it was harder to see how the charges were arranged. Because Rutherford only knew about electrons and the positively charged nucleus, he tried to balance the charges by assuming that the nucleus was made up of a mix of protons (positively charged particles that he discovered in 1918 by isolating the nuclei of hydrogen) and some electrons that partially neutralized the charge. The remaining electrons circled outside the nucleus in the familiar orbitals of quantum theory. Hydrogen, the

Three of a kind

Radioactive substances emit three types of radiation, called alpha, beta and gamma radiation. Alpha radiation consists of heavy helium nuclei comprising two protons and two neutrons bound together. Because they are heavy, alpha particles do not travel far before losing their energy in collisions and can be stopped easily, even by a piece of paper. A second type of radiation is carried by beta particles; these are high-speed electrons — very light and negatively charged. Beta particles can travel farther than alpha radiation but may be halted by metal, like an aluminium plate. Third are gamma rays, which are electromagnetic waves, associated with photons, and so carry no mass but a lot of energy. Gamma rays are pervasive and can be shielded only with dense blocks of concrete or lead. All three types of radiation are emitted by unstable atoms that we describe as radioactive.

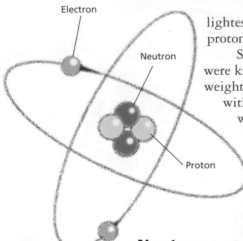

Electron

Neutron

Proton

lightest element, has a nucleus containing just one proton with one electron orbiting it.

Some other forms of elements with odd weights were known, called isotopes. Carbon usually has a weight of 12 atomic units, but is occasionally seen with a weight of 14 units. Carbon-14 is unstable with a half-life (the time it takes for half the atoms to decay by emitting a radioactive particle) of 5,730 years, emitting a beta particle to become nitrogen-14. This reaction is used in radiocarbon dating to measure the ages of archeological artefacts thousands of years old, such as wood or charcoal from fires.

Neutrons

In the early 1930s a new type of "radiation" was found that was heavy enough to free protons from paraffin but with no charge. Cambridge physicist James Chadwick showed that this new radiation was in fact a neutral particle with the same mass as the proton. It was named the neutron and the model of the atom was rearranged. Scientists realized that a carbon-12 atom, for instance, contains 6 protons and 6 neutrons in the nucleus (to give it a mass of 12 atomic units) and 6 orbiting electrons. Neutrons and protons are known as nucleons.

Strong force

The nucleus is absolutely tiny compared with the full extent of the atom and its veiling electrons. A hundred thousand times smaller than an atom, the nucleus is only a few femtometers (10–15 meters, or one ten million billionth of a meter) across. If the atom were scaled up to the diameter of the Earth, the nucleus at the center would be just 10 kilometers wide, or the length of Manhattan. The nucleus harbors practically all the mass of the atom in one tiny spot, and this can include many tens of protons. What holds all this positive charge together in such a small space so tightly? To overcome the electrostatic repulsion of the positive charges and bind the nucleus together, physicists had to invent a new type of force, called the strong nuclear force.

> **Nothing exists except atoms and empty space; everything else is opinion.**
> Democritus, 460–370 B.C.

If two protons are brought up close to one another, they initially repel because of their like charges (following Maxwell's inverse square law). But if they are pushed even closer the strong nuclear force locks them together. The strong force only appears at very small separations, but it is much greater than the electrostatic force. If the protons are pushed even closer together, they resist, acting as if they are hard spheres — so there is a firm limit to how close together they can go. This behavior means that the nucleus is tightly bound, very compact and rock hard.

In 1934, Hideki Yukawa proposed that the nuclear force was carried by special particles (called mesons), that act in a similar way to photons. Protons and neutrons are glued together by exchanging mesons. Even now it is a mystery why the strong nuclear force acts on such a specific distance scale — why it is so weak outside the nucleus and so strong at close range. It is as if it locks the nucleons together at a precise distance. The strong nuclear force is one of the four fundamental forces, along with gravity, electromagnetism and another nuclear force called the weak force.

53 **Antimatter**

Fictional spaceships are often powered by "antimatter drives," yet antimatter itself is real and has even been made artificially on Earth. A mirror image form of matter that has negative energy, antimatter cannot coexist with matter for long — both annihilate in a flash of energy if they come into contact. The very existence of antimatter hints at deep symmetries in particle physics.

Walking down the street you meet a replica of yourself. It is your antimatter twin. Do you shake hands? Antimatter was predicted in the 1920s and discovered in the 1930s by bringing together quantum theory and relativity. It is a mirror image form of matter, where particles' charges, energies and other quantum properties are all reversed in sign. So an anti-electron, called a positron, has the same mass as the electron but instead has a positive charge. Similarly, protons and other particles have opposite antimatter siblings.

> "For every one billion particles of antimatter there were one billion and one particles of matter. And when the mutual annihilation was complete, one billionth remained — and that's our present universe."
> **Albert Einstein, 1879–1955**

Negative energy

Creating an equation for the electron in 1928, British physicist Paul Dirac saw that it offered the possibility that electrons could have negative as well as positive energy. Just as the equation $x^2 = 4$ has the solutions $x=2$ and $x=-2$, Dirac had two ways of solving his problem: positive energy was expected, associated with a normal electron, but negative energy made no sense. But rather than ignore this confusing term, Dirac suggested that such particles might actually exist. This complementary state of matter is "anti-"matter.

Antiparticles

The hunt for antimatter began quickly. In 1932, Carl Anderson confirmed the existence of positrons experimentally. He was following the tracks of showers of particles produced by cosmic rays (energetic particles that crash into the atmosphere from space). He saw the track of a positively charged particle with the electron's mass, the positron. So antimatter was no longer just an abstract idea but was real.

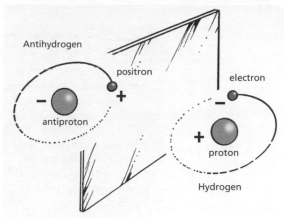

It took another two decades before the next antiparticle, the antiproton, was detected. Physicists built new particle-accelerating machines that used magnetic fields to increase the speeds of particles traveling through them. Such powerful beams of speeding protons produced enough energy to reveal the antiproton in 1955. Soon afterwards, the antineutron was also found.

With the antimatter equivalent building blocks in place, was it possible to build an anti-atom, or at least an anti-nucleus? The answer, shown in 1965, was yes. A heavy hydrogen (deuterium) anti-nucleus (an anti-deuteron), containing an antiproton and antineutron, was created by scientists at CERN in Europe and Brookhaven Laboratory in the U.S. Tagging on a positron to an antiproton to make a hydrogen anti-atom (anti-hydrogen) took a little longer, but was achieved in 1995. Today experimenters are testing whether anti-hydrogen behaves in the same way as normal hydrogen.

On Earth, physicists can create antimatter in particle accelerators, such as those at CERN in Switzerland or Fermilab near Chicago. When the beams of particles and antiparticles meet, they annihilate each other in a flash of pure energy. Mass is converted to energy according to Einstein's $E=mc^2$ equation. So if you met your antimatter twin it might not be such a good idea to throw your arms around them.

Universal asymmetries

If antimatter were spread across the universe, these annihilation episodes would be occurring all the time. Matter and antimatter would gradually destroy each other in little explosions, mopping each other up. Because we don't see this, there cannot be much antimatter around.

Paul Dirac 1902–84

Paul Dirac was a talented but shy British physicist. People joke that his vocabulary consisted of "Yes," "No," and "I don't know." He once said "I was taught at school never to start a sentence without knowing the end of it." What he lacked in verbosity he made up for in his mathematical ability. His PhD thesis is famous for being impressively short and powerful, presenting a new mathematical description of quantum mechanics. He partly unified the theories of quantum mechanics and relativity theory, but he also is remembered for his outstanding work on the magnetic monopole and in predicting antimatter. Awarded the 1933 Nobel Prize, Dirac's first thought was to reject it to avoid the publicity. But he gave in when told he would get even more publicity if he turned it down. Dirac did not invite his father to the ceremony, possibly because of strained relations after the suicide of Dirac's brother.

In fact normal matter is the only widespread form of particle we see, by a very large margin. So at the outset of the creation of the universe there must have been an imbalance such that more normal matter was created than its antimatter opposite.

> **In science one tries to tell people, in such a way as to be understood by everyone, something that no one ever knew before. But in poetry, it's the exact opposite.**
>
> Paul Dirac, 1902–84

Like all mirror images, particles and their antiparticles are related by different kinds of symmetry. One is time. Because of their negative energy, antiparticles are equivalent mathematically to normal particles moving backwards in time. So a positron can be thought of as an electron traveling from future to past. The next symmetry involves charges and other quantum properties, which are reversed, and is known as "charge conjugation." A third symmetry regards motion through space. Returning to Mach's principle, motions are generally unaffected if we change the direction of coordinates marking out the grid of space. A particle moving left to right looks the same as one moving right to left, or is unchanged whether spinning clockwise or counterclockwise. This "parity" symmetry is true of most particles, but there are a few for which it does not always hold. Neutrinos exist in only one form, as a left-handed neutrino, spinning in one direction; there is no such thing as a right-handed neutrino. The converse is true for antineutrinos which are all right-handed. So parity symmetry can sometimes be broken, although a combination of charge conjugation and parity is conserved, called charge–parity or CP symmetry for short.

> ❝ The opposite of a correct statement is a false statement. But the opposite of a profound truth may well be another profound truth. ❞
> Niels Bohr, 1885–1962

Just as chemists find that some molecules prefer to exist in one version, as a left-handed or right-handed structure, it is a major puzzle why the universe contains mostly matter and not antimatter. A tiny fraction — less than 0.01 percent — of the stuff in the universe is made of antimatter. But the universe also contains forms of energy, including a great many photons. So it is possible that a vast amount of both matter and antimatter was created in the big bang, but then most of it was annihilated shortly after. Only the tip of the iceberg now remains. A minuscule imbalance in favor of matter would be enough to explain its dominance now. To do this, only 1 in every 10,000,000,000 (10^{10}) matter particles needed to survive a split second after the big bang, the remainder being annihilated. The leftover matter was likely preserved via a slight asymmetry from CP symmetry violation.

The particles that may have been involved in this asymmetry are a kind of heavy boson, called X bosons, that have yet to be found. These massive particles decay in a slightly imbalanced way to give a slight overproduction of matter. X bosons may also interact with protons and cause them to decay, which would be bad news as it means that all matter will eventually disappear into a mist of even finer particles. But the good news is that the timescale for this happening is very long. That we are here and no one has ever seen a proton decay means that protons are very stable and must live for at least 10^{17}–10^{35} years, or billions of billions of billions of years, hugely longer than the lifetime of the universe so far. But this does raise the possibility that if the universe gets really old, then even normal matter might, one day, disappear.

54 **Nuclear fission**

The demonstration of nuclear fission is one of the great highs and lows of science. Its discovery marked a huge leap in our understanding of nuclear physics, and broke the dawn of atomic energy. But the umbrella of war meant this new technology was implemented almost immediately in nuclear weapons, devastating the Japanese cities of Hiroshima and Nagasaki and unleashing a proliferation problem that remains difficult to resolve.

At the start of the 20th century, the atom's inner world began to be revealed. Like a Russian doll, it contains many outer shells of electrons enveloping a hard kernel or nucleus. By the early 1930s, the nucleus itself was cracked, showing it to be a mix of positively charged protons and uncharged neutrons, both much heavier than the ephemeral electron, and bonded together by the strong nuclear force. Unlocking the energy glue of the nucleus became the holy grail for scientists.

Breakup

The first successful attempt to split the nucleus occurred in 1932. John Cockroft and Ernest Walton in Cambridge, England, fired very fast protons at metals. The metals changed their composition and released energy according to Einstein's $E=mc^2$. But these experiments needed more energy to be put into them than was created and so physicists didn't believe that it was possible to harness this energy for commercial use.

In 1938 German scientists Otto Hahn and Fritz Strassmann shot neutrons into the heavy element uranium, attempting to create new even heavier elements. What they found instead was that much lighter elements, some just half the mass of uranium, were given off. It was as if the nucleus sheared in half when bombarded by something less than half a percent of its mass; like a water melon splitting in two when hit by a cherry. Hahn wrote describing this to Lise Meitner, their exiled Austrian colleague who had just fled fascist Germany for Sweden. Meitner was equally puzzled and discussed it with her physicist nephew, Otto Frisch, who was visiting from Copenhagen. Meitner and Frisch realized that energy would be released as the nucleus split because the two halves took up less energy overall. On his return to

Denmark, Frisch could not contain his excitement and mentioned their idea to Niels Bohr. Embarking on a sea voyage to America, Bohr immediately set to work on an explanation, bringing the news to Italian Enrico Fermi at Columbia University.

Meitner and Frisch published their paper ahead of Bohr, introducing the word "fission" after the division of a biological cell. Back in New York, Fermi and Hungarian exile Léo Szilárd realized that this uranium reaction could produce spare neutrons that would produce more fission and so could go on to cause a nuclear chain reaction (a self-sustaining reaction). Fermi obtained the first chain reaction in 1942 at the University of Chicago, beneath the football stadium.

> " . . . gradually we came to the idea that perhaps one should not think of the nucleus being cleaved in half as with a chisel but rather that perhaps there was something in Bohr's idea that the nucleus was like a liquid drop. "
> Otto Frisch, 1967

Chain reaction

Fellow physicist Arthur Compton remembered the day: "On the balcony a dozen scientists were watching the instruments and handling the controls. Across the room was a large cubical pile of graphite and uranium blocks in which we hoped the atomic chain reaction would

Nuclear power

Subcritical chain reactions can be kept stable and used for nuclear power plants. Boron control rods regulate the flow of neutrons through the uranium fuel by absorbing spare neutrons. In addition, coolant is needed to reduce the heat from the fission reactions. Water is most common, but pressurized water, helium gas and liquid sodium can all be used. Today, France leads the world in using nuclear power, producing more than 70 percent of its total power compared with around 20 percent in the U.S. or UK.

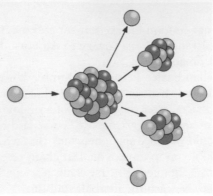

develop. Inserted into openings in this pile of blocks were control and safety rods. After a few preliminary tests, Fermi gave the order to withdraw the control rod another foot. We knew that that was going to be the real test. The geiger counters registering the neutrons from the reactor began to click faster and faster till their sound became a rattle. The reaction grew until there might be danger from the radiation up on the platform where we were standing. 'Throw in the safety rods,' came Fermi's order. The rattle of the counters fell to a slow series of clicks. For the first time, atomic power had been released. It had been controlled and stopped. Somebody handed Fermi a bottle of Italian wine and a little cheer went up."

Manhattan project

Szilárd was so concerned about German scientists copying their feat that he approached Albert Einstein and they presented a joint letter to warn President Roosevelt in 1939. However, not much happened until 1941 when physicists in the UK shared a calculation showing just how easy it was to build a nuclear weapon. That coincided with the Japanese attack at Pearl Harbor and Roosevelt soon began the U.S. nuclear bomb project, known as the Manhattan project. It was led by Berkeley physicist Robert Oppenheimer from a remote and secret base at Los Alamos in New Mexico.

In the summer of 1942, Oppenheimer's team designed mechanisms for the bomb. To set up the chain reaction leading to the explosion a critical mass of uranium was needed, but it should be split up before detonation. Two techniques were favored, a "gun" mechanism where a lump of uranium was shot into another with conventional explosives to complete the critical mass, and an "implosion" mechanism where conventional explosives caused a hollow sphere of uranium to implode onto a core of plutonium.

Nuclear waste

Fission reactors are efficient producers of energy, but they generate radioactive waste. The most toxic products include the remnants of the uranium fuels, which can remain radioactive for thousands of years, and heavier elements (such as plutonium) that can last for hundreds of thousands of years. These dangerous types of waste are made in only small quantities, but the extraction of uranium from its ore and other processes leaves a trail of lower level waste. How to dispose of this waste is a question that is still being considered worldwide.

Uranium comes in two types, or isotopes, hosting different numbers of neutrons in their nuclei. The most common isotope, uranium-238, is ten times more common than the other, uranium-235. It is uranium-235 that is most effective for a fission bomb, so raw uranium is enriched in uranium-235. When uranium-238 receives a neutron it becomes plutonium-239. Plutonium-239 is unstable and its breakdown produces even more neutrons per gram, so mixing in plutonium can trigger the chain reaction readily. The gun method was used with enriched uranium to build the first type of bomb, called "Little Boy." The spherical implosion bomb type, including plutonium, was also built and named "Fat Man."

On August 6 "Little Boy" was dropped on Hiroshima. Three days later, "Fat Man" destroyed Nagasaki. Each bomb released the equivalent of about 20,000 tons of dynamite, killing 70,000–100,000 people immediately, and twice that eventually.

> **I thought this day would go down as a black day in the history of mankind . . . I was also aware of the fact that something had to be done if the Germans get the bomb before we have it . . . They had the people to do it . . . We had no choice, or we thought we had no choice.**
> Léo Szilárd, 1898–1964

55 **Nuclear fusion**

All the elements around us, including those in our bodies, are the product of nuclear fusion. Fusion powers stars like the Sun, within which all the elements heavier than hydrogen are cooked up. We really are made of stardust. If we can harness the stars' power on Earth, fusion could even be the key to unlimited clean energy.

Nuclear fusion is the merging together of light atomic nuclei to form heavier ones. When pressed together hard enough, hydrogen nuclei can merge to produce helium, giving off energy — a great deal of energy — in the process. Gradually, by building up heavier and heavier nuclei through a series of fusion reactions, all the elements that we see around us can be created from scratch.

> I ask you to look both ways. For the road to a knowledge of the stars leads through the atom; and important knowledge of the atom has been reached through the stars.
>
> Sir Arthur Eddington, 1928

Tight squeeze
Fusing together even the lightest nuclei, such as hydrogen, is tremendously difficult. Enormous temperatures and pressures are needed, so fusion only happens naturally in extreme places, like the Sun and other stars. For two nuclei to merge, the forces that hold each one together must be overcome. Nuclei are made up of protons and neutrons locked together by the strong nuclear force. The strong force is dominant at the tiny scale of the nucleus, and is much weaker outside the nucleus. Because protons are positively charged, their electrical charges repel one another, so pushing each other apart slightly as well. But the strong force glue is more powerful so the nucleus holds together.

Because the strong nuclear force acts over such a short precise range, its combined strength is greater for small nuclei than for large ones. For a weighty nucleus, such as uranium, with 238 nucleons, the mutual attraction will not be as strong between nucleons on opposite sides of the nucleus. The electric repulsive force, on the other hand, is still felt at larger separations and so becomes stronger for larger nuclei because it can span the whole nucleus. It is also boosted by the greater numbers of positive charges they contain. The net effect of this balance is that the energy needed to bind the nucleus together, averaged per nucleon, increases with atomic weight up to the elements nickel and iron, which are very stable, and then drops off again for larger nuclei. So fission of large nuclei happens relatively easily as they can be disrupted by a minor knock.

For fusion, the energy barrier to overcome is least for hydrogen isotopes that contain just a single proton. Hydrogen comes in three types: "normal" hydrogen atoms contain one proton surrounded by a single electron; deuterium, or heavy hydrogen, has one proton, one electron and also one neutron; tritium has two neutrons added, so it is even heavier. The simplest fusion reaction therefore is the combination of hydrogen and deuterium to form tritium plus a lone neutron. Although it is the simplest, scorching temperatures of 800 million kelvins are needed to ignite even this reaction (which is why tritium is quite rare).

Fusion reactors

On Earth, physicists are trying to replicate these extreme conditions in fusion reactors to generate power. However, they are decades off from achieving this in practice. Even the most advanced fusion machines take in more energy than they give out, by many orders of magnitude.

Fusion power is the holy grail of energy production. Compared with fission technology, fusion reactions are relatively clean and, should they work, efficient. Very few atoms are needed to produce huge amounts of energy (from Einstein's $E=mc^2$ equation), there is very little waste and certainly nothing as nasty as the ultraheavy elements that come out of fission reactors. Fusion power does not produce greenhouse gases either, promising a self-contained, reliable source of energy assuming its fuel, hydrogen and deuterium, can be manufactured. But it is not perfect and will produce some radioactive by-products as neutrons are released in the main reactions and need to be mopped up.

At the high temperatures involved, controlling the scorching gases is the main difficulty, so although fusion has been achieved these monster machines only work for a few seconds at a time. To try and break through the next technology barrier, an international team of scientists is collaborating to build an even bigger fusion reactor in France, called the International Thermonuclear Experimental Reactor (ITER), that will test the feasibility of fusion being made to work commercially.

> ❝We are bits of stellar matter that got cold by accident, bits of a star gone wrong.❞
>
> Sir Arthur Eddington, 1882–1944

Stardust

Stars are nature's fusion reactors. German physicist Hans Bethe described how they shine by converting hydrogen nuclei (protons) into helium nuclei (two protons and two neutrons). Additional particles (positrons and neutrinos) are involved in the transfer, so that two of the original protons are turned into neutrons in the process.

Within stars, heavier elements are gradually built up in steps by fusion cookery, just like following a recipe. Larger and larger nuclei are constructed through a succession of "burning" first hydrogen, then helium, then other elements lighter than iron and, eventually, elements heavier than iron. Stars like the Sun shine because they are mostly fusing hydrogen into helium and this proceeds slowly enough that heavy elements are made in only small quantities. In bigger stars this reaction is sped up by the involvement of the elements carbon, nitrogen and oxygen in further reactions. So more heavy elements are made more quickly. Once helium is present, carbon can be made from it (three helium-4 atoms fuse, via unstable beryllium-8). Once some carbon is made it can combine with helium to make oxygen, neon and magnesium. These slow transformations take most of the life of the star. Elements heavier than iron are made in slightly different reactions, gradually building sequences of nuclei right up the periodic table.

First stars

Some of the first light elements were created not in stars but in the big bang fireball itself. At first the universe was so hot that not even atoms were stable. As it cooled hydrogen atoms condensed out first, along with a smattering of helium and lithium and a tiny amount of beryllium. These were the first ingredients for all stars and everything else. All elements heavier than this were created in and around stars and were then flung across space by exploding stars called supernovae.

Cold fusion

In 1989, the scientific world was rocked by a controversial claim. Martin Fleischmann and Stanley Pons reported they had performed nuclear fusion, not in a huge reactor, but in a test tube. By firing electric current through a beaker of heavy water (whose hydrogen atoms are replaced by deuterium), the pair believed they had created energy via "cold" fusion. They said their experiment gave out more energy than was put into it, due to fusion occurring. This caused uproar. Most scientists believed Fleischmann and Pons were mistaken in accounting for their energy budget, but even now this is not settled. Other disputed claims of lab-based fusion have occasionally cropped up. In 2002, Rusi Taleyarkhan proposed that fusion was behind so-called sono-luminescence, where bubbles in a fluid emit light when pulsed (and heated) rapidly by ultrasound waves. The jury is still out on whether fusion can really be made to work in a lab flask.

However, we still don't really understand how the first stars switched on. The very first star would not contain any heavy elements, just hydrogen, and so would not cool down quickly enough to collapse and switch on its fusion engine. The process of collapsing under gravity causes the hydrogen gas to heat up and swell, too much. Heavy elements can help it cool down by radiating light, so by the time that one generation of stars has existed and blown out all their by-products into space via supernovae, stars are easy to make. But forming the first star and its siblings quickly enough is still a challenge to the theorists.

Fusion is a fundamental power source across the universe. If we can tap it then our energy woes could be over. But it means harnessing the enormous power of the stars here on Earth, which isn't easy.

56 Standard model

Protons, neutrons and electrons are just the tip of the particle physics iceberg. Protons and neutrons are made up of even smaller quarks, electrons are accompanied by neutrinos, and forces are mediated by a whole suite of bosons, including photons. The "standard model" brings together the entire particle zoo in a single family tree.

To the ancient Greeks, atoms were the smallest components of matter. It was not until the end of the 19th century that even smaller ingredients, first electrons and then protons and neutrons, were etched out of atoms. So are these three particles the ultimate building blocks of matter?

Well, no. Even protons and neutrons are grainy. They are made up of even tinier particles called quarks. And that's not all. Just as photons carry electromagnetic forces, a myriad of other particles transmit the other fundamental forces. Electrons are indivisible, as far as we know, but they are paired with near massless neutrinos. Particles also have their antimatter doppelgangers. This all sounds pretty complicated, and it is, but this plethora of particles can be understood in a single framework called the standard model of particle physics.

> " Even if there is only one possible unified theory, it is just a set of rules and equations. What is it that breathes fire into the equations and makes a universe for them to describe? "
> **Stephen Hawking, 1988**

Excavation

In the early 20th century physicists knew that matter was made up of protons, neutrons and electrons. Niels Bohr had described how, due to quantum theory, electrons arranged themselves in a series of shells around the nucleus, like the orbits of planets around the Sun. The properties of the nucleus were even stranger. Despite their repelling positive charges, nuclei could host tens of protons alongside neutrons compressed into a tiny hard kernel, bound by the precise strong nuclear force. But as more was learned from radioactivity about how nuclei broke apart (via fission) or joined together (via fusion), it became clear that more phenomena needed to be explained.

First, the burning of hydrogen into helium in the Sun, via fusion, implicates another particle, the neutrino, which transforms protons into neutrons. In 1930, the neutrino's existence was inferred to explain the decay of a neutron into a proton and an electron — beta radioactive decay. The neutrino itself was not discovered until 1956, having virtually no mass. So, even in the 1930s there were many loose ends. Pulling on some of these dangling threads, in the 1940s and 50s other particles were sought and the collection grew.

Out of these searches evolved the standard model, which is a family tree of subatomic particles. There are three basic types of fundamental particle, "hadrons" made of "quarks," others called "leptons" that include electrons, and then particles (bosons) that transmit forces, like photons. Each of the quarks and leptons has a corresponding antiparticle as well.

Quarks

In the 1960s, by firing electrons at protons and neutrons physicists realized they hosted even smaller particles within them, called quarks. Quarks come in threes. They have three "colors": red, blue and green. Just as electrons and protons carry electric charge, quarks carry "color charge," which is conserved when quarks change from one type to another. Color charge is nothing to do with the visible colors of light — it is just physicists having to be inventive and finding an arbitrary way of naming the weird quantum properties of quarks.

Just as electric charges produce a force, so color charges (quarks) can exert forces on one another. The color force is transmitted by a force particle called a "gluon." The color force gets stronger the farther the

Quarks

Quarks were so named after a phrase used in James Joyce's *Finnegans Wake* to describe the cries of seagulls. He wrote that they gave "three quarks," or three cheers.

quarks are apart, so they stick together as if held by an invisible elastic band. Because the color force field tie is so strong, quarks cannot exist on their own and must always be locked together in combinations that are color neutral overall (exhibiting no color charge). Possibilities include threesomes called baryons, ("bary" means heavy) including normal protons and neutrons, or quark–antiquark pairs (called mesons).

As well as having color charge, quarks come in 6 types, or "flavors." Three pairs make up each generation of increasing mass. The lightest are the "up" and "down" quarks; next come "strange" and "charm" quarks; finally, the "top" and "bottom" quarks are the heaviest pair. The up, charm and top quarks have electric charges $+\frac{2}{3}$ and the down, strange and bottom quarks have charge $-\frac{1}{3}$. So quarks have fractional electric charge, compared with $+1$ for protons or -1 for electrons. So three quarks are needed to make up a proton (two ups and a down) or a neutron (two downs and an up).

Leptons

The second class of particles, the leptons, are related to and include electrons. Again there are three generations with increasing masses: electrons, muons and taus. Muons are 200 times heavier than an electron and taus 3,700 times heavier. Leptons all have single negative charge. They also have an associated particle called a neutrino (electron-, muon- and tau-neutrino) that has no charge. Neutrinos have almost no mass and do not interact much with anything. They can travel right through the Earth without noticing, so are difficult to catch. All leptons have antiparticles.

Interactions

Fundamental forces are mediated by the exchange of particles. Just as the electromagnetic wave can also be thought of as a stream of photons, the weak nuclear force can be thought of as being carried by W and Z particles while the strong nuclear force is transmitted via gluons. Like the photon, these other particles are bosons, which can all exist in the same quantum state simultaneously. Quarks and leptons are fermions and cannot.

Particle smashing

How do we know about all these subatomic particles? In the second half of the 20th century physicists exposed the inner workings of atoms and particles using brute force and smashing them apart. Particle physics has been described as taking an intricate Swiss watch and smashing it up with a hammer and looking at the shards to work out how it operates. Particle accelerators use giant magnets to accelerate particles to extremely high speeds and then smash those particle beams either into a target or into another oppositely directed beam. At modest speeds, the particles break apart a little and the lightest generations of particles are released. Because mass means energy, you need a higher-energy particle beam to release the later (heavier) generations of particles.

The particles produced in the atom smashers then need to be identified and particle physicists do this by photographing their tracks as they pass through a magnetic field. In the magnetic field, positive charged particles swerve one way and negative ones the other. The mass of the particle also dictates how fast it shoots through the detector and how much its path is curved by the magnetic field. So light particles barely curve and heavier particles may even spiral into loops. By mapping their characteristics in the detector, and comparing them with what they expect from their theories, particle physicists can tell what each particle is.

One thing that is not yet included in the standard model is gravity. The "graviton," or gravity force carrying particle, has been postulated but only as an idea. Unlike light, there's no evidence yet for any graininess in gravity. Some physicists are trying to put gravity into the standard model in what would be a grand unified theory (GUT). But we are a long way off yet.

57 Special relativity

Newton's laws of motion describe how most objects move, from cricket balls and cars to comets. But Albert Einstein showed in 1905 that strange effects happen when things move very quickly. Watching an object approach light speed, you'd see it become heavier, contract in length and age more slowly. That's because nothing can travel faster than the speed of light, so time and space themselves distort when approaching this universal speed limit.

Sound waves ring though air, but their vibrations cannot traverse empty space where there are no atoms. So it is true that "in space no one can hear you scream." But light is able to spread through empty space, as we know because we see the Sun and stars. Is space filled with a special medium, a sort of electric air, through which electromagnetic waves propagate? Physicists at the end of the 19th century thought so and believed that space was effused with a gas or "ether" through which light radiates.

> **The most incomprehensible thing about the world is that it is at all comprehensible.**
> Albert Einstein, 1879–1955

Light speed

In 1887, however, a famous experiment proved the ether did not exist. Because the Earth moves around the Sun, its position in space is always changing. If the ether were fixed then Albert Michelson and Edward Morley devised an ingenious experiment that would detect movement against it. They compared two beams of light traveling different paths, fired at right angles to one another and reflected back off identically faraway mirrors. Just as a swimmer takes less time to travel across a river from one bank to the other and back than to swim the same distance upstream against the current and downstream with it, they expected a similar result for light. The river current mimics the motion of the Earth through the ether. But

Twin paradox

Imagine if time dilation applied to humans. Well it could. If your identical twin was sent off into space on a rocket ship fast enough and for long enough, then they would age more slowly than you on Earth. On their return, they might find you to be elderly when they are still a sprightly youth. Although this seems impossible, it is not really a paradox because the space-faring twin would experience powerful forces that permit such a change to happen. Because of this time shift, events that appear simultaneous in one frame may not appear so in another. Just as time slows, so lengths contract also. The object or person moving at that speed would not notice either effect, it would just appear so to another viewer.

there was no such difference — the light beams returned to their starting points at exactly the same time. No matter which direction the light traveled, and how the Earth was moving, the speed of light remained unchanged. Light's speed was unaffected by motion. The experiment proved the ether did not exist — but it took Einstein to realize this.

Just like Mach's principle (see Chapter 34), this meant that there was no fixed background grid against which objects moved. Unlike water waves or sound waves, light appeared always to travel at the same speed. This was odd and quite different from our usual experience where velocities add together. If you are driving in a car at 50 km/h and another passes you at 65 km/h it is as if you are stationary and the other is traveling at 15 km/h past you. But even if you were rushing at hundreds of km/h light would still travel at the same speed. It is exactly 300 million meters per second whether you are shining a torch from your seat in a fast jet plane or the saddle of a bicycle.

It was this fixed speed of light that puzzled Albert Einstein in 1905, leading him to devise his theory of special relativity. Then an unknown Swiss patent clerk, Einstein worked out the equations from scratch in his spare moments. Special relativity was the biggest breakthrough since Newton and revolutionized physics. Einstein started with the assumption that the speed of light is a constant value, and appears the same for any observer no matter how fast they are moving. If the speed of light does not change then, reasoned Einstein, something else must change to compensate.

> **The introduction of a light-ether will prove to be superfluous since ... neither will a space in absolute rest endowed with special properties be introduced nor will a velocity vector be associated with a point of empty space in which electromagnetic processes take place.**
>
> **Albert Einstein, 1905**

Space and time

Following ideas developed by Edward Lorenz, George Fitzgerald and Henri Poincaré, Einstein showed that space and time must distort to accommodate the different viewpoints of observers traveling close to the speed of light. The three dimensions of space and one of time made up a four-dimensional world in which Einstein's vivid imagination worked. Speed is distance divided by time, so to prevent anything from exceeding the speed of light, distances must shrink and time slow down to compensate. So a rocket traveling away from you at near light speed looks shorter and experiences time more slowly than you do.

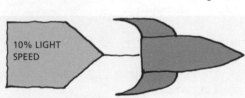

10% LIGHT SPEED

86.5% LIGHT SPEED

Einstein worked out how the laws of motion could be rewritten for observers traveling at different speeds. He ruled out the existence of a stationary frame of reference, such as the ether, and stated that all motion was relative with no privileged viewpoint. If you are sitting on a train and see the train next to you moving, you may not know whether it is your train or the other one pulling out. Moreover, even if you can see your train is stationary at the platform you cannot assume that you are immobile, just that you are not moving relative to that platform. We do not feel the motion of the Earth around the Sun; similarly, we never notice the Sun's path across our own Galaxy, or our Milky Way being pulled towards the huge Virgo cluster of galaxies beyond it. All that is experienced is relative motion, between you and the platform or the Earth spinning against the stars.

Einstein called these different viewpoints inertial frames. Inertial frames are spaces that move relative to one another at a constant speed, without experiencing accelerations or forces. So sitting in a car traveling at 50 km/h you are in one inertial frame, and you feel just the same as if you were in a train traveling at 100 km/h (another inertial frame) or a jet plane traveling at 500 km/h (yet another). Einstein stated that the laws of physics are the same in all inertial frames. If you dropped your pen in the car, train or plane, it would fall to the floor in the same way.

Slower and heavier

Turning next to understand relative motions near the speed of light, the maximum speed practically attainable by matter, Einstein predicted that time would slow down. Time dilation expressed the fact that clocks in different moving inertial frames may run at different speeds. This was proved in 1971 by sending four identical atomic clocks on scheduled flights twice around the world, two flying eastwards and two westwards. Comparing their times with a matched clock on the Earth's surface in the United States, the moving clocks had each lost a fraction of a second compared with the grounded clock, in agreement with Einstein's special relativity.

> It is impossible to travel faster than the speed of light, and certainly not desirable, as one's hat keeps blowing off.
>
> **Woody Allen, b.1935**

Another way that objects are prevented from passing the light speed barrier is that their mass grows, according to $E = mc^2$. An object would become infinitely large at light speed itself, making any further acceleration impossible. And anything with mass cannot reach the speed of light exactly, but only approach it, as the closer it gets the heavier and more difficult to accelerate it becomes. Light is made of mass-less photons so these are unaffected.

Einstein's special relativity was a radical departure from what had gone before. The equivalence of mass and energy was shocking, as were all the implications for time dilation and mass. Although Einstein was a scientific nobody when he published it, his ideas were read by Max Planck, and it is perhaps because of his adoption of Einstein's ideas that they became accepted and not sidelined. Planck saw the beauty in Einstein's equations, catapulting him to global fame.

58 General relativity

Incorporating gravity into his theory of special relativity, Einstein's theory of general relativity revolutionized our view of space and time. Going beyond Newton's laws, it opened up a universe of black holes, worm holes and gravitational lenses.

Imagine a person jumping off a tall building, or parachuting from a plane, being accelerated towards the ground by gravity. Albert Einstein realized that in this state of free fall they did not experience gravity. In other words they were weightless. Trainee astronauts today recreate the zero gravity conditions of space in just this way, by flying a passenger jet (attractively named the "vomit comet") in a path that mimics a rollercoaster. When the plane flies upwards the passengers are glued to their seats as they experience even stronger forces of gravity. But when the plane tips forwards and plummets downwards, they are released from gravity's pull and can float in the body of the aircraft.

Acceleration

Einstein recognized that this acceleration was equivalent to the force of gravity. So, just as special relativity describes what happens in reference frames, or inertial frames, moving at some constant speed relative to one another, gravity was a consequence of being in a reference frame that is accelerating. He called this the happiest thought of his life.

Over the next few years Einstein explored the consequences. Talking through his ideas with trusted colleagues and using the latest mathematical formalisms to encapsulate them, he pieced together the full theory of gravity that he called general relativity. The year 1915 when he published the work proved especially busy and almost immediately he revised it several times. His peers were astounded by his progress. The theory even produced bizarre testable predictions, including the idea that light could be bent by a gravitational field and also that Mercury's elliptical orbit would rotate slowly because of the gravity of the Sun.

Space–time

In general relativity theory, the three dimensions of space and one of time are combined into a four-dimensional space–time grid, or metric. Light's speed is still fixed, and nothing can exceed it. When moving and accelerating, it is this space–time metric that distorts to maintain the fixed speed of light.

General relativity is best imagined by visualizing space–time as a rubber sheet stretched across a hollow table top. Objects with mass are like weighted balls placed on the sheet. They depress space–time around them. Imagine you place a ball representing the Earth on the sheet. It forms a depression in the rubber plane in which it sits. If you then threw in a smaller ball, say as an asteroid, it would roll down the slope towards the Earth. This shows how it feels gravity. If the smaller ball was moving fast enough and the Earth's dip was deep enough, then just as a daredevil cyclist can ride around an inclined track, that body would maintain a moon-like circular orbit. You can think of the whole universe as a giant rubber sheet. Every one of the planets and stars and galaxies causes a depression that can attract or deflect passing smaller objects, like balls rolling over the contours of a golf course.

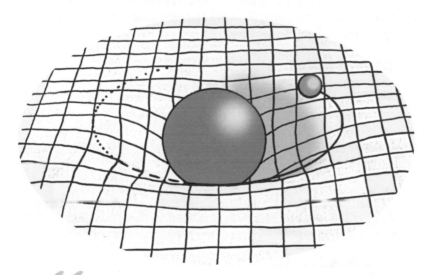

“ Time and space and gravitation have no separate existence from matter. **”**
Albert Einstein, 1915

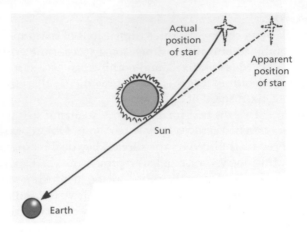

Actual position of star

Apparent position of star

Sun

Earth

Einstein understood that, because of this warping of space–time, light would be deflected if it passed near a massive body, such as the Sun. He predicted that the position of a star observed just behind the Sun would shift a little because light from it is bent as it passes the Sun's mass. On May 29, 1919 the world's astronomers gathered to test Einstein's predictions by observing a total eclipse of the Sun. It proved one of his greatest moments, showing that the theory some thought crazy was in fact close to the truth.

Warps and holes

The bending of light rays has now been confirmed with light that has traveled right across the universe. Light from very distant galaxies clearly flexes when it passes a very massive region such as a giant cluster

Gravity waves

Another aspect of general relativity is that waves can be set up in the space–time sheet. Gravitational waves can radiate, especially from black holes and dense spinning compact stars like pulsars. Astronomers have seen pulsars' spin decreasing so they expect that this energy will have been lost to gravity waves, but the waves have not yet been detected. Physicists are building giant detectors on Earth and in space that use the expected rocking of extremely long laser beams to spot the waves as they pass by. If gravity waves were detected then this would be another coup for Einstein's general relativity theory.

of galaxies or a really big galaxy. The background dot of light is smeared out into an arc. Because this mimics a lens the effect is known as "gravitational lensing". If the background galaxy is sitting right behind the heavy intervening object then its light is smeared out into a complete circle, called an Einstein ring. Many beautiful photographs of this spectacle have been taken with the Hubble Space Telescope.

Einstein's theory of general relativity is now widely applied to modeling the whole universe. Space–time can be thought of like a landscape, complete with hills, valleys and pot holes. General relativity has lived up to all observational tests so far. The regions where it is tested most are where gravity is extremely strong, or perhaps very weak.

> ❝ We shall therefore assume the complete physical equivalence of a gravitational field and the corresponding acceleration of the reference frame. This assumption extends the principle of relativity to the case of uniformly accelerated motion of the reference frame. ❞
> Albert Einstein, 1907

Black holes (see Chapter 59) are extremely deep wells in the space–time sheet. They are so deep and steep that anything that comes close enough can fall in, even light. They mark holes, or singularities, in space–time. Space–time may also warp into wormholes, or tubes, but no one has actually seen such a thing yet.

At the other end of the scale, where gravity is very weak it might be expected to break up eventually into tiny quanta, similar to light that is made up of individual photon building blocks. But no one has yet seen any graininess in gravity. Quantum theories of gravity are being developed but, without evidence to back it up, the unification of quantum theory and gravity is elusive. This hope occupied Einstein for the rest of his career but even he did not manage it and the challenge still stands.

59 Black holes

Falling into a black hole would not be pleasant, having your limbs torn asunder and all the while appearing to your friends to be frozen in time just as you fell in. Black holes were first imagined as frozen stars whose escape velocity exceeds that of light, but are now considered as holes or "singularities" in Einstein's space–time sheet. Not just imaginary, giant black holes populate the centers of galaxies, including our own, and smaller ones punctuate space as the ghosts of dead stars.

If you throw a ball up in the air, it reaches a certain height and then falls back down. The faster you fling it the higher it goes. If you hurled it fast enough it would escape the Earth's gravity and whiz off into space. The speed that you need to reach to do this, called the "escape velocity," is 11 km/s (or about 25,000 mph). A rocket needs to attain this speed if it is to escape the Earth. The escape velocity is lower if you are standing on the smaller Moon: 2.4 km/s would do. But if you were standing on a more massive planet then the escape velocity rises. If that planet was heavy enough, then the escape velocity could reach or exceed the speed of light itself, and so not even light could escape its gravitational pull. Such an object, that is so massive and dense that not even light can escape it, is called a black hole.

> " God not only plays dice, but also sometimes throws them where they cannot be seen. "
> **Stephen Hawking, 1977**

Event horizon

The black hole idea was developed in the 18th century by geologist John Michell and mathematician Pierre-Simon Laplace. Later, after Einstein had proposed his relativity theories, Karl Schwarzschild worked out what a black hole would look like. In Einstein's theory of general relativity, space and time are linked and behave together like a vast rubber sheet. Gravity distorts the sheet according to an object's mass. A heavy planet rests in a dip in space–time, and its gravitational pull is equivalent to the force felt as you roll into the dip, perhaps warping your path or even pulling you into orbit.

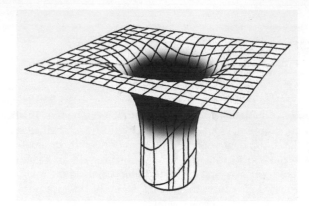

So what then is a black hole? It would be a pit that is so deep and steep that anything that comes close enough to it falls straight in and cannot return. It is a hole in the sheet of space–time, like a basketball net (from which you will never get your ball back).

If you pass far from a black hole, your path might curve towards it, but you needn't fall in. But if you pass too close to it, then you will spiral in. The same fate would even befall a photon of light. The critical distance that borders these two outcomes is called the "event horizon." Anything that falls within the event horizon plummets into the black hole, including light.

Falling into a black hole has been described as being "spaghetti-fied." Because the sides are so steep, there is a very strong gravity gradient within the black hole. If you were to fall into one feet first, and let's hope you never do, then your feet would be pulled more than your head and so your body would be stretched like being on a rack. Add to that any spinning motion and you would be pulled out like chewing gum into a scramble of spaghetti. Not a nice way to go. Some scientists have worried about trying to protect an unlucky person who might accidentally stumble into a black hole. One way you could protect yourself, apparently, is to don a leaden life-saver ring. If the ring was heavy and dense enough, it would counteract the gravity gradient and preserve your shape — and your life.

Frozen stars

The name "black hole" was coined in 1967 by John Wheeler as a catchier alternative to describe a frozen star. Frozen stars were predicted in the 1930s by Einstein and Schwarzschild's theories. Because of the weird behavior of space and time close to the event horizon, glowing matter falling in would seem to slow down as it does so, due to the

Evaporation

Strange as it may sound, black holes eventually evaporate. In the 1970s, Stephen Hawking suggested that black holes are not completely black but radiate particles due to quantum effects. Mass is gradually lost in this way and so the black hole shrinks until it disappears. The black hole's energy continually creates pairs of particles and their corresponding antiparticles. If this happens near the event horizon then sometimes one of the particles might escape even if the other falls in. To an outside eye the black hole seems to emit particles, called Hawking radiation. This radiated energy then causes the hole to diminish. This idea is still based in theory, and no one really knows what happens to a black hole. The fact that they are relatively common suggests that this process takes a long time, so black holes hang around.

light waves taking longer and longer to reach an observer looking on. As it passes the event horizon, this outside observer sees time actually stop so that the material appears to be frozen at the time it crosses the horizon. Hence, frozen stars, frozen in time just at the point of collapsing into the event horizon, were predicted. Astrophysicist Subrahmanyan Chandrasekhar predicted that stars more than 1.4 times the Sun's mass would ultimately collapse into a black hole; however, we now know that white dwarfs and neutron stars will prop themselves up by quantum pressure, which arises because the rules of quantum mechanics prevent some particles from taking the same energy and occupying the same space. So, to overcome this, black holes need more than three times the sun's mass to form. Evidence of these frozen stars or black holes was not discovered until the 1960s.

If black holes suck in light, how can we see them to know they are there? There are two ways. First, you can spot them because of the way they pull other objects towards them. And second, as gas falls into them it can heat up and glow before it disappears. The first method has been used to identify a black hole lurking in the center of our own Galaxy. Stars that pass close to it have been seen to whip past it and be flung out on elongated orbits. The Milky Way's black hole has a mass of a million Suns, squashed into a region of radius just 10 million kilometers (30 light seconds) or so. Black holes that lie in galaxies are called supermassive black holes. We don't know how they formed, but they seem to affect how galaxies grow so might have been there from day one or, perhaps, grew from millions of stars collapsing into one spot.

The second way to see a black hole is by the light coming from hot gas that is fired up as it falls in. Quasars, the most luminous things in the universe, shine due to gas being sucked into supermassive black holes in the centers of distant galaxies. Smaller black holes, just a few solar masses, can also be identified by x-rays shining from gas falling towards them.

> The black holes of nature are the most perfect macroscopic objects there are in the universe: the only elements in their construction are our concepts of space and time.
> Subrahmanyan Chandrasekhar, 1983

Wormholes

What lies at the bottom of a black hole in the space–time sheet? Supposedly they just end in a sharp point, or truly are holes, punctures in the sheet. But theorists have asked what might happen if they joined another hole. You can imagine that two nearby black holes might appear as long tubes dangling from the space–time sheet. If the tubes were joined together, then you could imagine a tube or wormhole being formed between the two mouths of the black holes. Armed with your "life-saver" you might be able to jump into one black hole and pop out of another. This idea has been used a lot in science fiction for transport across time and space. Perhaps the wormhole could flow through to an entirely different universe. The possibilities for rewiring the universe are endless, but don't forget your life-saver ring.

60 Olbers' paradox

Why is the night sky dark? If the universe were endless and had existed for ever then it should be as bright as the Sun, yet it is not. Looking up at the night sky you are viewing the entire history of the universe. The limited number of stars is real and implies that the universe has a limited size and age. Olbers' paradox paved the way for modern cosmology and the big bang model.

You might think that mapping the entire universe and viewing its history would be difficult and call for expensive satellites in space, huge telescopes on remote mountain tops, or a brain like Einstein's. But in fact if you go out on a clear night you can make an observation that is every bit as profound as general relativity. The night sky is dark. Although this is something we take for granted, the fact that it is dark and not as bright as the Sun tells us a lot about our universe.

Star light star bright

If the universe were infinitely big, extending for ever in all directions, then in every direction we look we would eventually see a star. Every sight line would end on a star's surface. Going farther away from the Earth, more and more stars would fill space. It is like looking through a forest of trees — nearby you can distinguish individual trunks, appearing larger the closer they are, but more and more distant trees fill your view. So, if the forest was really big, you could not see the landscape beyond. This is what would happen if the universe were infinitely big. Even though the stars are more widely spaced than the trees, eventually there would be enough of them to block the entire view.

If all the stars were like the Sun, then every point of sky would be filled with star light. Even though a single star far away is faint, there are more stars at that distance. If you add up all the light from those stars they provide as much light as the Sun, so the entire night sky should be as bright as the Sun.

Dark skies

The beauty of the dark night sky is becoming harder and harder to see due to the glow of lights from our cities. On clear nights throughout history people have been able to look upward and see a brightly lit backbone of stars, stretched across the heavens. This was christened the Milky Way, and we now know that when we gaze at it we are looking towards the central plane of our Galaxy. Even in large cities 50 years ago it was possible to see the brightest stars and the Milky Way's swath, but nowadays hardly any stars are visible from towns and even the countryside views of the heavens are washed out by yellow smog. The vista that has inspired generations before us is becoming obscured. Sodium street lights are the main culprit, especially ones that waste light by shining upwards as well as down. Groups worldwide, like the International Dark-Sky Association, which includes astronomers, are now campaigning for curbs on light pollution so that our view out to the universe is preserved.

Obviously this is not so. The paradox of the dark night sky was noted by Johannes Kepler in the 17th century, but only formulated in 1823 by German astronomer Heinrich Olbers. The solutions to the paradox are profound. There are several explanations, and each one has elements of truth that are now understood and adopted by modern astronomers. Nevertheless, it is amazing that such a simple observation can tell us so much.

Eureka!

Edgar Allan Poe, in his 1848 prose poem *Eureka*, observed:

"Were the succession of stars endless, then the background of the sky would present us an uniform luminosity, like that displayed by the Galaxy — since there could be absolutely no point, in all that background, at which would not exist a star. The only mode, therefore, in which, under such a state of affairs, we could comprehend the voids which our telescopes find in innumerable directions, would be by supposing the distance of the invisible background so immense that no ray from it has yet been able to reach us at all."

End in sight

The first explanation is that the universe is not infinitely big. It must stop somewhere. So there must be a limited number of stars in it and not all sightlines will find a star. Similarly, standing near the edge of the forest or in a small wood you can see the sky beyond.

Another explanation could be that the more distant stars are fewer in number, so they do not add together to give as much light. Because light travels with a precise speed, the light from distant stars takes longer to reach us than from nearby stars. It takes eight minutes for light to reach us from the Sun but four years for light from the next nearest star, Alpha Centauri to arrive, and as much as 100,000 years for light to reach us from stars on the other side of our own Galaxy. Light from the next nearest galaxy, Andromeda, takes 2 million years to reach us; it is the most distant object we can see with the naked eye. So as we peer further into the universe, we are looking back in time and distant stars look younger than the ones nearby. This could help us with Olbers' paradox if those youthful stars eventually become rarer than Sun-like stars nearby. Stars like the Sun live for about 10 billion years (bigger ones live for shorter times and smaller ones for longer), so the fact that stars have a finite lifetime could also explain the paradox. Stars cease to exist earlier than a certain time because they have not been born yet. So stars have not existed for ever.

Making distant stars fainter than the Sun is also possible through redshift. The expansion of the universe stretches light wavelengths causing the light from distant stars to appear redder. So stars a long way away will look a little cooler than stars nearby. This could also restrict the amount of light reaching us from the outermost parts of the universe.

Wackier ideas have been put forward such as the distant light being blocked out, by soot from alien civilizations, iron needles or weird gray dust. But any absorbed light would be re-radiated as heat and so would turn up elsewhere in the spectrum. Astronomers have checked the light in the night sky at all wavelengths, from radio waves to gamma rays, and they have seen no sign that the visible star light is blocked.

Middle of the road universe

So, the simple observation that the night sky is dark tells us that the universe is not infinite. It has only existed for a limited amount of time, it is restricted in size, and the stars in it have not existed forever.

Modern cosmology is based on these ideas. The oldest stars we see are around 13 billion years old, so we know the universe must have been formed before this time. Olbers' paradox suggests it cannot be very much ahead of this or we would expect to see many previous generations of stars and we do not.

Distant galaxies of stars are indeed redder than nearby ones, due to redshift, making them harder to see with optical telescopes and confirming that the universe is expanding. The most distant galaxies known today are so red they become invisible and can only be picked up at infrared wavelengths. So all this evidence supports the idea of the big bang, such that the universe grew out of a vast explosion some 14 billion years ago.

61 Hubble's law

Edwin Hubble was first to realize that galaxies beyond our own are all moving away from us together. The further away they are, the faster they recede, following Hubble's law. This galactic diaspora formed the first evidence that the universe is expanding, an astounding finding that changed our view of our entire universe and its destiny.

Copernicus' deduction in the 16th century that the Earth goes around the Sun caused major consternation. Humans no longer inhabited the exact center of the cosmos. But in the 1920s, American astronomer Edwin Hubble made telescope measurements that were even more unsettling. He showed the entire universe was not static but expanding. Hubble mapped out the distances to other galaxies and their relative speeds compared with our Milky Way; he found that they were all hurtling away from us. We were so cosmically unpopular that only a few close neighbors were inching towards us. The more distant the galaxy, the faster it was receding, with a speed proportional to its distance away (Hubble's law). The ratio of speed to distance is always the same number and is called the Hubble constant.

> **The history of astronomy is a history of receding horizons.**
> Edwin Hubble, 1938

Astronomers today have measured its value to be close to 75 kilometers per second per megaparsec (a megaparsec, or a million parsecs, is equivalent to 3,262,000 light years or 3×10^{22} m). So galaxies continually recede from us by this amount.

The great debate

Before the 20th century, astronomers barely understood our own Galaxy, the Milky Way. They had measured hundreds of stars within it but also noted it was marked with many faint smudges, called nebulae. Some of these nebulae were gas clouds associated with the births and deaths of stars. But some looked different. Some had spiral or oval shapes that suggested they were more regular than a cloud.

In 1920 two famous astronomers held a debate on the origin of these smudges. Harlow Shapley argued that everything in the sky was part of the Milky Way, which constituted the entire universe. On the other side, Heber Curtis proposed that some of these nebulae were separate "island universes" or external "universes" outside our own Milky Way. The term "galaxy" was coined only later to describe these nebulous universes. Both astronomers cited evidence to back up their own idea, and the debate was not settled on the day. Later work by Hubble showed that Curtis' view was correct. These spiral nebulae really were external galaxies and did not lie within the Milky Way. The universe had suddenly opened up into a vast canvas.

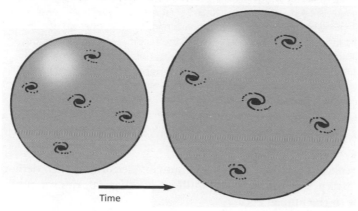

Time

Flying apart

Hubble used the 100-inch Hooker Telescope at Mount Wilson to measure the light from flickering stars in the Andromeda nebula, now known to be a spiral galaxy very similar to the Milky Way and also a sibling in the group of galaxies associated with us. These flickering stars are called Cepheid variable stars, after the prototype star found in the constellation Cepheus, and are even now invaluable probes of distance. The amount and timing of flickering scales with the intrinsic brightness of the star, so once you know how its light varies you know how bright it is. Knowing its brightness you can then work out how far away it is because it is dimmed when placed at a distance. It is analogous to seeing a light bulb placed a distance away, reading that its power is 100 Watts, and then working out how far away it is by comparing its brightness with a 100 Watt bulb in front of you.

In this way Hubble measured the distance to the Andromeda galaxy. It was much farther away than the size of our Milky Way, as given by Shapley, so it must lie outside. This simple fact was revolutionary. It meant that the universe was vast, and filled with other galaxies just like the Milky Way. If putting the Sun at the center of the

Hubble Space Telescope

The Hubble Space Telescope is surely the most popular satellite observatory ever. Its stunning photographs of nebulae, distant galaxies and disks around stars have graced the front pages of many newspapers for almost 20 years. Launched in 1990 from the space shuttle *Discovery*, the spacecraft is about the size of a double-decker bus, 13 m long, 4 m across and weighing 11,000 kg. It carries an astronomical telescope whose mirror is 2.4 m across and a suite of cameras and electronic detectors that are able to take crystal clear images, in visible and ultraviolet light and the infrared. Hubble's power lies in the fact it is located above the atmosphere — so its photographs are not blurred. Now getting old, Hubble's fate is uncertain. NASA may upgrade its instruments but that would require a manned shuttle crew, or it may terminate its program and either rescue the craft for posterity or crash it safely into the ocean.

universe annoyed the church and humans' sensibility then demoting the Milky Way to just one in millions of other galaxies was a bigger blow to the human ego.

Hubble then set about mapping distances to many other galaxies. He also found that the light from them was mostly redshifted by an amount that scaled with distance. The redshift is similar to the Doppler shift of a speeding object (see Chapter 43). Finding that frequencies of light, such as atomic transitions of hydrogen, were all redder than expected meant that these galaxies were all rushing away from us, like many ambulance sirens falling off in tone as they speed away. It was very strange that all the galaxies were rushing away, with only "local" ones moving towards us. The farther away you looked, the faster they receded. Hubble saw that the galaxies weren't simply receding from us, which would have made our place in the universe very privileged indeed. Instead, they were all hurtling away from each other. Hubble concluded that the universe itself was expanding, being inflated like a giant balloon. The galaxies are like spots on the balloon, getting further apart from one another as more air is added.

How far how fast?

Even today astronomers use Cepheid variable stars to map out the local universe's expansion. Measuring the Hubble constant accurately has been a major goal. To do so you need to know how far away something is and its speed or redshift. Redshifts are straightforward to measure from atomic spectra. The frequency of a particular atomic transition in star light can be checked against its known wavelength in the laboratory; the difference gives its redshift. Distances are harder to determine, because you need to observe something in the distant galaxy either whose true length is known or whose true brightness is known, a "standard candle."

> **We find them smaller and fainter, in constantly increasing numbers, and we know that we are reaching into space, farther and farther, until, with the faintest nebulae that can be detected with the greatest telescopes, we arrive at the frontier of the known universe.**
> Edwin Hubble, 1938

There are a variety of methods for inferring astronomical distances. Cepheid stars work for nearby galaxies when you can separate the individual stars. But farther away other techniques are needed. All the different techniques can be tied together one by one to build up a giant measuring rod, or "distance ladder." But because each method comes with peculiarities there are still many uncertainties in the accuracy of the extended ladder.

The Hubble constant is now known to an accuracy of about 10 percent, thanks largely to observations of galaxies with the Hubble Space Telescope and the cosmic microwave background radiation. The expansion of the universe began in the big bang, the explosion that created the universe, and galaxies have been flying apart ever since then. Hubble's law sets a limit on the age of the universe. Because it is continuously expanding, if you trace back the expansion to the beginning point, you can work out how long ago that was. It turns out to be around 14 billion years. This expansion rate is fortunately not enough to break apart the universe. The cosmos instead is finely balanced, in between completely blowing apart and containing enough mass to collapse back in on itself eventually.

62 The Big Bang

The birth of the universe in a phenomenal explosion created all space, matter and time as we know it. Predicted from the mathematics of general relativity, we see evidence for the big bang in the rush of galaxies away from our own, the quantities of light elements in the universe and the microwave glow that fills the sky.

The Big Bang is the ultimate explosion — the birth of the universe. Looking around us today, we see signs that our universe is expanding and infer it must have been smaller, and hotter, in the past. Taking this to its logical conclusion means that the entire cosmos could have originated from a single point. At the moment of ignition, space and time and matter were all created together in a cosmic fireball. Very gradually, over 14 billion years, this hot, dense cloud swelled and cooled. Eventually it fragmented to produce the stars and galaxies that dot the heavens today.

It's no joke

The "Big Bang" phrase itself was actually coined in ridicule. The eminent British astronomer Fred Hoyle thought it preposterous that the whole universe grew from a single seed. In a series of lectures first broadcast in 1949 he derided as far-fetched the proposition of Belgian mathematician Georges Lemaître who found such a solution in Einstein's equations of general relativity. Instead, Hoyle preferred to believe in a more sustainable vision of the cosmos. In his perpetual "steady state" universe, matter and space were being continually created and destroyed and so could have existed for an unlimited time. Even so, clues were already amassing and by the 1960s Hoyle's static picture had to give way, given the weight of evidence that favored the Big Bang.

The expanding universe

Three critical observations underpin the success of the Big Bang model. The first is Edwin Hubble's observation in the 1920s that most galaxies are moving away from our own. Looked at from afar, all galaxies tend to fly apart from one another as if the fabric of space–time is expanding and stretching, following Hubble's law. One consequence of the stretching is that light takes slightly longer to reach us when traveling across an expanding universe than one where distances are fixed. This effect is recorded as a shift in the frequency of the light, called the "redshift" because the received light appears redder than it was when it left the distant star or galaxy. Redshifts can be used to infer astronomical distances.

> Tune your television to any channel it doesn't receive, and about 1 percent of the dancing static you see is accounted for by this ancient remnant of the big bang. The next time you complain that there is nothing on, remember that you can always watch the birth of the universe.
>
> Bill Bryson, 2005

Light elements

Going back in time to the first hours of the newborn universe, just after the Big Bang, everything was packed close together in a seething superheated cauldron. Within the first second, the universe was so hot and dense that not even atoms were stable. As it grew and cooled a particle soup emerged first, stocked with quarks, gluons and other fundamental particles (see Chapter 56). After just a minute the quarks stuck together to form protons and neutrons. Then, within the first three minutes, cosmic chemistry mixed the protons and neutrons, according to their relative numbers, into atomic nuclei. This is when elements other than hydrogen were first formed by nuclear fusion. Once the universe cooled below the fusion limit, no elements heavier than beryllium could be made. So the universe initially was awash with the nuclei of hydrogen, helium and traces of deuterium (heavy hydrogen), lithium and beryllium created in the Big Bang itself.

In the 1940s Ralph Alpher and George Gamow predicted the proportions of light elements produced in the Big Bang, and this basic picture has been confirmed by even the most recent measurements in slow-burning stars and primitive gas clouds in our Milky Way.

BIG BANG TIMELINE

13.7 billion years
[after the big bang]
Now (temperature, T = 2.726 K)

200 million years "Reionization":
first stars heat and ionize hydrogen
gas (T = 50 K)

380 thousand years
"Recombination": hydrogen gas
cools down to form molecules
(T = 3,000 K)

10 thousand years End of the
radiation-dominated era
(T = 12,000 K)

1,000 seconds Decay of lone
neutrons (T = 500 million K)

180 seconds "Nucleosynthesis":
formation of helium and other
elements from hydrogen
(T = 1 billion K)

10 seconds Annihilation of
electron–positron pairs
(T = 5 billion K)

1 second Decoupling of neutrinos
(T ~ 10 billion K)

100 microseconds Annihilation of
pions (T ~ 1 trillion K)

50 microseconds "QCD phase
transition": quarks bound into
neutrons and protons
(T = 2 trillion K)

10 picoseconds "Electroweak phase
transition": electromagnetic and
weak force become different
(T ~ 1–2 quadrillion K)

Before this time the temperatures
were so high that our
knowledge of physics is
uncertain.

Time

Big bang

Microwave glow

Another pillar supporting the Big Bang model is the discovery in 1965 of the faint echo of the Big Bang itself. Arno Penzias and Robert Wilson were working on a radio receiver at Bell Labs in New Jersey when they were puzzled by a weak noise signal they could not get rid of. It seemed there was an extra source of microwaves coming from all over the sky, equivalent to something a few degrees in temperature.

After talking to astrophysicist Robert Dicke at nearby Princeton University, they realized that their signal matched predictions of the Big Bang afterglow. They had stumbled upon the cosmic microwave background radiation, a sea of photons left over from the very young hot universe. Dicke, who had built a similar radio antenna to look for the background radiation, was a little less jubilant: "Boys, we've been scooped," he quipped.

In Big Bang theory, the existence of the microwave background had been predicted in 1948 by George Gamow, Ralph Alpher and Robert Hermann. Although nuclei were synthesized within the first three minutes, atoms were not formed for 400,000 years. Eventually, negatively charged electrons paired with positively charged nuclei to make atoms of hydrogen and light elements. The removal of charged particles, which scatter and block the path of light, cleared the fog and made the universe transparent. From then onwards, light could travel freely across the universe, allowing us to see back that far.

Although the young universe fog was originally hot (some 3,000 kelvins), the expansion of the universe has redshifted the glow from it so that we see it today at a temperature of less than 3 K (three degrees above absolute zero). This is what Penzias and Wilson spotted. So with these three major foundations so far intact, Big Bang theory is widely accepted by most astrophysicists. A handful still pursue the steady state model that attracted Fred Hoyle, but it is difficult to explain all these observations in any other model.

Fate and past

What happened before the Big Bang? Because space–time was created in it, this is not really a very meaningful question to ask — a bit like "where does the Earth begin?" or "what is north of the north pole on Earth?" However, mathematical physicists do ponder the triggering of the Big Bang in multi-dimensional space (often 11 dimensions) through the mathematics of M-theory and string theory. These look at the physics and energies of strings, and membranes in these multi-dimensions and incorporate ideas of particle physics and quantum mechanics to try to trigger such an event. With parallels to quantum physics ideas, some cosmologists also discuss the existence of parallel universes.

In the Big Bang model, unlike the steady state model, the universe evolves. The cosmos' fate is dictated largely by the balance between the amount of matter pulling it together through gravity and other physical forces that pull it apart, including the expansion. If gravity wins, then the universe's expansion could one day stall and it could start to fall back in on itself, ending in a rewind of the Big Bang, known as the big crunch. Universes could follow many of these birth–death cycles. Alternatively, if the expansion and other repelling forces (such as dark energy) win, they will eventually pull all the stars and galaxies and planets apart and our universe could end up a dark desert of black holes and particles, a "big chill." Lastly there is the "Goldilocks universe," where the attractive and repellent forces balance and the universe continues to expand forever but gradually slows. It is this ending that modern cosmology is pointing to as being most likely. Our universe is just right.

> " There is a coherent plan in the universe, though I don't know what it's a plan for. "
>
> Fred Hoyle, 1915–2001

63 Cosmic inflation

Why does the universe look the same in all directions?
And why, when parallel light rays traverse space, do they remain parallel so we see separate stars? We think that the answer is inflation — the idea that the baby universe swelled up so fast in a split second that its wrinkles smoothed out and its subsequent expansion balanced gravity exactly.

The universe we live in is special. When we look out into it we see clear arrays of stars and distant galaxies without distortion. It could so easily be otherwise. Einstein's general relativity theory describes gravity as a warped sheet of space and time upon which light rays wend their way along curved paths (see Chapter 58). So, potentially, light rays could become scrambled, and the universe we look out onto could appear distorted like reflections in a hall of mirrors. But overall, apart from the odd deviation as they skirt a galaxy, light rays tend to travel more or less in straight lines right across the universe. Our perspective remains clear all the way to the visible edge.

> It is said that there's no such thing as a free lunch. But the universe is the ultimate free lunch.
>
> Alan Guth, b.1947

Flatness
Although relativity theory thinks of space–time as being a curved surface, astronomers sometimes describe the universe as "flat," meaning that parallel light rays remain parallel no matter how far they travel through space, just as they would do if traveling along a flat plain. Space–time can be pictured as a rubber sheet, where heavy objects weigh down the sheet and rest in dips in it, representing gravity. In reality, space–time has more dimensions (at least four: three of space and one of time) but it is hard to imagine those. The fabric is also continually expanding, following the big bang explosion. The universe's geometry is such that the sheet remains mostly flat, like a table top, give or take some small dips and lumps here or there due to

Geometry of the universe

From the latest observations of the microwave background, such as those of the Wilkinson Microwave Anisotropy Probe (WMAP) satellite in 2003 and 2006, physicists have been able to measure the shape of space–time right across the universe. By comparing the sizes of hot and cold patches in the microwave sky with the lengths predicted for them by big bang theory, they show that the universe is "flat." Even over a journey across the entire universe lasting billions of years, light beams that set out parallel will remain parallel.

the patterns of matter. So light's path across the universe is relatively unaffected, bar the odd detour around a massive body.

If there was too much matter, then everything would weigh the sheet down and it would eventually fold in on itself, reversing the expansion. In this scenario initially parallel light rays would eventually converge and meet at a point. If there were too little matter weighing it down, then the space–time sheet would stretch and pull itself apart. Parallel light rays would diverge as they crossed it. However, our real universe seems to be somewhere in the middle, with just enough matter to hold the universe's fabric together while expanding steadily. So the universe appears to be precisely poised (see box).

Sameness

Another feature of the universe is that it looks roughly the same in all directions. The galaxies do not concentrate in one spot, they are littered in all directions. This might not seem that surprising at first, but it is unexpected. The puzzle is that the universe is so big that its opposite edges should not be able to communicate — even at the speed of light. Having only existed for 14 billion years, the universe is more than 14 billion light years across in size. So light, even though it is traveling at the fastest speed attainable by any transmitted signal, has not had time to get from one side of the universe to the other. So how does one side of the universe know what the other side should look like? This is the "horizon problem," where the "horizon" is the farthest

10^{10} years — Steady expansion — Now

Inflation

10^{-35}s

Big bang

distance that light has traveled since the birth of the universe, marking an illuminated sphere. So there are regions of space that we cannot and will never see, because light from there has not had time to travel to us yet.

Smoothness

The universe is also quite smooth. Galaxies are spread quite uniformly across the sky. If you squint, they form a uniform glow rather than clumping in a few big patches. Again this need not have been the case. Galaxies have grown over time due to gravity. They started out as just a slightly overdense spot in the gas left over from the big bang. That spot started to collapse due to gravity, forming stars and eventually building up a galaxy. The original overdense seeds of galaxies were set up by quantum effects, miniscule shifts in the energies of particles in the hot embryonic universe. But they could well have amplified to make large galaxy patches, like a cow's hide, unlike the widely scattered sea that we see. There are many molehills in the galaxy distribution rather than a few giant mountain ranges.

Growth spurt

The flatness, horizon and smoothness problems of the universe can all be fixed with one idea: inflation. Inflation was developed as a solution in 1981 by American physicist, Alan Guth. The horizon problem, that the universe looks the same in all directions even though it is too large to know this, implies that the universe must at one time have been so

Microwave background

One observation that encompasses all these problems is that of the cosmic microwave background radiation. This background marks the afterglow of the big bang fireball, redshifted now to a temperature of 2.73 K. It is precisely 2.73 K all over the sky, with hot and cold patches differing from this by as little as one part in 100,000. To this day this temperature measurement remains the most accurate one made for any body at a single temperature. This uniformity is surprising because when the universe was very young, its distant regions could not communicate even at light speed. So it is puzzling that they nevertheless have exactly the same temperature. The tiny variations in temperature are the fossil imprints of the quantum fluctuations in the young universe.

> **It is rather fantastic to realize that the laws of physics can describe how everything was created in a random quantum fluctuation out of nothing, and how over the course of 15 billion years, matter could organize in such complex ways that we have human beings sitting here, talking, doing things intentionally.**
> Alan Guth, b.1947

small that light could communicate between all the regions in it. Because it is no longer like this, it must have then inflated quickly to the proportionately bigger universe we see now. But this period of inflation must have been extraordinarily rapid, much faster than the speed of light. The rapid expansion, doubling in size and doubling again and again in a split second, smeared out the slight density variations imprinted by quantum fluctuations, just like a printed pattern on an inflated balloon becomes fainter. So the universe became smooth. The inflationary process also fixed up the subsequent balance between gravity and the final expansion, proceeding at a much more leisurely pace thereafter. Inflation happened almost immediately after the big bang fireball (10^{-35} seconds after).

Inflation has not yet been proven and its ultimate cause is not well understood — there are as many models as theorists — but understanding it is a goal of the next generation of cosmology experiments, including the production of more detailed maps of the cosmic microwave background radiation and its polarization.

64 Dark matter

Ninety percent of the matter in the universe does not glow but is dark. Dark matter is detectable by its gravitational effect but hardly interacts with light waves or matter. Scientists think it may be in the form of MACHOs, failed stars and gaseous planets, or WIMPs, exotic subatomic particles — the hunt for dark matter is the wild frontier of physics.

Dark matter sounds exotic, and it may be, but its definition is quite down to Earth. Most of the things we see in the universe glow because they emit or reflect light. Stars twinkle by pumping out photons, and the planets shine by reflecting light from the Sun. Without that light, we simply would not see them. When the Moon passes into the Earth's shadow it is dark; when stars burn out they leave husks too faint to see; even a planet as big as Jupiter would be invisible if it was set free to wander far from the Sun. So it is, at first sight, perhaps not a big surprise that much of the stuff in the universe does not glow. It is dark matter.

Dark side

Although we cannot see dark matter directly, we can detect its mass through its gravitational pull on other astronomical objects and also light rays. If we did not know the moon was there, we could still infer its presence because its gravity would tug and shift the orbit of the Earth slightly. We have even used the gravity-induced wobble applied to a parent star to discover planets around distant stars.

In the 1930s, Swiss astronomer Fritz Zwicky realized that a nearby giant cluster of galaxies was behaving in a way that implied its mass was much greater than the weight of all the stars in all the galaxies within it. He inferred that some unknown dark matter accounted for 400 times as much material as luminous matter, glowing stars and hot gas, across the entire cluster. The sheer amount of dark matter was a big surprise, implying that most of the universe was not in the form of stars and gas but something else. So what is this dark stuff? And where does it hide?

Mass is also missing from individual spiral galaxies. Gas in the outer regions rotates faster than it should if the galaxy was only as heavy as the combined mass of stars within it. So such galaxies are more massive than expected by looking at the light alone. Again, the extra dark matter needs to be hundreds of times more abundant than the visible stars and gas. Dark matter is not only spread throughout galaxies but its mass is so great it dominates the motions of every star within them. Dark matter even extends beyond the stars, filling a spherical "halo" or bubble around every flattened spiral galaxy disk.

Weight gain

Astronomers have now mapped dark matter not only in individual galaxies but also in clusters of galaxies, containing thousands of galaxies bound together by mutual gravity, and superclusters of galaxies, chains of clusters in a vast web that stretches across all of space. Dark matter features wherever there is gravity at work, on every scale. If we add up all the dark matter, we find that there is a thousand times more dark stuff as luminous matter.

The fate of the entire universe depends on its overall weight. Gravity's attraction counterbalances the expansion of the universe following the big bang explosion. There are three possible outcomes.

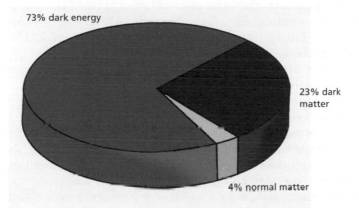

73% dark energy

23% dark matter

4% normal matter

❝The universe is made mostly of dark matter and dark energy, and we don't know what either of them is.❞
Saul Perlmutter, 1999

Energy budget

Today we know that only about 4 percent of the universe's matter is made up of baryons (normal matter comprising protons and neutrons). Another 23 percent is exotic dark matter. We do know that this isn't made up of baryons. It is harder to say what it is made from, but it could be particles such as WIMPs. The rest of the universe's energy budget consists of another thing entirely — dark energy.

Either the universe is so heavy that gravity wins and the universe eventually collapses back in on itself (a closed universe ending in a big crunch), or there is too little mass and it expands forever (an open universe), or the universe is precisely balanced and the expansion gradually slows by gravity, but over such a long time that it never ceases. The latter seems the best case for our universe, it has precisely the right amount of matter to slow — but never halt — the expansion.

WIMPs and MACHOs

What might dark matter be made of? First, it could be dark gas clouds, dim stars or unlit planets. These are called MACHOs, or MAssive Compact Halo Objects. Alternatively the dark matter could be new kinds of subatomic particles, called WIMPs, short for Weakly Interacting Massive Particles, which would have virtually no effect on other matter or light.

Astronomers have found MACHOs roaming within our own Galaxy. Because MACHOs are large, akin to the planet Jupiter, they can be spotted individually by their gravitational effect. If a large gas planet or failed star passes in front of a background star, its gravity bends the starlight around it. The bending focuses the light when the MACHO is right in front of the star, so the star appears much brighter for a moment as it passes. This is called "gravitational lensing."

In terms of relativity theory, the MACHO planet distorts space–time, like a heavy ball depressing a rubber sheet, which curves the light's wavefront around it (see Chapter 58). Astronomers have looked for this brightening of stars by the passage of a foreground MACHO against millions of stars in the background. They have found a few such flare-ups, but too few to explain all the missing mass of the Milky Way.

MACHOs are made of normal matter, or baryons, built of protons, neutrons and electrons. The tightest limit on the amount of baryons in the universe is given by tracking the heavy hydrogen isotope, deuterium. Deuterium was only produced in the big bang itself and is not formed by stars afterwards, although it can be burned within them. So, by measuring the amount of deuterium in pristine gas clouds in space, astronomers can estimate the total number of protons and neutrons that were made in the big bang, because the mechanism for making deuterium is precisely known. This turns out to be just a few percent of the mass of the entire universe. So the rest of the universe must be in some entirely different form, such as WIMPs.

The search for WIMPS is now the focus of attention. Because they are weakly interacting these particles are intrinsically difficult to detect. One candidate is the neutrino. In the last decade physicists have measured its mass and found it to be very small but not zero. Neutrinos make up some of the universe's mass, but again not all. So there is still room for other more exotic particles out there waiting to be detected, some new to physics such as axions and photinos. Understanding dark matter may yet light up the world of physics.

65 Cosmological constant

Einstein called adding his cosmological constant into the equations of general relativity his biggest blunder. The term allowed for the speeding up or slowing down of the rate of expansion of the universe to compensate gravity. Einstein did not need this number and abandoned it. However, new evidence in the 1990s required that it be reintroduced. Astronomers found that mysterious dark energy is causing the expansion of the universe to speed up, leading to the rewriting of modern cosmology.

Albert Einstein thought we lived in a steady state universe rather than one with a big bang. Trying to write down the equations for it, he ran into a problem. If you just had gravity, then everything in the universe would ultimately collapse into a point, perhaps a black hole. Obviously the real universe wasn't like that and appeared stable. So Einstein added another term to his theory to counterbalance gravity, a sort of repulsive "anti-gravity" term. He introduced this purely to make the equations look right, not because he knew of such a force. But this formulation was immediately problematic.

If there was a counterforce to gravity then, just as untrammeled gravity could cause collapse, an anti-gravity force could just as easily amplify to tear apart regions of the universe that were not held together by gravity's glue. Rather than allow such shredding of the universe, Einstein preferred to ignore his second repulsive term and admitted he had made a mistake in introducing it. Other physicists also preferred to exclude it, relegating it to history. Or so they thought. The term was not forgotten — it was preserved in the relativity equations — but its value, the cosmological constant, was set to zero to dismiss it.

Accelerating universe

In the 1990s, two groups of astronomers were mapping supernovae in distant galaxies to measure the geometry of space and found that distant supernovae appeared fainter than they should be. Supernovae, the brilliant explosions of dying stars, come in many types. Type Ia supernovae have a predictable brightness and so are useful for inferring distances. Just like the Cepheid variable stars that were used to measure the distances to galaxies to establish Hubble's law, the intrinsic brightness of Type Ia supernovae can be worked out from their light spectra so that it is possible to say how far away they must be. This all worked fine for supernovae that were quite nearby, but the more distant supernovae were too faint. It was as if they were farther away from us than they should be.

> " For 70 years, we've been trying to measure the rate at which the universe slows down. We finally do it, and we find out it's speeding up. "
> Michael S. Turner, 2001

As more and more distant supernovae were discovered, the pattern of the dimming with distance began to suggest that the expansion of the universe was not steady, as in Hubble's law, but was accelerating. This was a profound shock to the cosmology community, and one that is still being disentangled today.

The supernova results fitted well with Einstein's equations, but only once a negative term was included by raising the cosmological constant from zero to about 0.7. The supernova results, taken with other cosmological data, such as the cosmic microwave background radiation pattern, showed that a new repulsive force counteracting gravity was needed. But it was quite a weak force. It is still a puzzle today why it is so weak, as there is no particular reason why it did not adopt a much larger value and perhaps completely dominate space over gravity. Instead it is very close in strength to gravity so has a subtle effect on space–time as we see it now. This negative energy term has been named "dark energy."

Dark energy

Dark energy's origin is still elusive. All we know is that it is a form of energy associated with the vacuum of free space, causing a negative pressure in regions devoid of gravity-attracting matter. So it causes regions of empty space to inflate. We know its strength roughly from

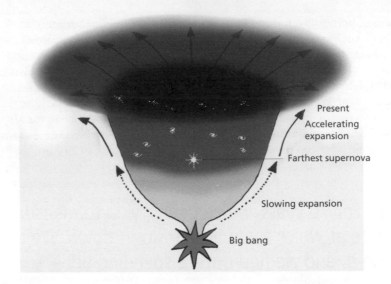

Present
Accelerating expansion

Farthest supernova

Slowing expansion

Big bang

> **It [dark energy] seems to be something that is connected to space itself and unlike dark matter which gravitates this has an effect which is sort of the opposite, counter to gravity, it causes the universe to be repulsed by itself.**
> Brian Schmidt, 2006

the supernova observations, but we do not know much more. We don't know if it truly is a constant — whether it always takes the same value right across the universe and for all time (as do gravity and the speed of light) — or whether its value changes with time so it may have had a different value just after the big bang compared with now or in the future. In its more general form it has also been called "quintessence" or the fifth force, encompassing all the possible ways its strength could change with time. But it is still not known how this elusive force manifests itself or how it arises within the physics of the big bang. It is a hot topic of study for physicists.

Nowadays we have a much better understanding of the geometry of the universe and what it is made up of. The discovery of dark energy has balanced the books of cosmology, making up the difference in the energy budget of the whole universe. So we now know that it is

4 percent normal baryonic matter, 23 percent exotic non-baryonic matter, and 73 percent dark energy. These numbers add up to about the right amount of stuff for the balanced "Goldilocks universe," close to the critical mass where it is neither open nor closed.

The mysterious qualities of dark energy however mean that even knowing the total mass of the universe, its future behavior is hard to predict because it depends on whether the influence of dark energy increases or not in the future. If the universe is accelerating then, at this point in time, dark energy is only just as significant as gravity in dominating the universe. But, at some point, the acceleration will pick up and the faster expansion will overtake gravity. So the universe's fate may well be to expand forever, faster and faster. Some scary scenarios have been proposed — once gravity is outstripped, then tenuously held together massive structures will disconnect and fly apart, eventually even galaxies themselves will break up, then stars will be evaporated into a mist of atoms. Ultimately the negative pressure could strip atoms, leaving only a grim sea of subatomic particles.

Nevertheless, although cosmology's jigsaw is fitting together now, and we have measured a lot of the numbers that describe the geometry of the universe, there are still some big unanswered questions. We just don't know what 95 percent of the stuff in the universe is, nor what this new force of quintessence really is. So it is not yet time to sit back and rest on our laurels. The universe has kept its mystery.

❝It is to be emphasized, however, that a positive curvature of space is given by our results, even if the supplementary term [cosmological constant] is not introduced. That term is necessary only for the purpose of making possible a quasi-static distribution of matter.❞

Albert Einstein, 1918

66 Fermi paradox

The detection of life elsewhere in the universe would be the greatest discovery of all time. Enrico Fermi wondered why, given the age and vastness of the universe, and the presence of billions of stars and planets that have existed for billions of years, we have not yet been contacted by any other alien civilizations. This was his paradox.

Chatting with his colleagues over lunch in 1950, physics professor Enrico Fermi supposedly asked "Where are they?" Our own Galaxy contains billions of stars and there are billions of galaxies in the universe, so that is trillions of stars. If just a fraction of those anchored planets, that's a lot of planets. If a fraction of those planets sheltered life, then there should be millions of civilizations out there. So why haven't we seen them? Why haven't they got in touch with us?

Drake equation

In 1961, Frank Drake wrote down an equation for the probability of a contactable alien civilization living on another planet in the Milky Way. This is known as the Drake equation. It tells us that there is a chance that we may coexist with another civilization but the probability is still quite uncertain. Carl Sagan once suggested that as many as a million alien civilizations could populate the Milky Way, but he later revised this down and others since have estimated that the value is just one, namely humans. More than half a century after Fermi asked the question, we have still heard nothing. Despite our communication systems, no one has called. The more we explore our local neighborhood, the lonelier it seems. No concrete signs of any life, not even the simplest bacteria, have been found on the Moon, Mars, asteroids, the outer solar system planets and moons. There are no signs of interference in the light from stars that could indicate giant orbiting machines harvesting energy from them. And it is not because no one has been looking. Given the stakes there is great attention given to searching for extraterrestrial intelligence.

Search for life

So, how would you go about looking for signs of life? The first way is to start looking for microbes within our solar system. Scientists have scrutinized rocks from the moon but they are inanimate basalt. Meteorites from Mars have been suggested to host the remnants of bacteria, but it is still not proven that the ovoid bubbles in those rocks hosted alien life and were not contaminated after having fallen to Earth or produced by natural geological processes. Even without rock samples, cameras on spacecraft and landers have scoured the surfaces of Mars, asteroids and now even a moon in the outer solar system — Titan, orbiting Saturn.

> ❝Who are we? We find that we live on an insignificant planet of a humdrum star lost in a galaxy tucked away in some forgotten corner of a universe in which there are far more galaxies than people.❞
>
> Werner von Braun, 1960

But the Martian surface is a dry desert of volcanic sand and rocks, not unlike the Atacama desert in Chile. Titan's surface is damp, drenched in liquid methane, but so far devoid of life. One of Jupiter's moons, Europa, has been touted as a popular target for future searches for life in the solar system, as it may host seas of liquid water beneath its frozen surface. Space scientists are planning a mission there that will drill through the ice crust and look below. Other moons in the outer solar system have been found to be quite geologically active, releasing heat as they are squeezed and pulled by the gravitational torques of their orbits around the giant gas planets. So liquid water may not be so rare a commodity in the outer solar system, raising expectations that one day life may be found. Spacecraft that go into this region are extensively sterilized to make sure that we do not contaminate them with foreign microbes from Earth.

But microbes are not going to call home. What about more sophisticated animals or plants? Now that individual planets are being detected around distant stars, astronomers are planning on dissecting the light from them to hunt for chemistry that could support or indicate life. Spectral hints of ozone or chlorophyll might be picked up, but these will need precise observations like those possible with the next generation of space missions such as NASA's Terrestrial Planet Finder. These missions might find us a sister Earth one day, but if they

Drake equation

$$N = N_* \times f_p \times n_e \times f_l \times f_i \times f_c \times f_L$$

where:

N is the number of civilizations in the Milky Way Galaxy whose electromagnetic emissions are detectable

N_* is the number of stars in the Galaxy

f_p is the fraction of those stars with planetary systems

n_e is the number of planets, per solar system, with an environment suitable for life

f_l is the fraction of suitable planets on which life actually appears

f_i is the fraction of life-bearing planets on which intelligent life emerges

f_c is the fraction of civilizations that develop a technology that releases detectable signs of their existence into space

f_L is the fraction of a planetary lifetime such civilizations release detectable signals into space (for earth this fraction is so far very small).

did, would it be populated with humans, fish or dinosaurs, or just contain empty lifeless continents and seas?

Contact

Life on other planets, even Earth-like ones, might have evolved differently to that on Earth. So it is not certain that aliens there would be able to communicate with us on Earth. Since radio and television began broadcasting, their signals have been spreading away from Earth, traveling outwards at the speed of light. So any TV fan on Alpha Centauri (four light years away) would be watching the Earth channels from four years ago, perhaps enjoying repeats of the film *Contact*. Black and white movies would be reaching the star Arcturus, and Charlie Chaplin could be starring at Aldebaran. So, Earth is giving off plenty of signals, if you have an antenna to pick them up. Wouldn't other advanced civilizations do the same? Radio astronomers are scouring nearby stars for signs of unnatural signals.

The radio spectrum is vast, so they are focusing on frequencies near key natural energy transitions, such as those of hydrogen, which should be the same anywhere in the universe. They are looking for transmissions that are regular or structured but not made by any known astronomical objects. In 1967, graduate student Jocelyn Bell got a fright in Cambridge, England, when she discovered regular pulses of radio waves coming from a star. Some thought this was indeed an alien Morse code, but in fact it was a new type of spinning neutron star now called a pulsar. Because this process of scanning thousands of stars takes a long time, a special program has been started in the USA called SETI (Search for ExtraTerrestrial Intelligence). Despite analyzing years of data, no odd signals have yet been picked up. Other radio telescopes search occasionally, but these too have seen nothing that does not have a more mundane origin.

> " Our sun is one of 100 billion stars in our galaxy. Our galaxy is one of billions of galaxies populating the universe. It would be the height of presumption to think that we are the only living things in that enormous immensity. "
>
> Carl Sagan, 1980

Out to lunch

So, given that we can think of many ways to communicate and detect signs of life, why might any civilizations not be returning our calls or sending their own? Why is Fermi's paradox still true? There are many ideas. Perhaps life only exists for a very short time in an advanced state where communication is possible. Why might this be? Perhaps intelligent life always wipes itself out quickly. Perhaps it is self-destructive and does not survive long, so the chances of being able to communicate and having someone nearby to communicate to are very low indeed. Or there are more paranoid scenarios. Perhaps aliens simply do not want to contact us and we are deliberately isolated. Or, perhaps they are just too busy and haven't got around to it yet.

67 Anthropic principle

The anthropic principle states that the universe is as it is because if it were different we would not be here to observe it. It is one explanation for why every parameter in physics takes the value that it does, from the size of the nuclear forces to dark energy and the mass of the electron. If any one of those varied even slightly then the universe would be uninhabitable.

If the strong nuclear force was a little different then protons and neutrons would not stick together to make nuclei and atoms could not form. Chemistry would not exist. Carbon would not exist and so biology and humans would not exist. If we did not exist, who would "observe" the universe and prevent it from existing only as a quantum soup of probability?

Equally, even if atoms existed and the universe had evolved to make all the structures we know today, then if dark energy were a little stronger galaxies and stars would already be being pulled apart. So, tiny changes in the values of the physical constants, in the sizes of forces or masses of particles, can have catastrophic implications. Put another way, the universe appears to be fine tuned. The forces are all "just right" for humanity to have evolved now. Is it a chance happening that we are living in a universe that is 14 billion years old, where dark energy and gravity balance each other out and the subatomic particles take the forms they do?

Just so

Rather than feel that humanity is particularly special and the entire universe was put in place just for us, perhaps a rather arrogant assumption, the anthropic principle explains that it is no surprise. If any of the forces were slightly different, then we simply would not be here to witness it. Just as the fact that there are many planets but as far as we know only one that has the right conditions for life, the universe could have been made in many ways but it is only in this one that we would come to exist. Equally, if my parents had never met, if the combustion engine had not been invented when it was and my father had not been able to travel north to meet my mother, then I would not be here. That does not mean that the entire universe evolved thus just so that I could exist. But the fact that I exist ultimately required, amongst other things, that the engine was invented beforehand, and so narrows the range of universes that I might be found in.

> ❝ The observed values of all physical and cosmological quantities are not equally probable but they take on values restricted by the requirement that there exist sites where carbon-based life can evolve and . . . that the Universe be old enough for it to have already done so. ❞
> John Barrow and Frank Tipler, 1986

The anthropic principle was used as an argument in physics and cosmology by Robert Dicke and Brandon Carter, although its argument is familiar to philosophers. One formulation, the weak anthropic principle, states that we would not be here if the parameters were different, so the fact that we exist restricts the properties of inhabitable physical universes that we could find ourselves in. Another stronger version emphasizes the importance of our own existence, such that life is a necessary outcome for the universe coming into being. For example, observers are needed to make a quantum universe concrete by observing it. John Barrow and Frank Tipler also suggested yet another version, whereby information processing is a fundamental purpose of the universe and so its existence must produce creatures able to process information.

Many worlds

To produce humans, you need the universe to be old, so that there's enough time for carbon to be made in earlier generations of stars, and the strong and weak nuclear forces must be "just so" to allow nuclear physics and chemistry. Gravity and dark energy must also be in balance to make stars rather than rip apart the universe. Further, stars need to be long lived to let planets form, and large enough so that we can find

We can avoid the anthropic dilemma if many parallel, or bubble, universes accompany the one we live in. Each bubble universe can take on slightly different parameters of physics. These govern how each universe evolves and whether a given one provides a nice niche in which life can form. As far as we know, life is fussy and so will choose only a few universes. But since there are so many bubble universes, this is a possibility and so our existence is not so improbable.

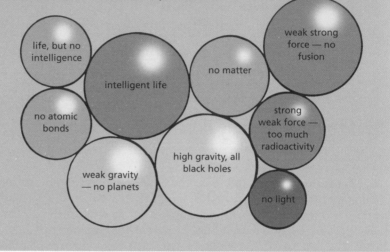

ourselves on a nice suburban temperate planet that has water, nitrogen, oxygen and all the other molecules needed to seed life.

Because physicists can imagine universes where these quantities are different, some have suggested that those universes can be created just as readily as one like our own. They may exist as parallel universes, or multi-verses, such that we only exist in one realization.

The idea of parallel universes fits in with the anthropic principle in allowing other universes also to exist where we cannot. These may exist in multiple dimensions and are split off along the same lines as quantum theory requires for observations to trigger outcomes (see Chapter 51).

On the other hand

The anthropic principle has its critics. Some think it is a truism — it is like this because it is like this — and is not telling us much that's new.

Others are unhappy that we have just this one special universe to test, and prefer to search the mathematics for ways of automatically tuning our universe to fall out of the equations simply because of the physics. The multiverse idea comes close to this by allowing an infinite number of alternatives. Yet other theorists, including string theorists and proponents of M-theory, are trying to go beyond the big bang to fine tune the parameters. They look at the quantum sea that preceded the big bang as a sort of energy landscape and ask where a universe is most likely to end up if you let it roll and unfold. For instance, if you roll a ball down a ridged hill, then the ball is more likely to end up in some places than others, such as in valley floors. So in trying to minimize its energy, the universe may well seek out certain combinations of parameters quite naturally, irrespective of whether we are a product of it billions of years later.

> **In order to make an apple pie from scratch, you must first create the universe.**
> Carl Sagan, 1980

Proponents of the anthropic principle and others, who pursue more mathematical means of ending up with the universe we know, disagree about how we got to be where we are and whether that is even an interesting question to ask. Once we get beyond the big bang and the observable universe, into the realms of parallel universes and pre-existing energy fields, we are really on philosophical ground. But whatever triggered the universe to appear in its current garb, we are lucky that it has turned out this way billions of years hence. It is understandable that it takes time to cook up the chemistry needed for life. But why we should be living here at a particular time in the universe's history when dark energy is relatively benign and balancing out gravity is more than lucky.

MATHEMA

Mathematics is far removed from its popular image as an unchanging body of knowledge handed down on tablets of stone. While the basics like 2 + 2 = 4 remain true for everyday transactions, the concepts of mathematics are constantly undergoing revision and updating. In different epochs, and in our time too, dramatic developments have taken place and major problems have been solved. But with every success new questions arise to spur on future generations of mathematicians.

The use of mathematics: connections with the world

The connection of mathematics with its applications has never been keener. The traffic is not only one-way: mathematics helps all areas of scientific activity but science also suggests new mathematics. Mathematics is firmly established in the engine room of physics and is a growing presence in the relatively young science of genetics.

We have to look no further than Sir Isaac Newton's *Mathematical Principles of Natural Philosophy* (1687) to realize a strong affiliation with physics (the modern name for Natural Philosophy). Newton believed that all laws of physics were intrinsically mathematical (see Chapter 35). In the 19th century James Clerk Maxwell wrote down his famous four equations which sought to capture the link between

30,000 B.C. Paleolithic peoples in Europe make number marks on bones

2,000 B.C. The Babylonians use symbols for numbers; they observe that π is approximately equal to 3

c.300 B.C. Euclid's *Elements* is compiled. Euclid proves there are infinitely many prime numbers

A.D. 1202 Leonardo of Pisa publishes the *Liber Abaci* and introduces the sequence of Fibonacci numbers

1650s Christiaan Huygens and Blasie Pascal discuss the foundations of a theory of probability

TICS

magnetism and electricity (see Chapter 45). But mathematics does not adopt a passive role and it can even suggest scientific progress: the astronomer John Couch Adams sat at his desk and discovered the planet Neptune "at the tip of his pen."

Inspired by Darwin's theory of evolution in the natural world, R. A. Fisher pioneered population genetics, his output of books including *The Genetical Theory of Natural Selection* (1930). Concepts in genetics are made precise by mathematics and Fisher's work is framed in terms of probability and statistics (see Chapter 18). A different link with genetics was made by the detection of the structure of DNA in the 1950s. Just as in the 17th century Johannes Kepler discovered that the planets moved in ellipses so it was found by Francis Crick and James Watson in the 20th that the structure of life is bound up with two spiral shapes winding around each other (see Chapter 6).

The basics of mathematics: number systems

At the basis of mathematical applications to genetics and physics is the evolution of a sophisticated number system. Our system, based on the Hindu–Arabic system of past centuries, has the power to

1660s–1670s Isaac Newton and Gottfried Leibniz take first steps in the Differential and Integral Calculus

1733 Abraham De Moivre publishes work on the normal distribution as an approximation to the binomial distribution of probabilities

1735 Leonhard Euler solves the Königsberg bridge problem and prepares the way for a theory of graphs

1780 Leonhard Euler explores the notion of a Latin square, the forerunner of modern Sudoku

1801 Carl Friedrich Gauss publishes *Disquitiones Arithmeticae* and includes the construction of the regular 17-sided polygon by ruler and compass

express both the smallest of numbers and also the largest. It can handle the number of electrons in nuclear reactions and at the same time deal with nanotechnology, the science of the very small.

Not that we can completely escape from previous systems. Vestiges of the Babylonian system based on the number 60 are still with us in the measurement of time (60 minutes in an hour, 60 seconds in a minute). But we have something the ancients never had. With all their achievements, neither the Ancient Greeks nor the Romans had the concept of zero and this absence placed a severe limitation on their ability to make calculations. Zero occupies a pivotal place in our system. We are so used to it that it passes unnoticed, silently performing all kinds of services. We might even wonder if it is necessary, like the anatomical appendix, something we can do without. This would be a mistake.

New problems arose when zero was introduced in India around the seventh century A.D.: how could it be integrated into the existing number systems? How could we add, subtract, multiply and divide by zero? Special rules had to be brought in but a widespread knowledge of these took centuries to assimilate. Zero is essential and complements the number signs 1, 2, 3, 4, 5, 6, 7, 8, 9 to give us our present decimal system based on 10. The superiority of the decimal system can be contrasted with the Babylonian system or the Roman system based on symbols I, V, X, L, C, D, M. With these systems, the number range is severely limited and calculations are only achieved with difficulty (try multiplying CXIII by XXIX for example).

Fractions have a tangled history. The Egyptians of around 1650 B.C. based their system of fractions on hieroglyphs denoting unit fractions and a few privileged exceptions like $\frac{2}{3}$. A unit fraction

1829–31 Nikolai Ivanovich Lobachevsky and János Bolyai independently publish work on "non-Euclidean" hyperbolic geometry

1832 Évariste Galois proposes the idea of groups of permutations in connection with the theory of equations

1847 George Boole publishes *The Mathematical Analysis of Logic* and paves the way for the language of the modern programmable computer

1858 Arthur Cayley publishes his definitive paper on the "Theory of Matrices," a theory in algebra with far-reaching consequences

1859 Bernard Riemann puts forward the famous "Riemann hypothesis" as a conjecture — not yet settled as a mathematical fact

is one where the top number (the numerator) is unity, like $\frac{1}{7}$, for example. Their system depended on expressing general fractions in terms of unit fractions, so for example $\frac{5}{7} = \frac{1}{2} + \frac{1}{7} + \frac{1}{14}$. Curiously the general task of expressing fractions in this way still poses unsolved mathematical problems.

The lure of mathematics: the emerging theory

Though the ancient Greeks lacked an efficient number system, they were brilliant mathematicians. They made advances in mathematics irrespective of any application to the real world. One discovery of great importance was the concept of irrational numbers. The Greeks initially believed that all numbers were either whole numbers or fractions, but the consideration of square roots challenged this viewpoint. The square root of a number, say 4, is that (positive) number which when multiplied by itself gives us 4 (the square root of 4 is actually 2 because 2 x 2 = 4). But what is the square root of 2 itself? What number multiplied by itself gives us 2? Expressed in decimals 1.414 is close because using our calculator we find that 1.414 x 1.414 = 1.999396. While this is nearly 2, it is not *exactly* 2. The number 1.41421356 is even closer, but it turns out that no matter how many decimal places we take, the answer will never be exactly 2. The ancient Greeks gave a famous mathematical proof of this fact — the upshot being that ordinary numbers are split into two separate camps: those that are fractions or whole numbers (rational numbers) and those like the square root of 2 (irrational numbers). The famous numbers π and e which start this section of the book are both irrational numbers.

1874 Georg Cantor introduces a controversial notion of infinity, and specifies different orders of infinity

1880s Francis Galton, a cousin of Charles Darwin, introduces ideas on regression and correlation in statistics

1882 Ferdinand von Lindemann proves that π is a transcendental number and consequently "squaring the circle" is impossible

1889 Henri Poincaré encounters chaos in his entry for a competition in honor of Sweden's King Oscar II

1890 Giuseppe Peano constructs his famous space filling curve prompting an overhaul of the theory of curves

While numbers form the bedrock of mathematics there is much more to the subject. Geometry is a major aspect of mathematics. This ancient craft, literally earth (*geo*) measuring (*metry*), has expanded vastly in modern times. Its organization started with the ancient Greeks in the third century B.C. Euclid's organization of geometry published in the famous *Elements* dealt with a "common sense geometry" built on self-evident axioms. The compendium was the most influential mathematical text ever written but new types of "non-Euclidean" geometries were studied in the 19th century. Although these appeared strange, they pushed the study of geometry forward and provided the tools for a proper understanding of Einstein's theory of relativity (see Chapter 58).

The promise of mathematics: the modern expanse

New theories in mathematics mushroomed in the 20th century and the first decade of the 21st. Amongst them, the advent of the geometry of fractals has provided a valuable method for modeling physical behavior such as cloud formations. Fractals are shapes which step outside the confines of dimensions 1, 2 and 3 of traditional geometry whether Euclidean or non-Euclidean. The novelty of fractals is that they can have dimension *between* these whole numbers. A fractal may well have dimension 1.262 for example (see Chapter 85).

In dealing with risk and uncertainty nowadays, the theory of probability is the appropriate vehicle. Gamblers in the 17th century used it to analyze games of chance but the theory has evolved and combines with the theory of statistics to provide tools for addressing

1900 David Hilbert suggests 23 famous problems as an agenda for mathematicians of the 20th century

1930 Bartel van der Waerden publishes his famous *Moderne Algebra*, which sets out the method of the new algebra

1931 Kurt Gödel proves that it is possible for a formal mathematical system to contain undecidable statements

1946 ENIAC the first general-purpose electronic computer comes into operation

1947 George Dantzig outlines a theory of linear programing and formulates the Simplex algorithm

a wide range of practical situations very different from narrowly focused gambling problems (see Chapter 89).

Mathematics has expanded on all fronts. The well-known handbooks that summarize the subject have nowadays to consist of thousands of pages. There have been so many areas of expansion: theories of sets, groups, graphs, probability, matrices, topology, alternative logics and many others. It is remarkable that many of these began in pure mathematics but have since proven their worth in applications to the real world.

The selection of topics in this section of the book has been chosen to give an understanding of leading mathematical ideas and the role of mathematics in science. We wend our way from π, touching on topics which have excited mathematicians throughout the ages ending up at the problem which has stumped the most brilliant mathematicians, the celebrated Riemann hypothesis. The challenge of finding a proof of this brings us back to the basic building blocks of mathematics. It involves the prime numbers — those whole numbers divisible only by themselves and unity. This problem, and so many others, beckon and cajole. Like genetics and physics, mathematics does not stand still.

1975 Benoît Mandelbrot introduces fractals into geometry and indicates their widespread applications

1983 The classification of finite simple groups is achieved and the so-called "Enormous theorem" is finally accomplished

1994 John F. Nash is awarded the Nobel Prize for Economics for his work on game theory

1994 Andrew Wiles proves "Fermat's last theorem" and collects the Wolfskehl prize

2002 Grigori Perelman completes the proof of the Poincaré Conjecture by establishing the result for dimension 3

68 **Pi**

Pi is the most famous number in mathematics. Forget all the other constants of nature, Pi will always come at the top of the list. If there were Oscars for numbers, Pi would get an award every year.

Pi, or π, is the length of the outside of a circle (the circumference) divided by the length across its center (the diameter). Its value, the ratio of these two lengths, does not depend on the size of the circle. Whether the circle is big or small, π is indeed a mathematical constant. The circle is the natural habitat for π but it occurs everywhere in mathematics, and in places not remotely connected with the circle.

Archimedes of Syracuse

The ratio of the circumference to the diameter of a circle was a subject of ancient interest. Around 2000 B.C. the Babylonians made the observation that the circumference was roughly three times as long as its diameter.

It was Archimedes of Syracuse who made a real start on the mathematical theory of π, in around 225 B.C. Archimedes is right up there with the greats. Mathematicians love to rate their co-workers and they place him on a level with Carl Friedrich Gauss (the "Prince of Mathematicians") and Sir Isaac Newton. Whatever the merits of this judgment it is clear that Archimedes would be in any mathematics Hall of Fame. He was hardly an ivory tower figure though — as well as his contributions to astronomy, mathematics and physics, he also designed weapons of war, such as catapults, levers and "burning mirrors," all used to help keep the Romans at bay. But by all accounts he did have something of the absent-mindedness of the professor, for what else would induce him to leap from his bath and run naked down the street shouting "Eureka" at discovering the law of buoyancy in hydrostatics? How he celebrated his work on π is not recorded.

Given that π is defined as the ratio of its circumference to its diameter, what does it have to do with the area of a circle? It is a deduction that the area of a circle of radius r is πr^2, though this is probably better known than the circumference/diameter definition of π. The fact that π does double duty for area and circumference is remarkable.

How can this be shown? The circle can be split up into a number of narrow equal triangles with base length b whose height is approximately the radius r. These form a polygon inside the circle which approximates the area of the circle. Let's take 1,000 triangles for a start. The whole process is an exercise in approximations. We can join together each adjacent pair of these triangles to form a rectangle (approximately) with area $b \times r$ so that the total area of the polygon will be $500 \times b \times r$. As $500 \times b$ is about half the circumference it has length πr, the area of the polygon is $\pi r \times r = \pi r^2$. The more triangles we take the closer will be the approximation and in the limit we conclude the area of the circle is πr^2.

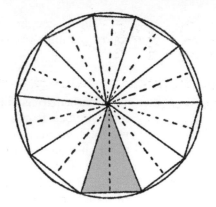

Archimedes estimated the value of π as bounded between $^{223}/_{71}$ and $^{220}/_{70}$. And so it is to Archimedes that we owe the familiar approximation 22/7 for the value of π. The honor for designating the actual symbol π goes to the little known William Jones, a Welsh mathematician who became Vice President of the Royal Society of London in the 18th century. It was the mathematician and physicist Leonhard Euler who popularized π in the context of the circle ratio.

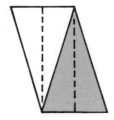

The exact value of π

We can never know the *exact* value of π because it is an irrational number, a fact proved by Johann Lambert in 1768. The decimal expansion is infinite with no predictable pattern. The first 20 decimal places are 3.14159265358979323846... The value of $\sqrt{10}$ used by the Chinese mathematicians is 3.16227766016837933199 and this was adopted around A.D. 500 by Brahmagupta. This value is in fact little better than the crude value of 3 and it differs in the second decimal place from π.

π can be computed from a series of numbers. A well known one is

$$\frac{\pi}{4} = 1 - \frac{1}{3} + \frac{1}{5} - \frac{1}{7} + \frac{1}{9} - \frac{1}{11} + \mathbf{l} \ \cdots$$

though this is painfully slow in its convergence on π and quite hopeless for calculation. Euler found a remarkable series that converges to π:

$$\frac{\pi^2}{6} = 1 + \frac{1}{2^2} + \frac{1}{3^2} + \frac{1}{4^2} + \frac{1}{5^2} + \frac{1}{6^2} + \cdots$$

The self-taught genius Srinivasa Ramanujan devised some spectacular approximating formulae for π. One involving only the square root of 2 is:

$$\frac{9801}{4412}\sqrt{2} = 3.14159273001330566603139961890\ldots$$

Mathematicians are fascinated by π. While Lambert had proved it could not be a fraction, in 1882 the German mathematician Ferdinand von Lindemann solved the most outstanding problem associated with π. He showed that π is "transcendental;" that is, π cannot be the solution of an algebraic equation (an equation which only involves powers of x). By solving this "riddle of the ages" Lindemann concluded the problem of "squaring the circle." Given a circle the challenge was to construct a square of the same area using only a pair of compasses and a straight edge. Lindemann proved conclusively that it cannot be done. Nowadays the phrase squaring the circle is the equivalent of an impossibility.

The actual calculation of π continued apace. In 1853, William Shanks claimed a value correct to 607 places (actually correct up to only 527). In modern times the quest for calculating π to more and more decimal places gained momentum through the modern computer. In 1949, π was calculated to 2,037 decimal places, which took 70 hours to do on an ENIAC computer. By 2002, π had been computed to a staggering 1,241,100,000,000 places, but it is an ever growing tail.

π in poetry

If you really want to remember the first values in the expansion of π perhaps a little poetry will help. Following the tradition of teaching mathematics in the "mnemonic way" there is a brilliant variation of Edgar Allan Poe's poem "The Raven" by Michael Keith.

The real poem by Poe begins
The Raven E.A. Poe

Once upon a midnight dreary, while I pondered weak and weary,
Over many a quaint and curious volume of forgotten lore,

Keith's variant for π begins
Poe, E. Near A Raven

Midnights so dreary, tired and weary.
Silently pondering volumes extolling all by-now obsolete lore.

The letter count of each successive word in Keith's version provides the first 740 digits of π.

If we stood on the equator and started writing down the expansion of π, Shanks' calculation would take us a full 14 meters, but the length of the 2002 expansion would take us about 62 laps around the world!

Various questions about π have been asked and answered. Are the digits of π random? Is it possible to find a predetermined sequence in the expansion? For instance, is it possible to find the sequence 0123456789 in the expansion? In the 1950s this seemed unknowable. No one had found such a sequence in the 2,000 known digits of π. L.E.J. Brouwer, a leading Dutch mathematician, said the question was devoid of meaning since he believed it could not be experienced. In fact these digits were found in 1997 beginning at the position 17,387,594,880, or, using the equator metaphor, about 3,000 miles before one lap is completed. You will find ten sixes in a row before you have completed 600 miles but will have to wait until one lap has been completed and gone a further 3,600 miles to find ten sevens in a row.

The importance of π

What is the use of knowing π to so many places? After all, most calculations only require a few decimal places, probably no more than ten places are needed for any practical application, and Archimedes' approximation of 22/7 is good enough for most. But the extensive calculations are not just for fun. They are used to test the limits of computers, besides exerting a fascination on the group of mathematicians who have called themselves the "friends of pi."

Perhaps the strangest episode in the story of π was the attempt in the Indiana State Legislature to pass a bill that would fix its value. This occurred at the end of the 19th century when a medical doctor, Dr. E.J. Goodwin, introduced the bill to make π "digestible." A practical problem encountered in this piece of legislation was the proposer's inability to fix the value he wanted. Happily for Indiana, the folly of legislating on π was realized before the bill was fully ratified. Since that day, politicians have left π well alone.

For a circle of diameter *d* and radius *r*:
circumference $= \pi d = 2\pi r$ area $= \pi r^2$

For a sphere of diameter d and radius *r*:
surface area $= \pi d^2 = 4 \pi r^2$

volume $= \frac{4}{3} \pi r^3$

69 Euler's number

Euler's number, or e, is the new kid on the block when compared with its only rival π. While π is more august and has a grand past dating back to the Babylonians, e is not so weighed down by the barnacles of history. The constant e is youthful and vibrant and is ever present when "growth" is involved. Whether it's populations, money or other physical quantities, growth invariably involves e.

e is the number whose approximate value is 2.71828. So why is that so special? It isn't a number picked out at random, but is one of the great mathematical constants. It came to light in the early 17th century when several mathematicians put their energies into clarifying the idea of a logarithm, the brilliant invention that allowed the multiplication of large numbers to be converted into addition.

But the story really begins with some 17th-century e-commerce. Jacob Bernoulli was one of the illustrious Bernoullis of Switzerland, a family which made it their business to supply a dynasty of mathematicians to the world. Jacob set to work in 1683 with the problem of compound interest.

Money, money, money

Suppose we consider a one-year time period, an interest rate of a whopping 100 percent, and an initial deposit (called a "principal" sum) of $1. Of course we rarely get 100 percent on our money but this figure suits our purpose and the concept can be adapted to realistic interest rates. Likewise, if we have greater principal sums like $10,000 we can multiply everything we do by 10,000.

At the end of the year at 100 percent interest, we will have the principal and the amount of interest earned which in this case is also $1. So we shall have the princely sum of $2. Now we suppose that the interest rate is halved to 50 percent but is applied for each half-year separately. For the first half-year we gain an interest of 50 cents and our principal has grown to $1.50 by the end of the first half-year. So, by the end of the full year we would have this amount and the 75 cents interest on this sum. Our $1 has grown to $2.25 by the end of the year!

By compounding the interest each half-year we have made an extra 25 cents. It may not seem much but if we had $10,000 to invest, we would have $2,250 interest instead of $2,000. By compounding every half-year we gain an extra $250.

But if compounding every half-year means we gain on our savings, the bank will also gain on any money we owe — so we must be careful! Suppose now that the year is split into four quarters and 25 percent is applied to each quarter. Carrying out a similar calculation, we find that our $1 has grown to $2.44141. Our money is growing and with our $10,000 it would seem to be advantageous if we could split up the year and apply the smaller percentage interest rates to the smaller time intervals.

Will our money increase beyond all bounds and make us millionaires? If we keep dividing the year up into smaller and smaller units, as shown in the table, this "limiting process" shows that the amount appears to be settling down to a constant number. Of course, the only realistic compounding period is per day (and this is what banks do). The mathematical message is that this

Compounding each ...	Accrued sum
year	$2.00000
half-year	$2.25000
quarter	$2.44141
month	$2.61304
week	$2.69260
day	$2.71457
hour	$2.71813
minute	$2.71828
second	$2.71828

limit, which mathematicians call e, is the amount $1 grows to if compounding takes place continuously. Is this a good thing or a bad thing? You know the answer: if you are saving, "yes"; if you owe money, "no." It's a matter of "e-learning."

The exact value of e

Like π, e is an irrational number so, as with π, we cannot know its exact value. To 20 decimal places, the value of e is 2.71828182845904523536...

Using only fractions, the best approximation to the value of e is 87/32 if the top and bottom of the fraction are limited to two-digit numbers. Curiously, if the top and bottom are limited to three-digit numbers the best fraction is 878/323. This second fraction is a sort of palindromic extension of the first one — mathematics has a habit of offering these little surprises. A well-known series expansion for e is given by:

$$e = 1 + \frac{1}{1} + \frac{1}{2 \times 1} + \frac{1}{3 \times 2 \times 1} + \frac{1}{4 \times 3 \times 2 \times 1} + \frac{1}{5 \times 4 \times 3 \times 2 \times 1} \cdots$$

The factorial notation using an exclamation mark is handy here. In this, for example, $5! = 5 \times 4 \times 3 \times 2 \times 1$. Using this notation, e takes the more familiar form

$$e = 1 + \frac{1}{1!} + \frac{1}{2!} + \frac{1}{3!} + \frac{1}{4!} + \frac{1}{5!} + \cdots$$

So the number e certainly seems to have some pattern. In its mathematical properties, e appears more "symmetric" than π.

If you want a way of remembering the first few places of e, try this: "We attempt a mnemonic to remember a strategy to memorize this count...," where the letter count of each word gives the next number of e. If you know your American history then you might remember that e is "2.7 Andrew Jackson Andrew Jackson," because Andrew Jackson ("Old Hickory"), the seventh president of the United States, was elected in 1828. There are many such devices for remembering e but their interest lies in their quaintness rather than any mathematical advantage.

That e is irrational (not a fraction) was proved by Leonhard Euler in 1737. In 1840, French mathematician Joseph Liouville showed that e was not the solution of any quadratic equation and in 1873, in a path-breaking work, his countryman Charles Hermite, proved that e is transcendental (it cannot be the solution of any algebraic equation). What was important here was the method Hermite used. Nine years later, Ferdinand von Lindemann adapted Hermites' method to prove that π was transcendental, a problem with a much higher profile.

One question was answered but new ones appeared. Is e raised to the power of e transcendental? It is such a bizarre expression, how could this be otherwise? Yet this has not been proved rigorously and, by the strict standards of mathematics, it must still be classified as a conjecture. Mathematicians have inched towards a proof, and have proved it is impossible for both it and e raised to the power of e^2 to be transcendental. Close, but not close enough.

The connections between π and e are fascinating. The values of e^{π} and π^e are close but it is easily shown (without actually calculating their values) that $e^{\pi} > \pi^e$. If you "cheat" and have a look on your calculator, you will see that approximate values are $e^{\pi} = 23.14069$ and $\pi^e = 22.45916$.

The number e^{π} is known as Gelfond's constant (named after the Russian mathematician Aleksandr Gelfond) and has been shown to be a transcendental. Much less is known about π^e; it has not yet been proved to be irrational — if indeed it is.

Is e important?

The chief place where e is found is in growth. Examples are economic growth and the growth of populations. Connected with this are the curves depending on e used to model radioactive decay.

The number e also occurs in problems not connected with growth. Pierre Montmort investigated a probability problem in the 18th century and it has since been studied extensively. In the simple version a group of people go to lunch and afterwards pick up their hats at random. What is the probability that no one gets their own hat?

It can be shown that this probability is ¼ (about 37 percent) so that the probability of at least one person getting their own hat is 1 − ¼ (about 63 percent). This application in probability theory is one of many. The Poisson distribution which deals with rare events is another. These were early instances but by no means isolated ones: James Stirling achieved a remarkable approximation to the factorial value n! involving e (and π); in statistics the familiar "bell curve" of the normal distribution involves e; and in engineering the curve of a suspension bridge cable depends on e. The list is endless.

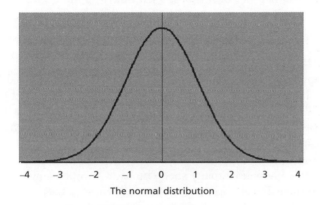

The normal distribution

An earth-shattering identity

The prize for the most remarkable formula of all mathematics involves e. When we think of the famous numbers of mathematics we think of 0, 1, π, e and the imaginary number $i = \sqrt{-1}$. How could it be that

$$e^{i\pi} + 1 = 0$$

It is! This is a result attributed to Euler.

Perhaps e's real importance lies in the mystery by which it has captivated generations of mathematicians. All in all, e is unavoidable. Just why an author like E.V. Wright should put himself through the effort of writing an e-less novel — presumably he had a pen name too — is hard to fathom, but his *Gadsby* is just that. It is hard to imagine a mathematician setting out to write an e-less textbook, or being able to do so.

70 Infinity

How big is infinity? The short answer is that ∞ (the symbol for infinity) is very big. Think of a straight line with larger and larger numbers lying along it and the line stretching "off to infinity." For every huge number produced, say 10^{1000}, there is always a bigger one, such as $10^{1000} + 1$.

This is a traditional idea of infinity, with numbers marching on forever. Mathematics uses infinity in any which way, but care has to be taken in treating infinity like an ordinary number. It is not.

Counting

The German mathematician Georg Cantor gave us an entirely different concept of infinity. In the process, he single-handedly created a theory which has driven much of modern mathematics. The idea on which Cantor's theory depends has to do with a primitive notion of counting, simpler than the one we use in everyday affairs.

Imagine a farmer who didn't know about counting with numbers. How would he know how many sheep he had? Simple — when he lets his sheep out in the morning he can tell whether they are all back in the evening by pairing each sheep with a stone from a pile at the gate of his field. If there is a sheep missing there will be a stone left over. Even without using numbers, the farmer is being very mathematical. He is using the idea of a one-to-one correspondence between sheep and stones. This primitive idea has some surprising consequences.

Cantor's theory involves *sets* (a set is simply a collection of objects). For example $N = \{1, 2, 3, 4, 5, 6, 7, 8, \ldots\}$ means the set of (positive) whole numbers. Once we have a set, we can talk about subsets, which are smaller sets within the larger set. The most obvious subsets connected with our example N are the subsets $O - \{1, 3, 5, 7, \ldots\}$ and $E = \{2, 4, 6, 8, \ldots\}$, which are the sets of the odd and even numbers respectively. If we were to ask "is there the same number of odd numbers as even numbers?" what would be our answer? Though we cannot do this by counting the elements in each set and comparing answers, the answer would still surely be "yes." What is this confidence based on? — probably something like "half the whole numbers are odd and half are even." Cantor would agree with the answer, but would give a different reason. He would say that every time we have an odd number, we have an even "mate" next to it. The idea that both sets **O** and **E** have the same number of elements is based on the pairing of each odd number with an even number:

$$
\begin{array}{ccccccccccccc}
\text{O:} & 1 & 3 & 5 & 7 & 9 & 11 & 13 & 15 & 17 & 19 & 21\ldots \\
& \updownarrow & \updownarrow & \updownarrow & \updownarrow & \updownarrow & \updownarrow & \updownarrow & \updownarrow & \updownarrow & \updownarrow & \updownarrow \\
\text{E:} & 2 & 4 & 6 & 8 & 10 & 12 & 14 & 16 & 18 & 20 & 22\ldots
\end{array}
$$

If we were to ask the further question "is there the same number of whole numbers as even numbers?" the answer might be "no," the argument being that the set **N** has twice as many numbers as the set of even numbers on its own.

The notion of "more" though, is rather hazy when we are dealing with sets with an indefinite number of elements. We could do better with the one-to-one correspondence idea. Surprisingly, there is a one-to-one correspondence between **N** and the set of even numbers **E**:

$$
\begin{array}{cccccccccccc}
\text{N:} & 1 & 2 & 3 & 4 & 5 & 6 & 7 & 8 & 9 & 10 & 11\ldots \\
& \updownarrow & \updownarrow & \updownarrow & \updownarrow & \updownarrow & \updownarrow & \updownarrow & \updownarrow & \updownarrow & \updownarrow & \updownarrow \\
\text{E:} & 2 & 4 & 6 & 8 & 10 & 12 & 14 & 16 & 18 & 20 & 22\ldots
\end{array}
$$

We make the startling conclusion that there is the "same number" of whole numbers as even numbers! This flies right in the face of the "common notion" declared by the ancient Greeks; the beginning of Euclid of Alexandria's *Elements* text says that "the whole is greater than the part."

Cardinality

The number of elements in a set is called its "cardinality." In the case of the sheep, the cardinality recorded by the farmer's accountants is 42. The cardinality of the set $\{a, b, c, d, e\}$ is 5 and this is written as *card*$\{a, b, c, d, e\}$ = 5. So cardinality is a measure of the "size" of a set. For the cardinality of the whole numbers **N**, and any set in a one-to-one correspondence with **N**, Cantor used the symbol $\aleph_0$ ($\aleph$ or "aleph" is from the Hebrew alphabet; the symbol $\aleph_0$ is read as "aleph nought"). So, in mathematical language, we can write *card*(**N**) = *card*(**O**) = *card*(**E**) = $\aleph_0$.

$$
\begin{array}{ccccccc}
1 & -1 & 2 & -2 & 3 & -3 & 4\dots \\
\tfrac{1}{2} & \tfrac{-1}{2} & \tfrac{3}{2} & \tfrac{-3}{2} & \tfrac{5}{2} & \tfrac{-5}{2} & \tfrac{7}{2}\dots \\
\tfrac{1}{3} & \tfrac{-1}{3} & \tfrac{2}{3} & \tfrac{-2}{3} & \tfrac{4}{3} & \tfrac{-4}{3} & \tfrac{5}{3}\dots \\
\tfrac{1}{4} & \tfrac{-1}{4} & \tfrac{3}{4} & \tfrac{-3}{4} & \tfrac{5}{4} & \tfrac{-5}{4} & \tfrac{7}{4}\dots \\
\tfrac{1}{5} & \tfrac{-1}{5} & \tfrac{2}{5} & \tfrac{-2}{5} & \tfrac{3}{5} & \tfrac{-3}{5} & \tfrac{4}{5}\dots \\
\vdots & \vdots & \vdots & \vdots & \vdots & \vdots &
\end{array}
$$

Any set which can be put into a one-to-one correspondence with **N** is called a "countably infinite" set. Being countably infinite means we can write the elements of the set down in a list. For example, the list of odd numbers is simply 1, 3, 5, 7, 9, . . . and we know which element is first, which is second, and so on.

Are the fractions countably infinite?

The set of fractions **Q** is a larger set than **N** in the sense that **N** can be thought of as a subset of **Q**. Can we write all the elements of **Q** down in a list? Can we devise a list so that every fraction (including negative ones) is somewhere in it? The idea that such a big set could be put in a one-to-one correspondence with **N** seems impossible. Nevertheless it can be done.

The way to begin is to think in two-dimensional terms. To start, we write down a row of all the whole numbers, positive and negative alternately. Beneath that we write all the fractions with 2 as a denominator but we omit those which appear in the row above (like $^{6}/_{2} = 3$). Below this row we write those fractions which have 3 as denominator, again omitting those which have already been recorded. We continue in this fashion, of course never ending, but knowing exactly where every fraction appears in the diagram. For example, $^{209}/_{67}$ is in the 67th row, around 200 places to the right of $^{1}/_{67}$.

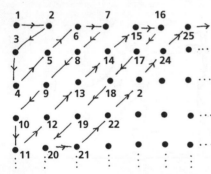

By displaying all the fractions in this way, potentially at least, we can construct a one-dimensional list. If we start on the top row and move to the right at each step we will never get to the second row. However, by choosing a devious zig-zagging route, we can be successful. Starting at 1, the promised linear list begins: 1, –1, ½, ⅓, –½, 2, –2, and follows the arrows. Every fraction, positive or negative, is somewhere in the linear list and conversely its position gives its "mate" in the two-dimensional list of fractions. So we can conclude that the set of fractions **Q** is countably infinite and write $card(\mathbf{Q}) = \aleph_0$.

Listing the real numbers

While the set of fractions accounts for many elements on the real number line there are also real numbers like $\sqrt{2}$, e and π which are *not* fractions. These are the irrational numbers — they "fill in the gaps" to give us the real number line **R**.

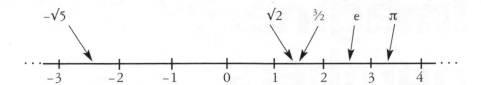

With the gaps filled in, the set **R** is referred to as the "continuum." So, how could we make a list of the real numbers? In a move of sheer brilliance, Cantor showed that even an attempt to put the real numbers *between 0 and 1* into a list is doomed to failure. This will undoubtedly come as a shock to people who are addicted to list-making, and they may indeed wonder how a set of numbers cannot be written down one after another.

Suppose you did not believe Cantor. You know that each number between 0 and 1 can be expressed as an extending decimal, for example, $\frac{1}{2} = 0.500000000000000000\ldots$ and $\frac{1}{\pi} = 0.31830988618379067153\ldots$ and you would have to say to Cantor, "here is my list of *all* the numbers between 0 and 1," which we'll call $r_1, r_2, r_3, r_4, r_5, \ldots$ If you could not produce one then Cantor would be correct.

Imagine Cantor looks at your list and he marks in bold the numbers on the diagonal:

$$r_1:\; 0.\boldsymbol{a_1}a_2a_3a_4a_5\ldots$$
$$r_2:\; 0.b_1\boldsymbol{b_2}b_3b_4b_5\ldots$$
$$r_3:\; 0.c_1c_2\boldsymbol{c_3}c_4c_5\ldots$$
$$r_4:\; 0.d_1d_2d_3\boldsymbol{d_4}d_5\ldots$$

Cantor would have said, "OK, but where is the number $x = x_1x_2x_3x_4x_5\ldots$ where x_1 differs from a_1, x_2 differs from b_2, x_3 differs from c_3 working our way down the diagonal?" His x differs from every number in your list in one decimal place and so it cannot be there. Cantor is right.

In fact, no list is possible for the set of real numbers **R**, and so it is a "larger" infinite set, one with a "higher order of infinity," than the infinity of the set of fractions **Q**. Big just got bigger.

71 Imaginary numbers

We can certainly imagine numbers. Sometimes I imagine my bank account is a million pounds in credit and there's no question that would be an "imaginary number." But the mathematical use of imaginary is nothing to do with this daydreaming.

The label "imaginary" is thought to be due to the philosopher and mathematician René Descartes, in recognition of curious solutions of equations which were definitely not ordinary numbers. Do imaginary numbers exist or not? This was a question chewed over by philosophers as they focused on the word imaginary. For mathematicians the existence of imaginary numbers is not an issue. They are as much a part of everyday life as the number 5 or π. Imaginary numbers may not help with your shopping trips, but go and ask any aircraft designer or electrical engineer and you will find they are vitally important. And by adding a real number and an imaginary number together we obtain what's called a "complex number," which immediately sounds less philosophically troublesome. The theory of complex numbers turns on the square root of *minus* 1. So what number, when squared, gives −1?

If you take any non-zero number and multiply it by itself (square it) you always get a positive number. This is believable when squaring positive numbers but is it true if we square negative numbers? We can use −1 × −1 as a test case. Even if we have forgotten the school rule

Engineering √−1

Even engineers, a very practical breed, have found uses for complex numbers. When Michael Faraday discovered alternating current in the 1830s, imaginary numbers gained a physical reality. In this case the letter j is used to represent √−1 instead of i because i stands for electrical current.

that "two negatives make a positive" we may remember that the answer is either -1 or $+1$. If we thought -1×-1 equaled -1 we could divide each side by -1 and end up with the conclusion that $-1 = 1$, which is nonsense. So we must conclude $-1 \times -1 = 1$, which is positive. The same argument can be made for other negative numbers besides -1, and so, when any real number is squared the result can *never* be negative.

This caused a sticking point in the early years of complex numbers in the 16th century. When this was overcome, the answer liberated mathematics from the shackles of ordinary numbers and opened up vast fields of inquiry undreamed of previously. The development of complex numbers is the "completion of the real numbers" to a naturally more perfect system.

The square root of –1

We have already seen that, restricted to the real number line,

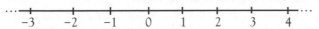

there is no square root of -1 as the square of any number cannot be negative. If we continue to think of numbers only on the real number line, we might as well give up, continue to call them imaginary numbers, go for a cup of tea with the philosophers, and have nothing more to do with them. Or we could take the bold step of accepting $\sqrt{-1}$ as a new entity, which we denote by i.

By this single mental act, imaginary numbers do exist. What they are we do not know, but we believe in their existence. At least we know $i^2 = -1$. So in our new system of numbers we have all our old friends like the real numbers $1, 2, 3, 4, \pi, e, \sqrt{2}$ and $\sqrt{3}$, with some new ones involving i such as $1 + 2i, -3 + i, 2 + 3i, 1 + i\sqrt{2}, \sqrt{3} + 2i, e + \pi i$ and so on.

This momentous step in mathematics was taken around the beginning of the 19th century, when we escaped from the one dimensional number line into a strange new two-dimensional number plane.

Adding and multiplying

Now that we have complex numbers in our mind, numbers with the form $a + bi$, what can we do with them? Just like real numbers, they can be added and multiplied together. We add them by adding their respective parts. So $2 + 3i$ added to $8 + 4i$ gives $(2 + 8) + (3 + 4)i$ with the result $10 + 7i$.

Multiplication is almost as straightforward. If we want to multiply $2 + 3i$ by $8 + 4i$ we first multiply each pair of symbols together

$$(2 + 3i) \times (8 + 4i) = (2 \times 8) + (2 \times 4i) + (3i \times 8) + (3i \times 4i)$$

and add the resulting terms, 16, $8i$, $24i$ and $12i^2$ (in this last term, we replace i^2 by -1), together. The result of the multiplication is therefore $(16 - 12) + (8i + 24i)$ which is the complex number $4 + 32i$.

With complex numbers, all the ordinary rules of arithmetic are satisfied. Subtraction and division are always possible (except by the complex number $0 + 0i$, but this was not allowed for zero in real numbers either). In fact the complex numbers enjoy all the properties of the real numbers save one. We cannot split them into positive ones and negative ones as we could with the real numbers.

The Argand diagram

The two-dimensionality of complex numbers is clearly seen by representing them on a diagram. The complex numbers $-3 + i$ and $1 + 2i$ can be drawn on what we call an Argand diagram. This way of picturing complex numbers was named after Jean Robert Argand, a Swiss mathematician, though others had a similar notion at around the same time.

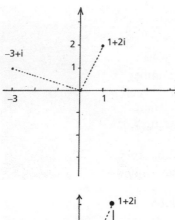

Every complex number has a "mate" officially called its "conjugate." The mate of $1 + 2i$ is $1 - 2i$ found by reversing the sign in front of the second component. The mate of $1 - 2i$, by the same token, is $1 + 2i$, so that is true mateship.

Adding and multiplying mates together always produces a real number. In the case of adding $1 + 2i$ and $1 - 2i$ we get 2, and multiplying them we get 5. This multiplication is more interesting. The answer 5 is the square of the "length" of the complex number $1 + 2i$ and this equals the length of its mate. Put the other way, we could define the length of a complex number as:

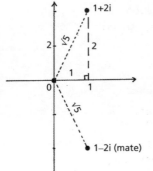

$$length\ of\ w = \sqrt{(w \times mate\ of\ w)}$$

Checking this for $-3 + i$, we find that length of $(-3 + i) = \sqrt{(-3 + i \times -3 - i)} = \sqrt{(9 + 1)}$ and so the length of $(-3 + i) = \sqrt{10}$.

The separation of the complex numbers from mysticism owes much to Sir William Rowan Hamilton, Ireland's premier mathematician in the 19th century. He recognized that i wasn't actually needed for the theory. It only acted as a placeholder and could be thrown away. Hamilton considered a complex number as an "ordered pair" of real numbers (a, b), bringing out their two-dimensional quality and making no appeal to the mystical $\sqrt{-1}$. Shorn of i, addition becomes

$$(2, 3) + (8, 4) = (10, 7)$$

and, a little less obviously, multiplication is

$$(2, 3) \times (8, 4) = (4, 32)$$

The completeness of the complex number system becomes clearer when we think of what are called "the nth roots of unity" (for mathematicians "unity" means "one"). These are the solutions of the equation $z^n = 1$. Let's take $z^6 = 1$ as an example. There are the two roots $z = 1$ and $z = -1$ on the real number line (because $1^6 = 1$ and $(-1)^6 = 1$), but where are the others when surely there should be six? Like the two real roots, all of the six roots have unit length and are found on the circle centered at the origin and of unit radius.

More is true. If we look at $w = \frac{1}{2} + \frac{\sqrt{3}}{2} i$ which is the root in the first quadrant, the successive roots (moving in a counterclockwise direction) are w^2, w^3, w^4, w^5, $w^6 = 1$ and lie at the vertices of a regular hexagon. In general the n roots of unity will each lie on the circle and be at the corners or "vertices" of a regular n-sided shape or polygon.

Extending complex numbers

Once mathematicians had complex numbers they instinctively sought generalizations. Complex numbers are two-dimensional, but what is special about two? For years, Hamilton sought to construct three-dimensional numbers and work out a way to add and multiply them but he was only successful when he switched to four dimensions. Soon afterwards these four-dimensional numbers were themselves generalized to eight dimensions (called Cayley numbers). Many wondered about 16-dimensional numbers as a possible continuation of the story — but 50 years after Hamilton's momentous feat, they were proved impossible.

72 **Primes**

Mathematics is such a massive subject, criss-crossing all avenues of human enterprise, that at times it can appear overwhelming. Occasionally we have to go back to basics. This invariably means a return to the counting numbers, 1, 2, 3, 4, 5, 6, 7, 8, 9, 10, 11, 12, . . . Can we get more basic than this?

Well, $4 = 2 \times 2$ and so we can break it down into primary components. Can we break up any other numbers? Indeed, here are some more: $6 = 2 \times 3$, $8 = 2 \times 2 \times 2$, $9 = 3 \times 3$, $10 = 2 \times 5$, $12 = 2 \times 2 \times 3$. These are composite numbers for they are built up from the very basic ones 2, 3, 5, 7, . . . The "unbreakable numbers" are the numbers 2, 3, 5, 7, 11, 13, . . . These are the prime numbers, or simply primes. A prime is a number which is only divisible by 1 and itself. You might wonder then if 1 itself is a prime number. According to this definition it should be, and indeed many prominent mathematicians in the past have treated 1 as a prime, but modern mathematicians start their primes with 2. This enables theorems to be elegantly stated. For us, too, the number 2 is the first prime.

For the first few counting numbers, we can underline the primes: 1, $\underline{2}$, $\underline{3}$, 4, $\underline{5}$, 6, $\underline{7}$, 8, 9, 10, $\underline{11}$, 12, $\underline{13}$, 14, 15, 16, $\underline{17}$, 18, $\underline{19}$, 20, 21, 22, $\underline{23}$, . . . Studying prime numbers takes us back to the very basics of the basics. Prime numbers are important because they are the "atoms" of mathematics. Like the basic chemical elements from which all other chemical compounds are derived, prime numbers can be built up to make mathematical compounds.

The mathematical result which consolidates all this has the grand name of the "prime-number decomposition theorem." This says that every whole number greater than 1 can be written by multiplying prime numbers in exactly one way. We saw that $12 = 2 \times 2 \times 3$ and there is no other way of doing it with prime components. This is often written in the power notation: $12 = 2^2 \times 3$. As another example, 6,545,448 can be written, $2^3 \times 3^5 \times 7 \times 13 \times 37$.

Discovering primes

Unhappily there are no set formulae for identifying primes, and there seems to be no pattern in their appearances among the whole numbers.

One of the first methods for finding them was developed by a younger contemporary of Archimedes who spent much of his life in Athens, Erastosthenes of Cyrene. His precise calculation of the length of the equator was much admired in his own time. Today he's noted for his sieve for finding prime numbers. Erastosthenes imagined the counting numbers stretched out before him. He underlined 2 and struck out all multiples of 2. He then moved to 3,

0	1	2	3	4	5	6	7	8	9
10	11	12	13	14	15	16	17	18	19
20	21	22	23	24	25	26	27	28	29
30	31	32	33	34	35	36	37	38	39
40	41	42	43	44	45	46	47	48	49
50	51	52	53	54	55	56	57	58	59
60	61	62	63	64	65	66	67	68	69
70	71	72	73	74	75	76	77	78	79
80	81	82	83	84	85	86	87	88	89
90	91	92	93	94	95	96	97	98	99

underlined it and struck out all multiples of 3. Continuing in this way, he sieved out all the composites. The underlined numbers left behind in the sieve were the primes.

So we can predict primes, but how do we decide whether a given number is a prime or not? How about 19,071 or 19,073? Except for the primes 2 and 5, a prime number must end in a 1, 3, 7 or 9 but this requirement is not enough to make that number a prime. It is difficult to know whether a large number ending in 1, 3, 7 or 9 is a prime or not without trying possible factors. By the way, $19,071 = 3^2 \times 13 \times 163$ is not a prime, but 19,073 is.

Another challenge has been to discover any patterns in the distribution of the primes. Let's see how many primes there are in each segment of 100 between 1 and 1,000.

Range	1–100	101–200	201–300	301–400	401–500	501–600	601–700	701–800	801–900	901–1,000	1–1,000
Number of primes	25	21	16	16	17	14	16	14	15	14	168

In 1792, when only 15 years old, Carl Friedrich Gauss suggested a formula $P(n)$ for estimating the number of prime numbers less than a given number n (this is now called the prime number theorem). For $n = 1,000$ the formula gives the approximate value of 172. The actual number of primes, 168, is less than this estimate. It had always been assumed this was the case for any value of n, but the primes often have surprises in store and it has been shown that for $n = 10^{371}$ (a huge number written longhand as a 1 with 371 trailing 0s) the actual number of primes *exceeds* the estimate. In fact, in some regions of the counting numbers the difference between the estimate and the actual number oscillates between less and excess.

How many?

There are infinitely many prime numbers. Euclid stated in his *Elements* (Book 9, Proposition 20) that "prime numbers are more than any assigned multitude of prime numbers." Euclid's beautiful proof goes like this:

> Suppose that P is the largest prime, and consider the number $N = (2 \times 3 \times 5 \times \ldots \times P) + 1$. Either N is prime or it is not. If N is prime we have produced a prime greater than P which is a contradiction to our supposition. If N is not a prime it must be divisible by some prime, say p, which is one of 2, 3, 5, ..., P. This means that p divides $N - (2 \times 3 \times 5 \times \ldots \times P)$. But this number is equal to 1 and so p divides 1. This cannot be since all primes are greater than 1. Thus, whatever the nature of N, we arrive at a contradiction. Our original assumption of there being a largest prime P is therefore false. *Conclusion*: the number of primes is limitless.

Though primes "stretch to infinity" this fact has not prevented people striving to find the largest known prime. One which has held the record recently is the enormous Mersenne prime $2^{24036583} - 1$, which is approximately $10^{7235732}$ or a number starting with 1 followed by 7,235,732 trailing 0s.

The unknown

Outstanding unknown areas concerning primes are the "twin primes problem" and the famous "Goldbach conjecture."

Twin primes are pairs of consecutive primes separated only by an even number. The twin primes in the range from 1 to 100 are 3, 5; 5, 7; 11, 13; 17, 19; 29, 31; 41, 43; 59, 61; 71, 73. On the numerical front, it is known that there are 27,412,679 twins less than 10^{10}. This means the even numbers with twins, like 12 (having twins 11, 13), constitute only 0.274 percent of the numbers in this range. Are there an infinite number of twin primes? It would be curious if there were not, but no one has so far been able to write down a proof of this.

Christian Goldbach conjectured that:

Every even number greater than 2 is the sum of two prime numbers.

For instance, 42 is an even number and we can write it as 5 + 37. The fact that we can also write it as 11 + 31, 13 + 29 or 19 + 23 is beside the point — all we need is one way. The conjecture is true for a huge

The number of the numerologist

One of the most challenging areas of number theory concerns "Waring's problem." In 1770 Edward Waring, a professor at Cambridge, England posed problems involving writing whole numbers as the addition of powers. In this setting the magic arts of numerology meet the clinical science of mathematics in the shape of primes, sums of squares and sums of cubes. In numerology, take the unrivaled cult number 666, the "number of the beast" in the biblical book of Revelation, and which has some unexpected properties. It is the sum of the squares of the first 7 primes:

$$666 = 2^2 + 3^2 + 5^2 + 7^2 + 11^2 + 13^2 + 17^2$$

Numerologists will also be keen to point out that it is the sum of palindromic cubes and, if that is not enough, the keystone 63 in the center is shorthand for 6 × 6 × 6:

$$666 = 1^3 + 2^3 + 3^3 + 4^3 + 5^3 + 6^3 + 5^3 + 4^3 + 3^3 + 2^3 + 1^3$$

The number 666 is truly the "number of the numerologist."

range of numbers — but it has never been proved in general. However, progress has been made, and some have a feeling that a proof is not far off. The Chinese mathematician Chen Jingrun made a great step. His theorem states that every sufficiently large even number can be written as the sum of two primes *or* the sum of a prime and a semi-prime (a number which is the multiplication of two primes).

The great number theorist Pierre de Fermat proved that primes of the form $4k + 1$ are expressible as the sum of two squares in exactly one way (e.g. $17 = 1^2 + 4^2$), while those of the form $4k + 3$ (like 19) cannot be written as the sum of two squares at all. Joseph Lagrange also proved a famous mathematical theorem about square powers: *every* positive whole number is the sum of four squares. So, for example, $19 = 1^2 + 1^2 + 1^2 + 4^2$. Higher powers have been explored and books filled with theorems, but many problems remain.

We described the prime numbers as the "atoms of mathematics." But "surely," you might say, "physicists have gone beyond atoms to even more fundamental units, like quarks. Has mathematics stood still?" If we limit ourselves to the counting numbers, 5 is a prime number and will always be so. But Gauss made a far-reaching discovery, that for some primes, like 5, $5 = (1 - 2i) \times (1 + 2i)$ where $i = \sqrt{-1}$ of the imaginary number system. As the product of two Gaussian integers, 5 and numbers like it are not as unbreakable as was once supposed.

73 Perfect numbers

In mathematics the pursuit of perfection has led its aspirants to different places. There are perfect squares, but here the term is not used in an esthetic sense. It's more to warn you that there are imperfect squares in existence. In another direction, some numbers have few divisors and some have many. But, like the story of the three bears, some numbers are "just right." When the addition of the divisors of a number equals the number itself it is said to be perfect.

The ancient Greek philosopher Speusippus, who took over the running of the Academy from his uncle Plato, declared that the Pythagoreans believed that 10 had the right credentials for perfection. Why? Because the number of prime numbers between 1 and 10 (namely 2, 3, 5, 7) equaled the non-primes (4, 6, 8, 9) and this was the smallest number with this property. Some people have a strange idea of perfection.

It seems the Pythagoreans actually had a richer concept of a perfect number. The mathematical properties of perfect numbers were delineated by Euclid in the *Elements* and studied in depth by Nicomachus 400 years later, leading to amicable numbers and even sociable numbers. These categories were defined in terms of the relationships between them and their divisors. At some point they came up with the theory of superabundant and deficient numbers and this led them to their concept of perfection.

Whether a number is superabundant is determined by its divisors and makes a play on the connection between multiplication and addition. Take the number 30 and consider its divisors — that is, all the numbers which divide into it exactly and which are *less* than 30. For such a small number as 30 we can see the divisors are 1, 2, 3, 5, 6, 10 and 15. Totaling up these divisors we get 42. The number 30 is superabundant because the addition of its divisors (42) is bigger than the number 30 itself.

A number is deficient if the opposite is true — if the sum of its divisors is less than itself. So the number 26 is deficient because its divisors 1, 2 and 13 add up to only 16, which is less than 26. Prime numbers are very deficient because the sum of their divisors is always just 1.

Rank	1	2	3	4	5	6	7
Perfect number	6	28	496	8,128	33,550,336	8,589,869,056	137,438,691,328

The first few perfect numbers

A number that is neither superabundant nor deficient is perfect. The addition of the divisors of a perfect number equal the number itself. The first perfect number is 6. Its divisors are 1, 2, 3 and when we add them up, we get 6. The Pythagoreans were so enchanted with the number 6 and the way its parts fitted together that they called it "marriage, health and beauty." There is another story connected with 6 told by St. Augustine (354–430). He believed that the perfection of 6 existed before the world came into existence and that the world was created in 6 days *because* the number was perfect.

The next perfect number is 28. Its divisors are 1, 2, 4, 7 and 14 and, when we add them up, we get 28. These first two perfect numbers, 6 and 28, are rather special in perfect number lore for it can be proved that every even perfect number ends in a 6 or a 28. After 28, you have to wait until 496 for the next perfect number. It is easy to check it really is the sum of its divisors: $496 = 1 + 2 + 4 + 8 + 16 + 31 + 62 + 124 + 248$. For the next perfect numbers we have to start going into the numerical stratosphere. The first five were known in the 16th century, but we still don't know if there is a largest one, or whether they go marching on without limit. The balance of opinion suggests that they, like the primes, go on for ever.

The Pythagoreans were keen on geometrical connections. If we have a perfect number of beads, they can be arranged around a hexagonal necklace. In the case of 6 this is the simple hexagon with beads placed at its corners, but for higher perfect numbers we have to add in smaller sub-necklaces within the large one.

Mersenne numbers

The key to constructing perfect numbers is a collection of numbers named after Father Marin Mersenne, a French monk who studied at a Jesuit college with René Descartes. Both men were interested in finding perfect numbers. Mersenne numbers are constructed from powers of 2, the doubling numbers 2, 4, 8, 16, 32, 64, 128, 256, . . ., and then subtracting a single 1. A Mersenne number is a number of the form $2^n - 1$. While they are always odd, they are not always prime. But it is those Mersenne numbers that are also prime that can be used to construct perfect numbers.

The power	Result	Take away 1 (Mersenne number)	Prime number?
2	4	3	prime
3	8	7	prime
4	16	15	not prime
5	32	31	prime
6	64	63	not prime
7	128	127	prime
8	256	255	not prime
9	512	511	not prime
10	1,024	1,023	not prime
11	2,048	2,047	not prime
12	4,096	4,095	not prime
13	8,192	8,191	prime
14	16,384	16,383	not prime
15	32,768	32,767	not prime

Mersenne knew that if the power was *not* a prime number, then the Mersenne number could not be a prime number either, accounting for the non-prime powers 4, 6, 8, 9, 10, 12, 14 and 15 in the table. The Mersenne numbers could only be prime if the power was a prime number, but was that enough? For the first few cases, we do get 3, 7, 31 and 127, all of which are prime. So is it generally true that a Mersenne number formed with a prime power should be prime as well?

Many mathematicians of the ancient world up to about the year 1500 thought this was the case. But primes are not constrained by simplicity, and it was found that for the power 11 (a prime number), $2^{11} - 1 = 2,047 = 23 \times 89$ and consequently it is not a prime number. There seems to be no rule. The Mersenne numbers $2^{17} - 1$ and $2^{19} - 1$ are both primes, but $2^{23} - 1$ is not a prime, because

$$2^{23} - 1 = 8,388,607 = 47 \times 178,481$$

Construction work

A combination of Euclid and Euler's work provides a formula which enables even perfect numbers to be generated: n is an even perfect number if and only if $n = 2^p - 1(2^p - 1)$ where $2^p - 1$ is a Mersenne prime.

Just good friends

The hard-headed mathematician is not usually given to the mystique of numbers but numerology is not yet dead. The amicable numbers came after the perfect numbers though they may have been known to the Pythagoreans. Later they became useful in compiling romantic horoscopes where their mathematical properties translated themselves into the nature of the ethereal bond. The two numbers 220 and 284 are amicable numbers. Why so? Well, the divisors of 220 are 1, 2, 4, 5, 10, 11, 20, 22, 44, 55 and 110 and if you add them up you get 284. You've guessed it. If you figure out the divisors of 284 and add them up, you get 220. That's true friendship.

Mersenne primes

Finding Mersenne primes is not easy. Many mathematicians over the centuries have added to the list, which has a chequered history built on a combination of error and correctness. The great Leonhard Euler contributed the eighth Mersenne prime, $2^{31} - 1 = 2,147,483,647$, in 1732. Finding the 23rd Mersenne prime, $2^{11213} - 1$, in 1963 was a source of pride for the mathematics department at the University of Illinois, who announced it to the world on their university postage stamp. But with powerful computers the Mersenne prime industry had moved on and in the late 1970s high school students Laura Nickel and Landon Noll jointly discovered the 25th Mersenne prime, and Noll the 26th Mersenne prime. To date 45 Mersenne primes have been discovered.

For example, $6 = 2^1(2^2 - 1)$, $28 = 2^2(2^3 - 1)$ and $496 = 2^4(2^5 - 1)$. This formula for calculating even perfect numbers means we can generate them if we can find Mersenne primes. The perfect numbers have challenged both people and machines and will continue to do so in a way which earlier practitioners had not envisaged. Writing at the beginning of the 19th century, the table maker Peter Barlow thought that no one would go beyond the calculation of Euler's perfect number

$$2^{30}(2^{31} - 1) = 2,305,843,008,139,952,128$$

as there was little point. He could not foresee the power of modern computers or mathematicians' insatiable need to meet new challenges.

Odd perfect numbers

No one knows if an odd perfect number will ever be found. Descartes did not think so but experts can be wrong. The English mathematician James Joseph Sylvester declared the existence of an odd perfect number "would be little short of a miracle" because it would have to satisfy so many conditions. It's little surprise Sylvester was dubious. It is one of the oldest problems in mathematics, but if an odd perfect number does exist quite a lot is already known about it. It would need to have at least eight distinct prime divisors, one of which is greater than a million, while it would have to be at least 300 digits long.

74 **Fibonacci numbers**

In *The Da Vinci Code*, **the author Dan Brown** made his murdered curator Jacques Saunière leave behind the first eight terms of a sequence of numbers as a clue to his fate. It required the skills of cryptographer Sophie Neveu to reassemble the numbers 13, 3, 2, 21, 1, 1, 8 and 5 to see their significance. Welcome to the most famous sequence of numbers in all of mathematics.

The Fibonacci sequence of whole numbers is:

1, 1, 2, 3, 5, 8, 13, 21, 34, 55, 89, 144, 233, 377, 610, 987, 1597, 2584, . . .

The sequence is widely known for its many intriguing properties. The most basic — indeed the characteristic feature which defines them — is that every term is the addition of the previous two. For example $8 = 5 + 3$, $13 = 8 + 5$, . . ., $2584 = 1587 + 987$, and so on. All you have to remember is to begin with the two numbers 1 and 1 and you can generate the rest of the sequence on the spot. The Fibonacci sequence is found in nature as the number of spirals formed from the number of seeds in the spirals in sunflowers (for example, 34 in one direction, 55 in the other), and the room proportions and building proportions designed by architects. Classical musical composers have used it as an inspiration, with Bartók's *Dance Suite* believed to be connected to the sequence. In contemporary music Brian Transeau (aka BT) has a track in his album *This Binary Universe* called 1.618 as a salute to the ultimate ratio of the Fibonacci numbers, a number we shall discuss a little later.

Origins

The Fibonacci sequence occurred in the *Liber Abaci* published by Leonardo of Pisa (Fibonacci) in 1202, but these numbers were probably known in India before that. Fibonacci posed the following problem of rabbit generation:

Mature rabbit pairs generate young rabbit pairs each month. At the beginning of the year there is one young rabbit pair. By the end of the first month they will have matured, by the end of the second month the mature pair is still there and they will have generated a young rabbit pair. The process of maturing and generation continues. Miraculously none of the rabbit pairs die.

○ = young pair

● = mature pair

Fibonacci wanted to know how many rabbit pairs there would be at the end of the year. The generations can be shown in a "family tree." Let's look at the number of pairs at the end of May (the fifth month). We see the number of pairs is 8. In this layer of the family tree the left-hand group

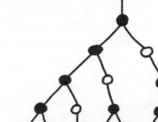

● ○ ● ○

is a duplicate of the whole row above, and the right-hand group

● ○ ●

is a duplicate of the row above that. This shows that the birth of rabbit pairs follows the basic Fibonacci equation:

The rabbit population

$$\text{number after } n \text{ months} = \text{number after } (n-1) \text{ month}$$
$$+ \text{ number after } (n-2) \text{ months}$$

Properties

Let's see what happens if we add the terms of the sequence:

$$1 + 1 = 2$$
$$1 + 1 + 2 = 4$$
$$1 + 1 + 2 + 3 = 7$$
$$1 + 1 + 2 + 3 + 5 = 12$$
$$1 + 1 + 2 + 3 + 5 + 8 = 20$$
$$1 + 1 + 2 + 3 + 5 + 8 + 13 = 33$$
$$\ldots$$

The result of each of these sums will form a sequence as well, which we can place under the original sequence, but shifted along:

Fibonacci   1   1   2   3   5   8 13 21 34 55 89...

Addition   2   4   7 12 20 33 54 88...

The addition of n terms of the Fibonacci sequence turns out to be 1 less than the next but one Fibonacci number. If you want to know the answer to the addition of $1 + 1 + 2 + \ldots + 987$, you just subtract 1 from 2584 to get 2583. If the numbers are added alternately by missing out terms, such as $1 + 2 + 5 + 13 + 34$, we get the answer 55, itself a Fibonacci number. If the other alternation is taken, such as $1 + 3 + 8 + 21 + 55$, the answer is 88 which is a Fibonacci number less 1.

The squares of the Fibonacci sequence numbers are also interesting. We get a new sequence by multiplying each Fibonacci number by itself and adding them.

Fibonacci	1	1	2	3	5	8	<u>13</u>	<u>21</u>	34	55 ...
Squares	1	1	4	9	25	64	169	441	1156	3025 ...
Addition of squares	1	2	6	15	40	104	<u>273</u>	714	1870	4895 ...

In this case, adding up all the squares up to the nth member is the same as multiplying the nth member of the original Fibonacci sequence by the next one to this. For example,

$$1 + 1 + 4 + 9 + 25 + 64 + 169 = 273 = 13 \times 21$$

Fibonacci numbers also occur when you don't expect them. Let's imagine we have a purse containing a mix of £1 and £2 coins. What if we want to count the number of ways the coins can be taken from the purse to make up a particular amount expressed in pounds. In this problem the order of actions is important. The value of £4, as we draw the coins out of the purse, can be any of the following ways, $1 + 1 + 1 + 1$; $2 + 1 + 1$; $1 + 2 + 1$; $1 + 1 + 2$; and $2 + 2$. There are five ways in all — and this corresponds to the fifth Fibonacci number. If you take out £20 there are 6,765 ways of taking the £1 and £2 coins out, corresponding to the 21st Fibonacci number! This shows the power of simple mathematical ideas.

The golden ratio

If we look at the ratio of terms formed from the Fibonacci sequence by dividing a term by its preceding term we find out another remarkable property of the Fibonacci numbers. Let's do it for a few terms 1, 1, 2, 3, 5, 8, 13, 21, 34, 55.

1/1	2/1	3/2	5/3	8/5	13/8	21/13	34/21	55/34
1.000	2.000	1.500	1.333	1.600	1.625	1.615	1.619	1.617

Pretty soon the ratios approach a value known as the "golden ratio", a famous number in mathematics, designated by the Greek letter ϕ. It takes its place amongst the top mathematical constants like π and e, and has the exact value

$$\phi = \frac{1+\sqrt{5}}{2}$$

and this can be approximated to the decimal 1.618033988. . . With a little more work we can show that each Fibonacci number can be written in terms of ϕ.

Despite the wealth of knowledge about the Fibonacci sequence, there are still many questions left to answer. The first few prime numbers in the Fibonacci sequence are 2, 3, 5, 13, 89, 233, 1597 — but we don't know if there are infinitely many primes in the Fibonacci sequence.

Family resemblances

The Fibonacci sequence holds pride of place in a wide ranging family of similar sequences. A spectacular member of the family is one we may associate with a cattle population problem. Instead of Fibonacci's rabbit pairs which transform in one month from young pairs to mature pairs which then start breeding, there is an intermediate stage in the maturation process as cattle pairs progress from young pairs to immature pairs and then to mature pairs. It is only the mature pairs which can reproduce. The cattle sequence is:

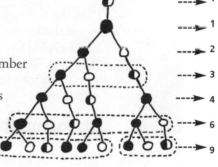

O = young pair

◐ = immature pair

● = mature pair

The cattle population

1, 1, 1, 2, 3, 4, 6, 9, 13, 19, 28, 41, 60, 88, 129, 189, 277, 406, 595, . . .

Thus the generation skips a value so for example, 41 = 28 + 13 and 60 = 41 + 19. This sequence has similar properties to the Fibonacci sequence. For the cattle sequence the ratios obtained by dividing a term by its preceding term approach the limit denoted by the Greek letter psi, written ψ, where

$$\psi = 1.46557123187676802665. . .$$

This is known as the "supergolden ratio."

75 **Golden rectangles**

Rectangles are all around us — buildings, photographs, windows, doors, even this book. Rectangles are present within the artists' community — Piet Mondrian, Ben Nicholson and others, who progressed to abstraction, all used one sort or another. So which is the most beautiful of all? Is it a long thin "Giacometti rectangle" or one that is almost a square? Or is it a rectangle in between these extremes?

Does the question even make sense? Some think so, and believe particular rectangles are more "ideal" than others. Of these, perhaps the golden rectangle has found greatest favor. Amongst all the rectangles one could choose for their different proportions — for that is what it comes down to — the golden rectangle is a very special one which has inspired artists, architects and mathematicians. Let's look at some other rectangles first.

International paper sizes as laid down by the International Standards Organization

Mathematical paper

This illustration is best demonstrated when looking at British paper sizes, in millimeters. If we take a piece of A4 paper, whose dimensions are a short side of 210 mm and a long side of 297 mm, the length-to-width ratio will be 297/210 which is approximately 1.4142. For any international A-size paper with short side equal to b, the longer side will always be $1.4142 \times b$. So for A4, $b = 210$ mm, while for A5, $b = 148$ mm. The A-formulae system used for paper sizes has a highly desirable property, one that does not occur for arbitrary paper sizes. If an A-size piece of paper is folded about the middle, the two smaller rectangles formed are directly in proportion to the larger rectangle. They are two smaller versions of the *same* rectangle.

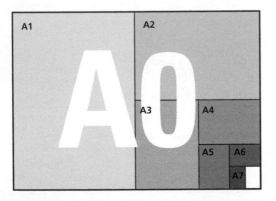

In this way, a piece of A4 folded into two pieces generates two pieces of A5. Similarly a piece of A5-size paper generates two pieces of A6. In the other direction, a sheet of A3 paper is made up of two pieces of A4. The smaller the number on the A-size the larger the piece of paper. How did we know that the particular number 1.4142 would do the trick? Let's fold a rectangle, but this time let's make it one where we don't know the length of its longer side. If we take the breadth of a rectangle to be 1 and we write the length of the longer side as x, then the length-to-width ratio is $x/1$. If we now fold the rectangle, the length-to-width ratio of the smaller rectangle is $1/\frac{1}{2}x$, which is the same as $2/x$. The point of A sizes is that our two ratios must stand for the same proportion, so we get an equation $x/1 = 2/x$ or $x^2 = 2$. The true value of x is therefore $\sqrt{2}$ which is approximately by 1.4142.

Mathematical gold

The golden rectangle is different, but only *slightly* different. This time the rectangle is folded along the line *RS* in the diagram so that the points *MRSQ* make up the corners of a *square*.

The key property of the golden rectangle is that the rectangle left over, *RNPS*, is proportional to the large rectangle — what is left over should be a mini-replica of the large rectangle.

As before, we'll say the breadth $MQ = MR$ of the large rectangle is 1 unit of length while we'll write the length of the longer side *MN* as x. The length-to-width ratio is again $x/1$. This time the breadth of the smaller rectangle *RNPS* is $MN - MR$, which is $x - 1$ so the length-to-width ratio of this rectangle is $1/(x-1)$. By equating them, we get the equation

$$\frac{x}{1} = \frac{1}{x-1}$$

which can be multiplied out to give $x^2 = x + 1$. An approximate solution is 1.618. We can easily check this. If you type 1.618 into a calculator and multiply it by itself you get 2.618 which is the same as $x + 1 = 2.618$. This number is the famous golden ratio and is designated by the Greek letter phi, written ϕ. Its definition and approximation is given by

$$\phi = \frac{1+\sqrt{5}}{2} = 1.61803398874989484820\ldots$$

and this number is related to the Fibonacci sequence and the rabbit problem (see Chapter 74).

Going for gold

Now let's see if we can build a golden rectangle. We'll begin with our square $MQSR$ with sides equal to 1 unit and mark the midpoint of QS as O. The length $OS = \frac{1}{2}$, and so by Pythagoras' theorem in the triangle ORS, $OR = \sqrt{\left(\frac{1}{2}\right)^2 + 1^2} = \frac{\sqrt{5}}{2}$

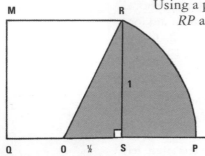

Using a pair of compasses centered on O, we can draw the arc RP and we'll find that $OP = OR = \sqrt{5}/2$. So we end up with

$$QP = \frac{1}{2} + \frac{\sqrt{5}}{2} = \phi$$

which is what we wanted: the "golden section" or the side of the golden rectangle.

History

Much is claimed of the golden ratio ϕ. Once its appealing mathematical properties are realized it is possible to see it in unexpected places, even in places where it is not. Added to this is the danger of claiming the golden ratio was there before the artefact — that musicians, architects and artists had it in mind at the point of creation. This foible is termed "golden numberism." The progress from numbers to general statements without other evidence is a dangerous argument to make.

Take the Parthenon in Athens. At its time of construction the golden ratio was certainly known but this does not mean that the Parthenon was based on it. Sure, in the front view of the Parthenon the ratio of the width to the height (including the triangular pediment) is 1.74 which is close to 1.618, but is it close enough to claim the golden ratio as a motivation? Some argue that the pediment should be left out of the calculation, and if this is done, the width-to-height ratio is actually the whole number 3.

In his 1509 book *De divina proportione*, Luca Pacioli "discovered" connections between characteristics of God and properties of the proportion determined by ϕ. He christened it the "divine proportion." Pacioli was a Franciscan monk who wrote influential books on mathematics. By some he is regarded as the "father of accounting" because he popularized the double-entry method of accounting used by Venetian merchants. His other claim to fame is that he taught mathematics to Leonardo da Vinci. In the Renaissance, the golden section achieved near mystical status — the astronomer Johannes Kepler described it as a mathematical "precious jewel." Later, Gustav Fechner, a German experimental psychologist, made thousands of measurements of rectangular shapes (playing cards, books, windows) and found the most commonly occurring ratio of their sides was close to ϕ.

Le Corbusier was fascinated by the rectangle as a central element in architectural design and by the golden rectangle in particular. He placed great emphasis on harmony and order and found this in mathematics. He saw architecture through the eyes of a mathematician. One of his planks was the "modulator" system, a theory of proportions. In effect this was a way of generating streams of golden rectangles, shapes he used in his designs. Le Corbusier was inspired by Leonardo da Vinci who, in turn, had taken careful notes on the Roman architect Vitruvius, who set store by the proportions found in the human figure.

Other shapes

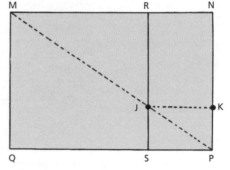

There is also a "supergolden rectangle" whose construction has similarities with the way the golden rectangle is constructed.

This is how we build the supergolden rectangle $MQPN$. As before, $MQSR$ is a square whose side is of length 1. Join the diagonal MP and mark the intersection on RS as the point J. Then make a line JK that's parallel to RN with K on NP. We'll say the length RJ is y and the length MN is x. For any rectangle, $RJ/MR = NP/MN$ (because triangles MRJ and MNP are similar), so $y/1 = 1/x$ which means $x \times y = 1$ and we say x and y are each other's "reciprocal." We get the supergolden rectangle by making the rectangle $RJKN$ proportional to the original rectangle $MQPN$, that is $y/(x - 1) = x/1$. Using the fact that $xy = 1$, we can conclude that the length of the supergolden rectangle x is found by solving the "cubic" equation $x^3 = x^2 + 1$, which is clearly similar to the equation $x^2 = x + 1$ (the equation that determines the golden rectangle). The cubic equation has one positive real solution ψ (replacing x with the more standard symbol ψ) whose value is

$$= 1.46557123187676802665\ldots$$

the number associated with the cattle sequence (see Chapter 74). Whereas the golden rectangle can be constructed by a straight edge and a pair of compasses, the supergolden rectangle cannot be made this way.

76 Pascal's triangle

The number 1 is important but what about 11? It is interesting too and so is 11 × 11 = 121, 11 × 11× 11 = 1331 and 11 × 11 × 11 × 11 = 14,641. Setting these out we get:

$$11$$
$$121$$
$$1,331$$
$$14,641$$

These are the first lines of Pascal's triangle. But where do we find it?

Throwing in $11^0 = 1$ for good measure, the first thing to do is forget the commas, and then introduce spaces between the numbers. So 14,641 becomes 1 4 6 4 1.

Pascal's triangle is famous in mathematics for its symmetry and hidden relationships. In 1653 Blaise Pascal thought so and remarked that he could not possibly cover them all in one paper. The many connections of Pascal's triangle with other branches of mathematics have made it into a venerable mathematical object, but its origins can be traced back much further than this. In fact Pascal didn't invent the triangle named after him — it was known to Chinese scholars of the 13th century.

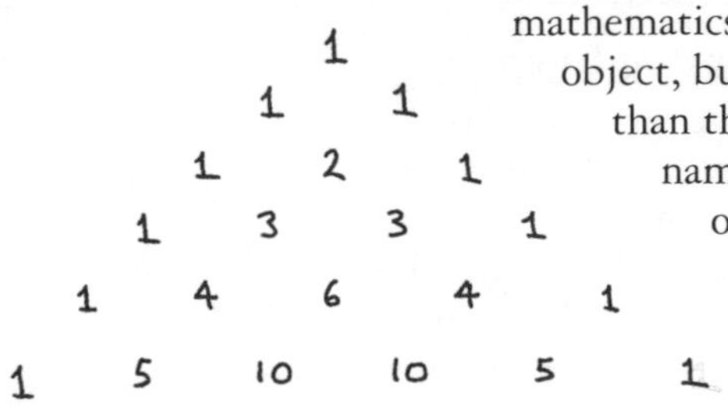

The Pascal pattern is generated from the top. Start with a 1 and place two 1s on either side of it in the next row down. To construct further rows we continue to place 1s on the ends of each row while the internal numbers are obtained by the sum of the two numbers immediately above. To obtain 6 in the fifth row, for example, we add 3 + 3 from the row above. The English mathematician G.H. Hardy said "a mathematician, like a painter or a poet, is a maker of patterns" and Pascal's triangle has patterns in spades.

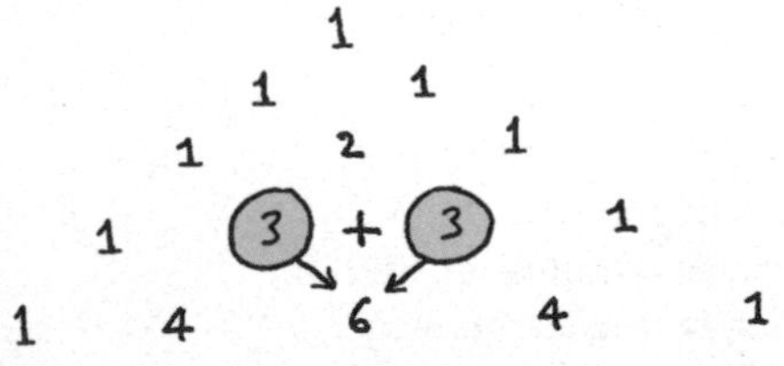

```
          1
        1   1
      1   2   1
    1   3   3   1
  1   4   6   4   1
1   5  10  10   5   1
```

Pascal's triangle

```
          1
        1   1
      1   2   1
    1  (3) + (3)  1
  1   4   6   4   1
```

Links with algebra

Pascal's triangle is founded on real mathematics. If we work out $(1 + x)$ $\times (1 + x) \times (1 + x) = (1 + x)^3$, for example, we get $1 + 3x + 3x^2 + x^3$. Look closely and you'll see the numbers in front of the symbols in this expression match the numbers in the corresponding row of Pascal's triangle. The scheme followed is:

$$(1 + x)^0 \qquad\qquad 1$$
$$(1 + x)^1 \qquad\qquad 1 \quad 1$$
$$(1 + x)^2 \qquad\qquad 1 \quad 2 \quad 1$$
$$(1 + x)^3 \qquad\quad 1 \quad 3 \quad 3 \quad 1$$
$$(1 + x)^4 \qquad 1 \quad 4 \quad 6 \quad 4 \quad 1$$
$$(1 + x)^5 \quad 1 \quad 5 \quad 10 \quad 10 \quad 5 \quad 1$$

If we add up the numbers in any row of Pascal's triangle we always obtain a power of 2. For example, in the fifth row down $1 + 4 + 6 + 4 + 1 = 16 = 2^4$. This can be obtained from the left-hand column above if we use $x = 1$.

Properties

The first and most obvious property of Pascal's triangle is its symmetry. If we draw a vertical line down through the middle, the triangle has "mirror symmetry" — it is the same to the left of the vertical line as to the right of it. This allows us to talk about plain "diagonals," because a northeast diagonal will be the same as a northwest diagonal. Under the diagonal made up of 1s we have the diagonal made up of the counting numbers 1, 2, 3, 4, 5, 6, . . . Under that there are the triangular numbers, 1, 3, 6, 10, 15, 21, . . . (the numbers which can be made up of dots in the form of triangles). In the diagonal under that we have the tetrahedral numbers, 1, 4, 10, 20, 35, 56, . . . These numbers correspond to tetrahedra ("three-dimensional triangles," or, if you like, the number of cannon balls which can be placed on triangular bases of increasing sizes). And what about the "almost diagonals"?

If we add up the numbers in lines across the triangle (which are not rows or true diagonals), we get the sequence 1, 2, 5, 13, 34, . . . Each number is three times the previous one with the one before that subtracted. For example $34 = 3 \times 13 - 5$. Based on this, the next number in the sequence will be $3 \times 34 - 13 = 89$. We have missed out the alternate "almost diagonals," starting with 1, 1 + 2 = 3, but these will give us the sequence 1, 3, 8, 21, 55, . . . and these are generated by the same "3 times minus 1" rule. We can therefore generate the next

Almost diagonals in Pascal's triangle

number in the sequence, as $3 \times 55 - 21 = \underline{144}$. But there's more. If we interleave these two sequences of "almost diagonals" we get the Fibonacci numbers:

$$1, \underline{1}, 2, \underline{3}, 5, \underline{8}, 13, \underline{21}, 34, \underline{55}, 89, \underline{144} \ldots$$

Pascal combinations

The Pascal numbers answer some counting problems. Think about seven people in a room. Let's call them **Alison, Catherine, Emma, Gary, John, Matthew** and **Thomas**. How many ways are there of choosing different groupings of three of them? One way would be **A, C, E**; another would be **A, C, T**. Mathematicians find it useful to write C(n,r) to stand for the

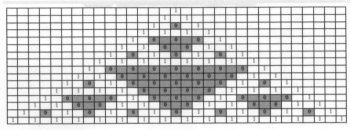

Even and odd numbers in Pascal's triangle

number in the nth row, in the rth position (counting from $r = 0$) of Pascal's triangle. The answer to our question is C(7,3). The number in the seventh row of the triangle, in the third position, is 35. If we choose one group of three we have automatically selected an "unchosen" group of four people. This accounts for the fact that C(7,4) = 35 too. In general, C(n,r) = C($n, n - r$) which follows from the mirror symmetry of Pascal's triangle.

0s and 1s

In Pascal's triangle, we see that the inner numbers form a pattern depending on whether they are even or odd. If we substitute 1 for the odd numbers and 0 for the even numbers we get a representation which is the same pattern as the remarkable fractal known as the Sierpiński gasket (see Chapter 85).

The Sierpiński gasket

Adding signs

We can write down the Pascal triangle that corresponds to the powers of $(-1 + x)$, namely $(-1 + x)^n$.

In this case the triangle is not completely symmetric about the vertical line, and instead of the rows adding to powers of 2, they add up to zero. However it is the diagonals which are interesting here. The southwestern diagonal 1, –1, 1, –1, 1, –1, 1, –1, . . . are the coefficients of the expansion

$$
\begin{array}{ccccccccc}
 & & & & 1 & & & & \\
 & & & -1 & & 1 & & & \\
 & & 1 & & -2 & & 1 & & \\
 & -1 & & 3 & & -3 & & 1 & \\
1 & & -4 & & 6 & & -4 & & 1 \\
1 & 5 & & -10 & & 10 & & -5 & 1
\end{array}
$$

Adding signs

$$(1 + x)^{-1} = 1 - x + x^2 - x^3 + x^4 - x^5 + x^6 - x^7 + \ldots$$

while the terms in the next diagonal along are the coefficients of the expansion

$$(1 + x)^{-2} = 1 - 2x + 3x^2 - 4x^3 + 5x^4 - 6x^5 + 7x^6 - 8x^7 + \ldots$$

The Leibniz harmonic triangle

The German polymath Gottfried Leibniz discovered a remarkable set of numbers in the form of a triangle. The Leibniz numbers have a symmetry relation about the vertical line. But unlike Pascal's triangle, the number in one row is obtained by adding the two numbers *below* it. For example $1/30 + 1/20 = 1/12$. To construct this triangle we can progress from the top and move from left to right by subtraction: we know $1/12$ and $1/30$ and so $1/12 - 1/30 = 1/20$, the number next to $1/30$. You might have spotted that the outside diagonal is the famous harmonic series

$$1 + \frac{1}{2} + \frac{1}{3} + \frac{1}{4} + \frac{1}{5} + \frac{1}{6} + \frac{1}{7} + \cdots$$

The Leibniz harmonic triangle

but the second diagonal is what is known as the Leibnizian series

$$\frac{1}{1 \times 2} + \frac{1}{2 \times 3} + \cdots + \frac{1}{n \times (n + 1)}$$

which by some clever manipulation turns out to equal $n/(n + 1)$. Just as we did before, we can write these Leibnizian numbers as $B(n,r)$ to stand for the nth number in the rth row. They are related to the ordinary Pascal numbers $C(n,r)$ by the formula:

$$B(n,r) \times C(n,r) = \frac{1}{n+1}$$

In the words of the old song, "the knee bone's connected to the thigh bone, the thigh bone's connected to the hip bone." So it is with Pascal's triangle and its intimate connections with so many parts of mathematics — modern geometry, combinatorics and algebra to name but three. More than this it is an exemplar of the mathematical trade — the constant search for pattern and harmony which reinforces our understanding of the subject itself.

77 **Algebra**

Algebra gives us a distinctive way of solving problems, a deductive method with a twist. That twist is "backwards thinking." For a moment consider the problem of taking the number 25, adding 17 to it, and getting 42. This is forwards thinking. We are given the numbers and we just add them together. But instead suppose we were given the answer 42, and asked a different question? We now want the number which when added to 25 gives us 42. This is where backwards thinking comes in. We want the value of x which solves the equation $25 + x = 42$ and we subtract 25 from 42 to give it to us.

Word problems which are meant to be solved by algebra have been given to schoolchildren for centuries:

My niece Michelle is 6 years of age, and I am 40.
When will I be three times as old as her?

We could find this by a trial and error method but algebra is more economical. In x years from now Michelle will be $6 + x$ years and I will be $40 + x$. I will be three times older than her when

$$3 \times (6 + x) = 40 + x$$

Multiply out the left-hand side of the equation and you'll get $18 + 3x = 40 + x$, and by moving all the xs over to one side of the equation and the numbers to the other, we find that $2x = 22$ which means that $x = 11$. When I am 51 Michelle will be 17 years old. Magic!

What if we wanted to know when I will be twice as old as her? We can use the same approach, this time solving

$$2 \times (6 + x) = 40 + x$$

to get $x = 28$. She will be 34 when I am 68. All the equations above are of the simplest type — they are called "linear" equations. They have no terms like x^2 or x^3, which make equations more difficult to solve. Equations with terms like x^2 are called "quadratic" and those with terms like x^3 are called "cubic" equations. In past times, x^2 was represented as a square and because a square has four sides the term quadratic was used; x^3 was represented by a cube.

Mathematics underwent a big change when it passed from the science of arithmetic to the science of symbols or algebra. To progress from numbers to letters is a mental jump but the effort is worthwhile.

Origins

Algebra was a significant element in the work of Islamic scholars in the ninth century. Al-Khwarizmi wrote a mathematical textbook which contained the Arabic word *al-jabr*. Dealing with practical problems in terms of linear and quadratic equations, al-Khwarizmi's "science of equations" gave us the word "algebra." Omar Khayyam is famed for writing the *Rubaiyat* and the immortal lines (in translation)

> *A Jug of Wine, a Loaf of Bread — and Thou*
> *Beside me singing in the Wilderness*

but in 1070, aged 22, he wrote a book on algebra in which he investigated the solution of cubic equations.

Girolamo Cardano's great work on mathematics, published in 1545, was a watershed in the theory of equations for it contained a wealth of results on the cubic equation and the quartic equation — those involving a term of the kind x^4. This flurry of research showed

The Italian connection

The theory of cubic equations was fully developed during the Renaissance. Unfortunately it resulted in an episode when mathematics was not always on its best behavior. Scipione Del Ferro found the solution to the various specialized forms of the cubic equation and, hearing of it, Niccolò Fontana — dubbed "Tartaglia" or "the stammerer" — a teacher from Venice, published his own results on algebra but kept his methods secret. Girolamo Cardano from Milan persuaded Tartaglia to tell him of his methods but was sworn to secrecy. The method leaked out and a feud between the two developed when Tartaglia discovered his work had been published in Cardano's 1545 book *Ars Magna*.

that the quadratic, cubic and quartic equations could all be solved by formulae involving only the operations $+, -, \times, \div, \sqrt[q]{}$ (the last operation means the qth root). For example, the quadratic equation $ax^2 + bx + c = 0$ can be solved using the formula:

$$x = \frac{-b \pm \sqrt{b^2 - 4ac}}{2a}$$

If you want to solve the equation $x^2 - 3x + 2 = 0$ all you do is feed the values $a = 1$, $b = -3$ and $c = 2$ into the formula.

The formulae for solving the cubic and quartic equations are long and unwieldy but they certainly exist. What puzzled mathematicians was that they could not produce a formula which was generally applicable to equations involving x^5, the "quintic" equations. What was so special about the power of five?

In 1826, the short-lived Niels Abel came up with a remarkable answer to this quintic equation conundrum. He actually proved a negative concept, nearly always a more difficult task than proving that something can be done. Abel proved there could not be a formula for solving all quintic equations, and concluded that any further search for this particular holy grail would be futile. Abel convinced the top rung of mathematicians, but news took a long time to filter through to the wider mathematical world. Some mathematicians refused to accept the result, and well into the 19th century people were still publishing work which claimed to have found the non-existent formula.

The modern world

For 500 years algebra meant "the theory of equations" but developments took a new turn in the 19th century. People realized that symbols in algebra could represent more than just numbers — they could represent "propositions" and so algebra could be related to the study of logic. They could even represent higher-dimensional objects such as those found in matrix algebra (see Chapter 94). And, as many non-mathematicians have long suspected, they could even represent nothing at all and just be symbols moved about according to certain (formal) rules.

A significant event in modern algebra occurred in 1843 when the Irishman Sir William Rowan Hamilton discovered the quaternions. Hamilton was seeking a system of symbols that would extend two-dimensional complex numbers to higher dimensions. For many years he tried three-dimensional symbols, but no satisfactory system resulted. When he came down for breakfast each morning his sons would ask him, "Well, Papa, can you *multiply* triplets?" and he was bound to answer that he could only add and subtract them.

Success came rather unexpectedly. The three-dimensional quest was a dead end — he should have gone for four-dimensional symbols. This flash of inspiration came to him as he walked with his wife along the Royal Canal to Dublin. He was ecstatic about the sensation of discovery. Without hesitation, the 38-year-old vandal, Astronomer Royal of Ireland and Knight of the Realm, carved the defining relations into the stone on Brougham Bridge — a spot that is acknowledged today by a plaque. With the date scored into his mind, the subject became Hamilton's obsession. He lectured on it year after year and published two heavyweight books on his "westward floating, mystic dream of four."

One peculiarity of quaternions is that when they are multiplied together, the order in which this is done is vitally important, contrary to the rules of ordinary arithmetic. In 1844 the German linguist and mathematician Hermann Grassmann published another algebraic system with rather less drama. Ignored at the time, it has turned out to be far-reaching. Today both quaternions and Grassmann's algebra have applications in geometry, physics and computer graphics.

The abstract

In the 20th century the dominant paradigm of algebra was the axiomatic method. This had been used as a basis for geometry by Euclid but it wasn't applied to algebra until comparatively recently.

Emmy Noether was the champion of the abstract method. In this modern algebra, the pervading idea is the study of structure where individual examples are subservient to the general abstract notion. If individual examples have the same structure but perhaps different notation they are called isomorphic.

The most fundamental algebraic structure is a group and this is defined by a list of axioms (see Chapter 93). There are structures with fewer axioms (such as groupoids, semi-groups and quasi-groups) and structures with more axioms (like rings, skew-fields, integral domains and fields). All these new words were imported into mathematics in the early 20th century as algebra transformed itself into an abstract science known as "modern algebra."

78 **Logic**

"If there are fewer cars on the roads the pollution will be acceptable. Either we have fewer cars on the road or there is road pricing, or both. If there is road pricing the summer will be unbearably hot. The summer is actually turning out to be quite cool. The conclusion is inescapable: pollution is acceptable."

Is this argument from the leader of a daily newspaper "valid" or is it illogical? We are not interested in whether it makes sense as a policy for road traffic or whether it makes good journalism. We are only interested in its validity as a rational argument. Logic can help us decide this question — for it concerns the rigorous checking of reasoning.

Two premises and a conclusion

As it stands the newspaper passage is quite complicated. Let's look at some simpler arguments first, going all the way back to the ancient Greek philosopher Aristotle of Stagira who is regarded as the founder of the science of logic. His approach was based on the different forms of the syllogism, a style of argument based on three statements: two premises and a conclusion. An example is

> All spaniels are dogs
> All dogs are animals
> _____
> All spaniels are animals

Above the line we have the premises, and below it, the conclusion. In this example, the conclusion has a certain inevitability about it whatever meaning we attach to the words "spaniels," "dogs" and "animals." The same syllogism, but using different words is

> All apples are oranges
> All oranges are bananas
> _____
> All apples are bananas

In this case, the individual statements are plainly nonsensical if we are using the usual connotations of the words. Yet both instances of the syllogism have the same structure and it is the structure which makes this syllogism valid. It is simply not possible to find an instance of As, Bs and Cs with this structure where the premises are true but the conclusion is false. This is what makes a valid argument useful.

A variety of syllogisms are possible if we vary the quantifiers such as "All," "Some" and "No" (as in No As are Bs). For example, another might be

All As are Bs
All Bs are Cs
―――――――――
All As are Cs

A valid argument

> Some As are Bs
> Some Bs are Cs
> ―――――――――
> Some As are Cs

Is this a valid argument? Does it apply to all cases of As, Bs and Cs, or is there a counterexample lurking, an instance where the premises are true but the conclusion false? What about making A spaniels, B brown objects, and C tables? Is the following instance convincing?

> Some spaniels are brown
> Some brown objects are tables
> ―――――――――――――――
> Some spaniels are tables

Our counterexample shows that this syllogism is not valid. There were so many different types of syllogism that medieval scholars invented mnemonics to help remember them. Our first example was known as BARBARA because it contains three uses of "All." These methods of analyzing arguments lasted for more than 2,000 years and held an important place in undergraduate studies in medieval universities. Aristotle's logic — his theory of the syllogism — was thought to be a perfect science well into the 19th century.

Propositional logic

Another type of logic goes further than syllogisms. It deals with propositions or simple statements and the combination of them. To analyze the newspaper leader we'll need some knowledge of this "propositional logic." It used to be called the "algebra of logic," which gives us a clue about its structure, since George Boole realized that it

a	b	a ∨ b
T	T	T
T	F	T
F	T	T
F	F	F

Or truth table

a	b	a ∧ b
T	T	T
T	F	F
F	T	F
F	F	F

And truth table

a	¬ a
T	F
F	T

Not truth table

a	b	a → b
T	T	T
T	F	F
F	T	T
F	F	T

Implies truth table

could be treated as a new sort of algebra. In the 1840s there was a great deal of work done in logic by such mathematicians as Boole and Augustus De Morgan.

Let's try it out and consider a proposition *a*, where *a* stands for "Freddy is a spaniel." The proposition *a* may be True or False. If I am thinking of my dog named Freddy who is indeed a spaniel then the statement is true (**T**) but if I am thinking that this statement is being applied to my cousin whose name is also Freddy then the statement is false (**F**). The truth or falsity of a proposition depends on its reference.

If we have another proposition **b** such as "Ethel is a cat" then we can combine these two propositions in several ways. One combination is written **a ∨ b**. The connective ∨ corresponds to "or" but its use in logic is slightly different from "or" in everyday language. In logic, **a ∨ b** is true if *either* "Freddy is a spaniel" is true or "Ethel is a cat" is true, *or* if both are true, and it is only false when *both a* and *b* are false. This conjunction of propositions can be summarized in a truth table.

We can also combine propositions using "and," written as **a ∧ b**, and "not," written as **¬a**. The algebra of logic becomes clear when we combine these propositions using a mixture of the connectives with **a**, **b** and **c** like **a ∧ (b ∨ c)**. We can obtain an equation we call an identity:

$$a \wedge (b \vee c) \equiv (a \wedge b) \vee (a \wedge c)$$

The symbol ≡ means equivalence between logical statements where both sides of the equivalence have the same truth table. There is a parallel between the algebra of logic and ordinary algebra because the symbols ∧ and ∨ act similarly to × and + in ordinary algebra, where we have $x \times (y + z) = (x \times y) + (x \times z)$. However, the parallel is not exact and there are exceptions.

Other logical connectives may be defined in terms of these basic ones. A useful one is the "implication" connective **a→b** which is defined to be equivalent to **¬a ∨ b** and has the truth table shown on the left.

Now if we look again at the newspaper leader, we can write it in symbolic form to give the argument in the margin:

C → P
C ∨ S
S → H
¬H
—————
P

C = fewer **C**ars on the roads
P = **P**ollution will be acceptable
S = there is a road pricing **S**cheme
H = summer will be unbearably **H**ot

Is the argument valid or not? Let's assume the conclusion **P** is false, but that all the premises are true. If we can show this forces a contradiction, it means the argument must be valid. It will then be impossible to have the premises true but the conclusion false. If **P** is false, then from the first premise **C** → **P**, **C** must be false. As **C** ∨ **S** is true, the fact that **C** is false means that S is true. From the third premise **S** → **H** this means that H is true. That is, ¬**H** is false. This contradicts the fact that ¬**H**, the last premise, was assumed to be true. The content of the statements in the newspaper leader may still be disputed, but the structure of the argument is valid.

Other logics

Gottlob Frege, C.S. Peirce and Ernst Schröder introduced quantification to propositional logic and constructed a "first-order predicate logic" (because it is predicated on variables). This uses the universal quantifier, ∀, to mean "for all," and the existential quantifier, ∃, to mean "there exists."

∨	or
∧	and
¬	not
→	implies
∀	for all
∃	there exists

    Another new development in logic is the idea of fuzzy logic. This suggests confused thinking, but it is really about a widening of the traditional boundaries of logic. Traditional logic is based on collections or sets. So we had the set of spaniels, the set of dogs, and the set of brown objects. We are sure what is included in the set and what is not in the set. If we meet a pure bred "Rhodesian ridgeback" in the park we are pretty sure it is not a member of the set of spaniels.

    Fuzzy set theory deals with what appear to be imprecisely defined sets. What if we had the set of heavy spaniels. How heavy does a spaniel have to be to be included in the set? With fuzzy sets there is a *gradation* of membership and the boundary as to what is in and what is out is left fuzzy. Mathematics allows us to be precise about fuzziness. Logic is far from being a dry subject. It has moved on from Aristotle and is now an active area of modern research and application.

79 **Proof**

Mathematicians attempt to justify their claims by proofs. The quest for cast iron rational arguments is the driving force of pure mathematics. Chains of correct deduction from what is known or assumed, lead the mathematician to a conclusion which then enters the established mathematical storehouse.

Proofs are not arrived at easily — they often come at the end of a great deal of exploration and false trails. The struggle to provide them occupies the center ground of the mathematician's life. A successful proof carries the mathematician's stamp of authenticity, separating the established theorem from the conjecture, bright idea or first guess.

Qualities looked for in a proof are rigor, transparency and, not least, elegance. To this add insight. A good proof is "one that makes us wiser" — but it is also better to have some proof than no proof at all. Progression on the basis of unproven facts carries the danger that theories may be built on the mathematical equivalent of sand.

Not that a proof lasts forever, for it may have to be revised in the light of developments in the concepts it relates to.

What is a proof?

When you read or hear about a mathematical result do you believe it? What would make you believe it? One answer would be a logically sound argument that progresses from ideas you accept to the statement you are wondering about. That would be what mathematicians call a proof, in its usual form a mixture of everyday language and strict logic. Depending on the quality of the proof you are either convinced or remain skeptical.

The main kinds of proof employed in mathematics are: the method of the counterexample; the direct method; the indirect method; and the method of mathematical induction.

The counterexample

Let's start by being skeptical — this is a method of proving a statement is incorrect. We'll take a specific statement as an example. Suppose you hear a claim that any number multiplied by itself results in an even number. Do you believe this? Before jumping in with an answer we should try a few examples. If we have a number, say 6, and multiply it by itself to get $6 \times 6 = 36$ we find that indeed 36 is an even number. But one swallow does not make a summer. The claim was for *any* number, and there are an infinity of these. To get a feel for the problem we should try some more examples. Trying 9, say, we find that $9 \times 9 = 81$. But 81 is an odd number. This means that the statement that *all* numbers when multiplied by themselves give an even number is false. Such an example runs counter to the original claim and is called a counterexample. A counterexample to the claim that "all swans are white," would be to see one black swan. Part of the fun of mathematics is seeking out a counterexample to shoot down a would-be theorem.

If we fail to find a counterexample we might feel that the statement is correct. Then the mathematician has to play a different game. A proof has to be constructed and the most straightforward kind is the direct method of proof.

The direct method

In the direct method we march forward with logical argument from what is already established, or has been assumed, to the conclusion. If we can do this we have a theorem. We cannot prove that multiplying any number by itself results in an even number because we have already disproved it. But we may be able to salvage something. The difference between our first example, 6, and the counterexample, 9, is that the first number is even and the counterexample is odd. Changing the hypothesis is something we can do. Our new statement is: if we multiply an even number by itself the result is an even number.

First we try some other numerical examples and we find this statement verified every time and we just cannot find a counterexample. Changing tack we try to prove it, but how can we start? We could begin with a general even number n, but as this looks a bit abstract we'll see how a proof might go by looking at a concrete number, say 6. As you know, an even number is one which is a multiple of 2, that is $6 = 2 \times 3$. As $6 \times 6 = 6 + 6 + 6 + 6 + 6 + 6$ or, written another way, $6 \times 6 = 2 \times 3 + 2 \times 3 + 2 \times 3 + 2 \times 3 + 2 \times 3 + 2 \times 3$ or, rewriting using brackets,

$$6 \times 6 = 2 \times (3 + 3 + 3 + 3 + 3 + 3)$$

This means 6×6 is a multiple of 2 and, as such, is an even number. But in this argument there is nothing which is particular to 6, and we could have started with $n = 2 \times k$ to obtain

$$n \times n = 2 \times (k + k + \ldots + k)$$

and conclude that $n \times n$ is even. Our proof is now complete. In translating Euclid's *Elements*, latter-day mathematicians wrote "QED" at the end of a proof to say job done — it's an abbreviation for the Latin *quod erat demonstrandum* (which was to be demonstrated). Nowadays they use a filled-in square ▪. This is called a halmos after Paul Halmos who introduced it.

The indirect method

In this method we pretend the conclusion is false and by a logical argument demonstrate that this contradicts the hypothesis. Let's prove the previous result by this method.

Our hypothesis is that n is even and we'll pretend $n \times n$ is odd. We can write $n \times n = n + n + \ldots + n$ and there are n of these. This means n cannot be even (because if it were $n \times n$ would be even). Thus n is odd, which contradicts the hypothesis ▪.

This is actually a mild form of the indirect method. The full-strength indirect method is known as the method of *reductio ad absurdum* (reduction to the absurd), and was much loved by the ancient Greeks. In the academy in Athens, Socrates and Plato loved to prove a debating point by wrapping up their opponents in a mesh of contradiction and out of it would come the point they were trying to prove. The classical proof that the square root of 2 is an irrational number is one of this form where we start off by assuming the square root of 2 is a rational number and derive a contradiction to this assumption.

The method of mathematical induction

Mathematical induction is a powerful way of demonstrating that a sequence of statements $P_1, P_2, P_3, \ldots$ are all true. This was recognized in the 1830s by Augustus De Morgan who formalized what had been known for hundreds of years. This specific technique (not to be confused with scientific induction) is widely used to prove statements involving *whole* numbers. It is especially useful in graph theory, number theory, and computer science generally. As a practical example, think of the problem of adding up the odd numbers. For instance, the addition of the first three odd numbers $1 + 3 + 5$ is 9, while the sum of the first four $1 + 3 + 5 + 7$ is 16. Now, 9 is $3 \times 3 = 3^2$ and 16 is $4 \times 4 = 4^2$, so could it be that the addition of the first n odd numbers is equal to n^2? If we try a randomly chosen value of n, say $n = 7$, we indeed find that the sum of the first seven is $1 + 3 + 5 + 7 + 9 + 11 + 13 = 49$, which is 7^2.

But is this pattern followed for *all* values of *n*? How can we be sure? We have a problem, because we cannot hope to check an infinite number of cases individually.

This is where mathematical induction steps in. Informally it is the domino method of proof. This metaphor applies to a row of dominos standing on their ends. If one domino falls it will knock the next one down. This is clear. All we need to make them *all* fall is the first one to fall. We can apply this thinking to the odd numbers problem. The statement P_n says that the sum of the first *n* odd numbers adds up to n^2. Mathematical induction sets up a chain reaction whereby P_1, P_2, P_3, . . . will *all* be true. The statement P_1 is trivially true because $1 = 1^2$. Next, P_2 is true because $1 + 3 = 1^2 + 3 = 2^2$, P_3 is true because $1 + 3 + 5 = 2^2 + 5 = 3^2$ and P_4 is true because $1 + 3 + 5 + 7 = 3^2 + 7 = 4^2$. We use the result at one stage to hop to the next one. This process can be formalized to frame the method of mathematical induction.

Difficulties with proof

Proofs come in all sorts of styles and sizes. Some are short and snappy, particularly those found in textbooks. Others detailing the latest research have taken up whole issues of journals and amount to thousands of pages. Very few people will have a grasp of the whole argument in these cases.

There are also foundational issues. For instance, a small number of mathematicians are unhappy with the *reductio ad absurdum* method of indirect proof where it applies to existence. If the assumption that a solution of an equation does not exist leads to a contradiction, is this enough to prove that a solution does exist? Opponents of this proof method would claim the logic is merely sleight of hand and doesn't tell us how actually to construct a concrete solution. They are called "Constructivists" (of varying shades) who say the proof method fails to provide "numerical meaning." They pour scorn on the classical mathematician who regards the *reductio* method as an essential weapon in the mathematical armoury. On the other hand the more traditional mathematician would say that outlawing this type of argument means working with one hand tied behind your back and, furthermore, denying so many results proved by this indirect method leaves the tapestry of mathematics looking rather threadbare.

80 **Sets**

Nicholas Bourbaki was a pseudonym for a self-selected group of French academics who wanted to rewrite mathematics from the bottom up in "the right way." Their bold claim was that everything should be based on the theory of sets. The axiomatic method was central and the books they produced were written in the rigorous style of "definition, theorem and proof." This was also the thrust of the modern mathematics movement of the 1960s.

Georg Cantor created set theory out of his desire to put the theory of real numbers on a sound basis. Despite initial prejudice and criticism, set theory was well established as a branch of mathematics by the turn of the 20th century.

What are sets?

A set may be regarded as a collection of objects. This is informal but gives us the main idea. The objects themselves are called "elements" or "members" of the set. If we write a set A which has a member a, we may write $a \in A$, as did Cantor. An example is $A = \{1, 2, 3, 4, 5\}$ and we can write $1 \in A$ for membership, and $6 \notin A$ for non-membership.

Sets can be combined in two important ways. If A and B are two sets then the set consisting of elements which are members of A *or B* (or both) is called the "union" of the two sets. Mathematicians write this as $A \cup B$. It can also be described by a Venn diagram, named after the Victorian logician the Rev. John Venn. Euler used diagrams like these even earlier.

The set $A \cap B$ consists of elements which are members of A *and B* and is called the "intersection" of the two sets.

If $A = \{1, 2, 3, 4, 5\}$ and $B = \{1, 3, 5, 7, 10, 21\}$, the union is $A \cup B = \{1, 2, 3, 4, 5, 7, 10, 21\}$ and the intersection is $A \cap B = \{1, 3, 5\}$. If we regard a set A as part of a universal set E, we can define the complement set $\neg A$ as consisting of those elements in E which are *not* in A.

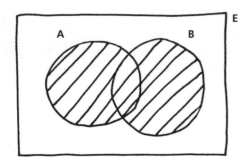

The union of A and B

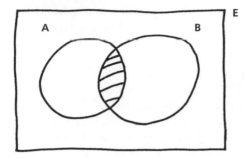

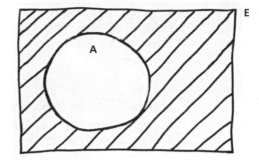

The intersection of A and B   The complement of A

The operations ∩ and ∪ on sets are analogous to × and + in algebra. Together with the complement operation ¬, there is an "algebra of sets." The Indian-born British mathematician Augustus De Morgan formulated laws to show how all three operations work together. In our modern notation, De Morgan's laws are:

$$\neg(A \cup B) = (\neg A) \cap (\neg B)$$

and

$$\neg(A \cap B) = (\neg A) \cup (\neg B)$$

The paradoxes

There are no problems dealing with finite sets because we can list their elements, as in $A = \{1, 2, 3, 4, 5\}$, but in Cantor's time, infinite sets were more challenging.

Cantor defined sets as the collection of elements with a specific property. Think of the set $\{11, 12, 13, 14, 15, \ldots\}$, all the whole numbers bigger than 10. Because the set is infinite, we can't write down all its elements, but we can still specify it because of the property that all its members have in common. Following Cantor's lead, we can write the set as $A = \{x: x$ is a whole number $> 10\}$, where the colon stands for "such that."

In primitive set theory we could also have a set of abstract things, $A = \{x: x$ is an abstract thing$\}$. In this case A is itself an abstract thing, so it is possible to have $A \in A$. But in allowing this relation, serious problems arise. The British philosopher Bertrand Russell hit upon the idea of a set S which contained *all* things which did *not* contain themselves. In symbols this is $S = \{x: x \notin x\}$.

He then asked the question "is $S \in S$?" If the answer is "Yes" then S must satisfy the defining sentence for S, and so $S \notin S$. On the other hand if the answer is "No" and $S \notin S$, then S does not satisfy the defining relation of $S = \{x : x \notin x\}$ and so $S \in S$. Russell's question ended with this statement, the basis of Russell's paradox,

$$S \in S \text{ if and only if } S \notin S$$

It is similar to the "barber paradox" where a village barber announces to the locals that he will only shave those who do not shave themselves. The question arises: should the barber shave himself? If he does not shave himself he should. If does shave himself he should not.

It is imperative to avoid such paradoxes, politely called "antinomies." For mathematicians it is simply not permissible to have systems that generate contradictions. Russell created a theory of types and only allowed $a \in A$ if a were of a lower type than A, so avoiding expressions such as $S \in S$.

Another way to avoid these antinomies was to formalize the theory of sets. In this approach we don't worry about the nature of sets themselves, but list formal axioms that specify rules for treating them. The ancient Greeks tried something similar with a problem of their own — they didn't have to explain what straight lines were, but only how they should be dealt with.

In the case of set theory, this was the origin of the Zermelo–Fraenkel axioms which prevented the appearance of sets in their system that were too "big." This effectively debarred such dangerous creatures as the set of all sets from appearing.

Gödel's theorem

Austrian mathematician Kurt Gödel dealt a knockout punch to those who wanted to escape from the paradoxes into formal axiomatic systems. In 1931, Gödel proved that even for the simplest of formal systems there were statements whose truth or falsity could not be deduced from within these systems. Informally, there were statements which the axioms of the system could not reach. They were undecidable statements. For this reason Gödel's theorem is paraphrased as "the incompleteness theorem." This result applied to the Zermelo–Fraenkel system as well as to other systems.

Cardinal numbers

The number of elements of a finite set is easy to count, for example $A = \{1, 2, 3, 4, 5\}$ has 5 elements, or we say its "cardinality" is 5 and write $card(A) = 5$. Loosely speaking, the cardinality measures the "size" of a set.

According to Cantor's theory of sets, the set of fractions **Q** and the real numbers **R** are very different. The set **Q** can be put in a list but the set **R** cannot (see Chapter 70). Although both sets are infinite, the set **R** has a higher order of infinity than **Q**. Mathematicians denote *card*(Q) by $\aleph_0$, the Hebrew "aleph nought" and *card*(R) = c. So this means $\aleph_0 < c$.

The continuum hypothesis

Brought to light by Cantor in 1878, the continuum hypothesis says that the next level of infinity after the infinity of **Q** is the infinity of the real numbers c. Put another way, the continuum hypothesis asserted there was no set whose cardinality lay strictly between $\aleph_0$ and c. Cantor struggled with it and though he believed it to be true he could not prove it. To disprove it would amount to finding a subset X of **R** with $\aleph_0 < card(X) < c$ but he could not do this either.

The problem was so important that German mathematician David Hilbert placed it at the head of his famous list of 23 outstanding problems for the next century, presented to the International Mathematical Congress in Paris in 1900.

Gödel emphatically believed the hypothesis to be false, but he did not prove it. He did prove (in 1938) that the hypothesis was compatible with the Zermelo–Fraenkel axioms for set theory. A quarter of a century later, Paul Cohen startled Gödel and the logicians by proving that the continuum hypothesis could not be deduced from the Zermelo–Fraenkel axioms. This is equivalent to showing the axioms and the negation of the hypothesis is consistent. Combined with Gödel's 1938 result, Cohen had shown that the continuum hypothesis was independent of the rest of the axioms for set theory.

This state of affairs is similar in nature to the way the parallel postulate in geometry (see Chapter 87) is independent of Euclid's other axioms. That discovery resulted in a flowering of the non-Euclidean geometries which, amongst other things, made possible the advancement of relativity theory by Einstein. In a similar way, the continuum hypothesis can be accepted or rejected without disturbing the other axioms for set theory. After Cohen's pioneering result a whole new field was created which attracted generations of mathematicians who adopted the techniques he used in proving the independence of the continuum hypothesis.

81 Calculus

A calculus is a way of calculating, so mathematicians sometimes talk about the "calculus of logic," the "calculus of probability," and so on. But all are agreed there is really only one Calculus, pure and simple, and this is spelled with a capital C.

Calculus is a central plank of mathematics. It would now be rare for a scientist, engineer or a quantitative economist not to have come across Calculus, so wide are its applications. Historically it is associated with Sir Isaac Newton and Gottfried Leibniz, who pioneered it in the 17th century. Their similar theories resulted in a priority dispute over who first discovered Calculus. In fact, both men came to their conclusions independently and their methods were quite different.

Since then Calculus has become a huge subject. Each generation bolts on techniques they think should be learned by the younger generation, and these days textbooks run beyond 1,000 pages and involve many extras. For all these add-ons, what is absolutely essential is *differentiation* and *integration*, the twin peaks of Calculus as set up by Newton and Leibniz. The words are derived from Leibniz's *differentialis* (taking differences or "taking apart") and *integralis* (the sum of parts, or "bringing together").

In technical language, differentiation is concerned with measuring *change* and integration with measuring *area*, but the jewel in the crown of Calculus is the "star result" that they are two sides of the same coin — differentiation and integration are the inverses of each other. Calculus is really one subject, and you need to know about both sides. No wonder that Gilbert and Sullivan's "very model of a modern Major General" in *The Pirates of Penzance* proudly proclaimed them both:

> *With many cheerful facts about the square of the hypotenuse.*
> *I'm very good at integral and differential calculus.*

Differentiation

Scientists are fond of conducting "thought experiments" — Einstein especially liked them. Imagine we are standing on a bridge high above a gorge and are about to let a stone drop. What will happen? The advantage of a thought experiment is that we do not actually have to be there in person. We can also do impossible things like stopping the stone in mid-air or watching it in slow motion over a short time interval.

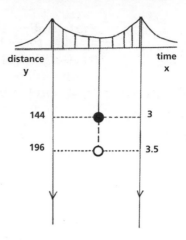

According to Newton's theory of gravity, the stone will fall. Nothing surprising in that; the stone is attracted to the earth and will fall faster and faster as the hand on our stopwatch ticks on. Another advantage of a thought experiment is that we can ignore complicating factors like air resistance.

What is the stone's speed at a given instant of time, say when the stopwatch reads *exactly* 3 seconds after it has been released? How can we work this out? We can certainly measure *average* speed but our problem is to measure *instantaneous* speed. As it's a thought experiment, why don't we stop the stone in mid-air and then let it move down a short distance by taking a fraction of a second more? If we divide this extra distance by the extra time we will have the average speed over the short time interval. By taking smaller and smaller time intervals the average speed will be closer and closer to the instantaneous speed at the place where we stopped the stone. This limiting process is the basic idea behind Calculus.

We might be tempted to make the small extra time equal to zero. But in our thought experiment, the stone has not moved at all. It has moved no distance and taken no time to do it! This would give us the average speed 0/0 which the Irish philosopher Bishop Berkeley famously described as the "ghosts of departed quantities." This expression cannot be determined — it is actually *meaningless*. By taking this route we are led into a numerical quagmire.

To go further we need some symbols. The exact formula connecting the distance fallen *y* and the time *x* taken to reach there was derived by Galileo:

$$y = 16 \times x^2$$

The factor "16" appears because feet and seconds are the chosen measurement units. If we want to know, say, how far the stone has dropped in 3 seconds we simply substitute $x = 3$ into the formula and calculate the answer $y = 16 \times 3^2 = 144$ feet. But how can we calculate the speed of the stone at time $x = 3$?

Let's take a further 0.5 of a second and see how far the stone has traveled between 3 and 3.5 seconds. In 3.5 seconds the stone has traveled $y = 16 \times 3.5^2 = 196$ feet, so *between* 3 and 3.5 seconds it has fallen $196 - 144 = 52$ feet. Since speed is distance divided by time, the average speed over this time interval is $52/0.5 = 104$ feet per second. This will be close to the instantaneous speed at $x = 3$, but you may well say that 0.5 seconds is not a small enough measure. Repeat the argument with a smaller time gap, say 0.05 seconds, and we see that the distance fallen is $148.84 - 144 = 4.84$ feet giving an average speed of $4.84/0.05 = 96.8$ feet per second. This indeed will be closer to the instantaneous speed of the stone at 3 seconds (when $x = 3$).

We must now take the bull by the horns and address the problem of calculating the average speed of the stone between x seconds and slightly later at $x + h$ seconds. After a little symbol shuffling we find this is

$$16 \times (2x) + 16 \times h$$

u	du/dx
x^2	$2x$
x^3	$3x^2$
x^4	$4x^3$
x^5	$5x^4$
$\cdots$	$\cdots$
x^n	nx^{n-1}

As we make h smaller and smaller, like we did in going from 0.5 to 0.05, we see that the first term is unaffected (because it does not involve h) and the second term itself becomes smaller and smaller. We conclude that

$$v = 16 \times (2x)$$

where v is the instantaneous velocity of the stone at time x. For example, the instantaneous velocity of the stone after 1 second (when $x = 1$) is $16 \times (2 \times 1) = 32$ feet per second; after 3 seconds it is $16 \times (2 \times 3)$ which gives 96 feet per second.

If we compare Galileo's distance formula $y = 16 \times x^2$ with the velocity formula $v = 16 \times (2x)$ the essential difference is the change x^2 to $2x$. This is the effect of differentiation, passing from $u = x^2$ to the *derivative* $\dot{u} = 2x$. Newton called $\dot{u} = 2x$ a "fluxion" and the variable x a fluent because he thought in terms of flowing quantities. Nowadays we frequently write $u = x^2$ and its *derivative* as $du/dx = 2x$. Originally introduced by Leibniz, this notation's continued use represents the success of the "d'ism of Leibniz over the dotage of Newton."

The falling stone was one example, but if we had other expressions that u stood for we could still calculate the derivative, which can be useful in other contexts. There is a pattern in this: the derivative is formed by multiplying by the previous power and subtracting 1 from it to make the new power.

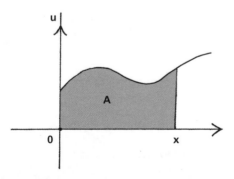

Integration

The first application of integration was to measure area. The measurement of the area under a curve is done by dividing it into approximate rectangular strips, each with width dx. By measuring the area of each and adding them up we get the "sum" and so the total area. The notation S standing for sum was introduced by Leibniz in an elongated form $\int$. The area of each of the rectangular strips is $u\,dx$, so the area A under the curve from 0 to x is

$$A = \int_0^x u\,dx$$

If the curve we're looking at is $u = x^2$, the area is found by drawing narrow rectangular strips under the curve, adding them up to calculate the approximate area, and applying a limiting process to their widths to gain the exact area. This answer gives the area

$$A = x^3/3$$

For different curves (and so other expressions for u) we could still calculate the integral. Like the derivative, there is a regular pattern for the integral of powers of x. The integral is formed by dividing by the "previous power $+1$" and adding 1 to it to make the new power.

The star result

If we differentiate the integral $A = x^3/3$ we actually get the original $u = x^2$. If we integrate the derivative $du/dx = 2x$ we also get the original $u = x^2$. Differentiation is the inverse of integration, an observation known as the "Fundamental Theorem of the Calculus" and one of the most important theorems in all mathematics.

Without Calculus there would be no satellites in orbit, no economic theory and statistics would be a very different discipline. Wherever change is involved, there we find Calculus.

u	$\int_0^x u\,dx$
x^2	$x^3/3$
x^3	$x^4/4$
x^4	$x^5/5$
x^5	$x^6/6$
$\ldots$	$\ldots$
x^n	$x^{n+1}/(n+1)$

82 **Curves**

It's easy to draw a curve. Artists do it all the time; architects lay out a sweep of new buildings in the curve of a crescent, or a modern avenue. A baseball pitcher throws a curveball. Soccer players make their way up the pitch in a curve, and when they shoot for goal, the ball follows a curve. But, if we were to ask "What is a curve?" the answer is not so easy to frame.

Mathematicians have studied curves for centuries and from many vantage points. It began with the ancient Greeks and the curves they studied are now called the "classical" curves.

Classical curves

The first family in the realm of the classical curves are what we call "conic sections." Members of this family are the circle, the ellipse, the parabola and the hyperbola. The conic is formed from the double cone, two ice-cream cones joined together where one is upside down. By slicing through this with a flat plane the curves of intersection will be a circle, an ellipse, a parabola or a hyperbola, depending on the tilt of the slicing plane to the vertical axis of the cone.

We can think of a conic as the projection of a circle onto a screen. The light rays from the bulb in a cylindrical table lamp form a double light cone where the light will throw out projections of the top and bottom circular rims. The image on the ceiling will be a circle but if we tip the lamp, this circle will become an ellipse. On the other hand the image against the wall will give the curve in two parts, the hyperbola.

The conics can also be described from the way points move in the plane. This is the "locus" method loved by the ancient Greeks, and unlike the projective definition it involves length. If a point moves so that its distance from one fixed point is always the

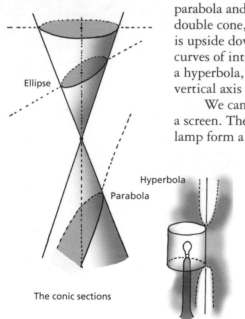

Ellipse

Hyperbola

Parabola

The conic sections

The parabola   The logarithmic spiral

same, we get a circle. If a point moves so that the sum of its distances from *two* fixed points (the foci) is a constant value we get an ellipse (where the two foci are the same, the ellipse becomes a circle). The ellipse was the key to the motion of the planets. In 1609, the German astronomer Johannes Kepler announced that the planets travel around the sun in ellipses, rejecting the old idea of circular orbits.

Not so obvious is the point which moves so that its distance from a point (the focus *F*) is the same as its perpendicular distance from a given line (the directrix). In this case we get a parabola. The parabola has a host of useful properties. If a light source is placed at the focus *F*, the emitted light rays are all parallel to *PM*. On the other hand, if TV signals are sent out by a satellite and hit a parabola-shaped receiving dish, they are gathered together at the focus and are fed into the TV set.

If a stick is rotated about a point any fixed point on the stick traces out a circle, but if a point is allowed to moved outwards along the stick in addition to it being rotated this generates a spiral. Pythagoras loved the spiral and much later Leonardo da Vinci spent ten years of his life studying their different types, while René Descartes wrote a treatise on them. The logarithmic spiral is also called the equiangular spiral because it makes the same angle with a radius and the tangent at the point where the radius meets the spiral.

Jacob Bernoulli of the famed mathematical clan from Switzerland was so enamoured with the logarithmic spiral that he wanted it carved on his tomb in Basle. The "Renaissance man" Emanuel Swedenborg regarded the spiral as the most perfect of shapes. A three-dimensional spiral which winds itself around a cylinder is called a helix. Two of these — a double helix — form the basic structure of DNA.

There are many classical curves, such as the limaçon, the lemniscate and the various ovals. The cardioid derives its name from

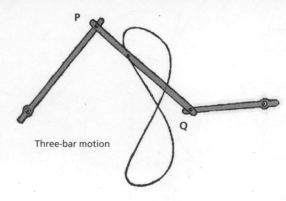
Three-bar motion

being shaped like a heart. The catenary curve was the subject of research in the 18th century and it was identified as the curve formed by a chain hanging between two points. The parabola is the curve seen in a suspension bridge hanging between its two vertical pylons.

One aspect of 19th-century research on curves was on those curves that were generated by mechanical rods. This type of question was an extension of the problem solved approximately by the Scottish engineer James Watt, who designed jointed rods to turn circular motion into linear motion. In the steam age this was a significant step forward.

The simplest of these mechanical gadgets is the three-bar motion, where the bars are jointed together with fixed positions at either end. If the "coupler bar" *PQ* moves in any which way, the locus of a point on it turns out to be a curve of degree six, a "sextic curve."

Algebraic curves

Following Descartes, who revolutionized geometry with the introduction of x, y and z coordinates and the Cartesian axes named after him, the conics could now be studied as algebraic equations. For example, the circle of radius 1 has the equation $x^2 + y^2 = 1$, which is an equation of the second degree, as *all* conics are. A new branch of geometry grew up called algebraic geometry.

In a major study Sir Isaac Newton classified curves described by algebraic equations of degree three, or cubic curves. Compared with the four basic conics, 78 types were found, grouped into five classes. The explosion of the number of different types continues for quartic curves, with so many different types that the full classification has never been carried out.

The study of curves as algebraic equations is not the whole story. Many curves such as catenaries, cycloids (curves traced out by a point on a revolving wheel) and spirals are not easily expressible as algebraic equations.

A definition

What mathematicians were after was a definition of a curve itself, not just specific examples. Camille Jordan proposed a theory of curves built on the definition of a curve in terms of variable points.

A simple closed
Jordan curve

Here's an example. If we let $x = t^2$ and $y = 2t$ then, for different values of t, we get many different points that we can write as coordinates (x, y). For example, if $t = 0$ we get the point $(0, 0)$, $t = 1$ gives the point $(1, 2)$, and so on. If we plot these points on the x–y axes and "join the dots" we will get a parabola. Jordan refined this idea of points being traced out. For him this was the definition of a curve.

Jordan's curves can be intricate, even when they are like the circle, in that they are "simple" (do not cross themselves) and "closed" (have no beginning or end). Jordan's celebrated theorem has meaning. It states that a simple closed curve has an inside and an outside. Its apparent "obviousness" is a deception.

In Italy, Giuseppe Peano caused a sensation when, in 1890, he showed that, according to Jordan's definition, a filled-in square is a curve. He could organize the points on a square so that they could *all* be "traced out" and at the same time conform to Jordan's definition. This was called a space-filling curve and blew a hole in Jordan's definition — clearly a square is not a curve in the conventional sense.

Examples of space-filling curves and other pathological examples caused mathematicians to go back to the drawing board once more and think about the foundations of curve theory. The whole question of developing a better definition of a curve was raised. At the start of the 20th century this task took mathematics into the new field of topology.

83 **Topology**

Topology is the branch of geometry that deals with the properties of surfaces and general shapes but is unconcerned with the measurement of lengths or angles. High on the agenda are qualities which do not change when shapes are transformed into other shapes. We are allowed to push and pull the shape in any direction and for this reason topology is sometimes described as "rubber sheet geometry." Topologists are people who cannot tell the difference between a donut and a coffee cup!

A donut is a surface with a single hole in it. A coffee cup is the same, where the hole takes the form of the handle. Here's how a donut can be transformed into a coffee cup.

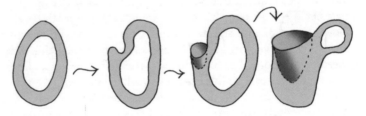

Classifying polyhedra

The most basic shapes studied by topologists are polyhedra ("poly" means "many" and "hedra" means "faces"). An example of a polyhedron is a cube, with 6 square faces, 8 vertices (points at the junction of the faces) and 12 edges (the lines joining the vertices). The cube is a *regular* polyhedron because:

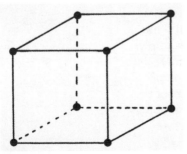

- all the faces are the same regular shape
- all the angles between edges meeting at a vertex are equal

Topology is a relatively new subject, but it can still be traced back to the ancient Greeks, and indeed the culminating result of Euclid's *Elements* is to show that there are *exactly* five regular polyhedra. These are the Platonic solids:

Tetrahedron

Cube

Octahedron

Dodecahedron

Icosahedron

Truncated
icosahedron

- tetrahedron (with 4 triangular faces)
- cube (with 6 square faces)
- octahedron (with 8 triangular faces)
- dodecahedron (with 12 pentagonal faces)
- icosahedron (with 20 triangular faces)

If we drop the condition that each face be the same, we are in the realm of the Archimedean solids which are semi-regular. Examples can be generated from the Platonic solids. If we slice off (truncate) some corners of the icosahedron we have the shape used as the design for the modern soccer ball. The 32 faces that form the panels are made up of 12 pentagons and 20 hexagons. There are 90 edges and 60 vertices. It is also the shape of buckminsterfullerene molecules, named after the visionary Richard Buckminster Fuller, creator of the geodesic dome. These "bucky balls" are a newly discovered form of carbon, C_{60}, with a carbon atom found at each vertex.

Euler's formula

Euler's formula is that the number of vertices V, edges E and faces F, of a polyhedron are connected by the formula

$$V - E + F = 2$$

For example, for a cube, $V = 8$, $E = 12$ and $F = 6$ so $V - E + F = 8 - 12 + 6 = 2$ and, for buckminsterfullerene, $V - E + F = 60 - 90 + 32 = 2$. This theorem actually challenges the very notion of a polyhedron.

If a cube has a "tunnel" through it, is it a real polyhedron? For this shape, $V = 16$, $E = 32$, $F = 16$ and $V - E + F = 16 - 32 + 16 = 0$. Euler's formula does not work. To reclaim the correctness of the formula, the type of polyhedron could be limited to those without tunnels. Alternatively, the formula could be generalized to include this peculiarity.

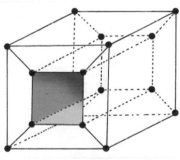

The cube with a
tunnel

Classification of surfaces

A topologist might regard the donut and the coffee cup as identical but what sort of surface is *different* from the donut? A candidate here is the rubber ball. There is no way of transforming the donut into a ball since the donut has a hole but the ball does not. This is a fundamental difference between the two surfaces. So a way of classifying surfaces is by the number of holes they contain.

Let's take a surface with *r* holes and divide it into regions bounded by edges joining vertices planted on the surface. Once this is done, we can count the number of vertices, edges and faces. For any division, the Euler expression $V - E + F$ always has the same value, called the Euler characteristic of the surface:

$$V - E + F = 2 - 2r$$

If the surface has no holes ($r = 0$) as was the case with ordinary polyhedra, the formula reduces to Euler's $V - E + F = 2$. In the case of one hole ($r = 1$), as was the case with the cube with a tunnel, $V - E + F = 0$.

One-sided surfaces

Ordinarily a surface will have two sides. The outside of a ball is different from the inside and the only way to cross from one side to the other is to drill a hole in the ball — a cutting operation which is not allowed in topology (you can stretch but you cannot cut). A piece of paper is another example of a surface with two sides. The only place where one side meets the other side is along the bounding curve formed by the edges of the paper.

Möbius strip

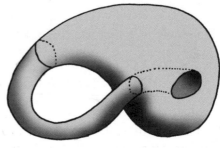

Klein bottle

The idea of a one-sided surface seems far-fetched. Nevertheless, a famous one was discovered by the German mathematician and astronomer August Möbius in the 19th century. The way to construct such a surface is to take a strip of paper, give it one twist and then stick the ends together. The result is a "Möbius strip," a one-sided surface with a boundary curve. You can take your pencil and start drawing a line along its middle. Before long you are back where you started!

It is even possible to have a one-sided surface that does not have a boundary curve. This is the "Klein bottle" named after the German mathematician Felix Klein. What's particularly impressive about this bottle is that it does not intersect itself. However, it is not possible to make a model of the Klein bottle in three-dimensional space without a physical intersection, for it properly lives in four dimensions where it would have no intersections.

Both these surfaces are examples of what topologists call "manifolds" — geometrical surfaces that look like pieces of two-dimensional paper when small portions are viewed by themselves. Since the Klein bottle has no boundary it is called a "closed" 2-manifold.

The Poincaré conjecture

For more than a century, an outstanding problem in topology was the celebrated Poincaré conjecture, named after Henri Poincaré. The conjecture centers on the connection between algebra and topology.

The part of the conjecture that remained unsolved until recently applied to closed 3-manifolds. These can be complicated — imagine a Klein bottle with an extra dimension. Poincaré conjectured that certain closed 3-manifolds which had all the algebraic hallmarks of being three-dimensional spheres actually had to be spheres. It was as if you walked around a giant ball and all the clues you received indicated it was a sphere but because you could not see the big picture you wondered if it really was a sphere.

No one could prove the Poincaré conjecture for 3-manifolds. Was it true or was it false? It had been proven for all other dimensions but the 3-manifold case was obstinate. There were many false proofs, until in 2002 when it was recognized that Grigori Perelman of the Steklov Institute in St. Petersburg had finally proved it. Like the solution to other great problems in mathematics, the solution techniques for the Poincaré conjecture lay outside its immediate area, in a technique related to heat diffusion.

84 **Dimension**

Leonardo da Vinci wrote in his notebook: "The science of painting begins with the point, then comes the line, the plane comes third, and the fourth the body in its vesture of planes." In da Vinci's hierarchy, the point has dimension zero, the line is one-dimensional, the plane is two-dimensional and space is three-dimensional. What could be more obvious? It is the way the point, line, plane and solid geometry had been propagated by the Greek geometer Euclid, and Leonardo was following Euclid's presentation.

That physical space is three-dimensional has been the view for millennia. In physical space we can move *out* of this page along the x-axis, or *across* it horizontally along the y-axis or vertically *up* the z-axis, or any combination of these. Relative to the origin (where the three axes meet) every point has a set of spatial coordinates specified by values of x, y and z and written in the form (x, y, z).

A cube plainly has these three dimensions and so does everything else which has solidity. At school we are normally taught the geometry of the plane which is two-dimensional and we then move up to three dimensions — to "solid geometry" — and stop there.

Around the beginning of the 19th century, mathematicians began to dabble in four dimensions and in even higher n-dimensional mathematics. Many philosophers and mathematicians began to ask whether higher dimensions existed.

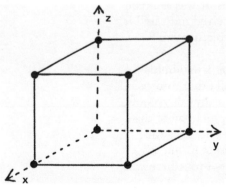

The space of three dimensions

Higher physical dimensions

Many leading mathematicians in the past thought that four dimensions could not be imagined. They queried the reality of four dimensions, and it became a challenge to explain this.

A common way to explain why four dimensions could be possible was to fall back to two dimensions. In 1884, an English schoolmaster and theologian, Edwin Abbott, published a highly popular book about "Flatlanders" who lived in the two-dimensional plane. They could not see triangles, squares or circles which existed in Flatland because they could not go out into the third dimension to view them. Their vision was severely limited. They had the same problems thinking about a third dimension that we do thinking of a fourth. But reading Abbott puts us into the frame of mind to accept the fourth dimension.

The need to contemplate the actual existence of a four-dimensional space became more urgent when Einstein came on the scene. Four-dimensional geometry became more plausible, and even understandable, because the extra dimension in Einstein's model is *time*. Unlike Newton, Einstein conceived time as bound together with space in a four-dimensional space–time continuum. Einstein decreed that we live in a four-dimensional world with four coordinates (x, y, z, t) where t designates time.

Nowadays the four-dimensional Einsteinian world seems quite tame and matter of fact. A more recent model of physical reality is based on "strings." In this theory, the familiar subatomic particles like electrons are the manifestations of extremely tiny vibrating strings. String theory suggests a replacement of the four-dimensional space–time continuum by a higher-dimensional version. Current research suggests that the dimension of the accommodating space–time continuum for string theory should be either 10, 11 or 26, depending on further assumptions and differing points of view.

A huge 2,000-tonne magnet at CERN near Geneva, Switzerland, designed to engineer collisions of particles at high speeds, might help to resolve the issue. It is intended to uncover the structure of matter and, as a by-product, may point to a better theory and the "correct" answer on dimensionality. The smart money seems to be on the theory that we're living in an 11-dimensional universe.

Hyperspace

Unlike higher physical dimensions, there is absolutely no problem with a *mathematical space* of more than three dimensions. Mathematical space can be any number of dimensions. Since the early 19th century mathematicians have habitually used n variables in their work. George Green, a miller from Nottingham, England, who explored the mathematics of electricity, and pure mathematicians A.L. Cauchy, Arthur Cayley and Hermann Grassmann, all described their mathematics in terms of n-dimensional hyperspace. There seemed no good reason to limit the mathematics and everything to be gained in elegance and clarity.

The idea behind n dimensions is merely an extension of three-dimensional coordinates (x, y, z) to an unspecified number of variables. A circle in two dimensions has an equation $x^2 + y^2 = 1$, a sphere in three dimensions has an equation $x^2 + y^2 + z^2 = 1$, so why not a hypersphere in four dimensions with equation $x^2 + y^2 + z^2 + w^2 = 1$.

The eight corners of a cube in three dimensions have coordinates of the form (x, y, z) where each of the x, y, z are either 0 or 1. The cube has six faces, each of which is a square, and there are $2 \times 2 \times 2 = 8$ corners. What about a four-dimensional cube? It will have coordinates of the form (x, y, z, w) where each of the x, y, z and w are either 0 or 1. So there are $2 \times 2 \times 2 \times 2 = 16$ possible corners for the four-dimensional cube, and eight faces, each of which is a cube. We cannot actually see this four-dimensional cube but we can create an artist's impression of it on this sheet of paper. This shows a projection of the four-dimensional cube which exists in the mathematician's imagination. The cubic faces can just about be perceived.

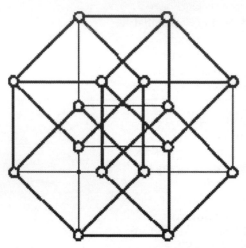

The four-dimensional cube

A mathematical space of many dimensions is quite a common occurrence for pure mathematicians. No claim is made for its actual existence though it may be assumed to exist in an ideal Platonic world. In the great problem of the classification of groups, for instance (see Chapter 93), the "monster group" is a way of measuring symmetry in a mathematical space of 196,883 dimensions. We cannot "see" this space in the same way as we can in the ordinary three-dimensional space, but it can still be imagined and dealt with in a precise way by modern algebra.

The mathematician's concern for dimension is entirely separate from the meaning the physicist attaches to dimensional analysis. The common units of physics are measured in terms of mass (M), length (L), and time (T). So, using their dimensional analysis a physicist can check whether equations make sense since both sides of an equation must have the same dimensions.

It is no good having force = velocity. A dimensional analysis gives velocity as meters per second so it has dimension of length divided by time or L/T, which we write as LT^{-1}. Force is mass times acceleration, and as acceleration is meters per second per second, the net result is that force will have dimensions MLT^{-2}.

Coordinated people

Human beings themselves are many-dimensioned things.
A human being has many more "coordinates" than three. We could use
(*a, b, c, d, e, f, g, h*), for age, height, weight, gender, shoe size, eye color,
hair color, nationality and so on. In place of geometrical points we might
have people. If we limit ourselves to this eight-dimensional "space" of
people, John Doe might have coordinates like (43 years, 165 cm, 83 kg,
male, 9, blue, blond, Danish) and Mary Smith's coordinates might be
(26 years, 157 cm, 56 kg, female, 4, brown, brunette, British).

Topology

Dimension theory is part of general topology. Other concepts of
dimension can be defined independently in terms of abstract
mathematical spaces. A major task is to show how they relate to each
other. Leading figures in many branches of mathematics have delved
into the meaning of dimension, including Henri Lebesgue, L.E.J.
Brouwer, Karl Menger, Paul Urysohn and Leopold Vietoris (who died
in 2002 aged 110).

   The pivotal book on the subject was *Dimension Theory*. Published
in 1948 by Witold Hurewicz and Henry Wallman — it is still seen as
a watershed in our understanding of the concept of dimension.

Dimension in all its forms

From the three dimensions introduced by the ancient Greeks the
concept of dimension has been critically analyzed and extended.

   The *n* dimensions of mathematical space were introduced quite
painlessly, while physicists have based theories on space–time (of
dimension four) and recent versions of string theory which demand,
10, 11 and 26 dimensions. There have been forays into fractional
dimensions with fractal shapes (see Chapter 85) with several different
measures being studied. Hilbert introduced an infinite-dimensional
mathematical space that is now a basic framework for pure
mathematicians. Dimension is so much more than the one, two, three
of Euclidean geometry.

85 **Fractals**

In March 1980, the state-of-the-art mainframe computer at the IBM research center at Yorktown Heights, New York State, was issuing its instructions to an ancient Tektronix printing device. It dutifully struck dots in curious places on a white page, and when it had stopped its clatter the result looked like a handful of dust smudged across the sheet. Benoît Mandelbrot rubbed his eyes in disbelief. He saw it was important, but what was it? The image that slowly appeared before him was like the black and white print emerging from a photographic developing bath. It was a first glimpse of that icon in the world of fractals — the Mandelbrot set.

This was experimental mathematics par excellence, an approach to the subject in which mathematicians had their laboratory benches just like the physicists and chemists. They too could now do experiments. New vistas opened up — literally. It was a liberation from the arid climes of "definition, theorem, proof," though a return to the rigors of rational argument would have to come later.

The downside of this experimental approach was that the visual images preceded a theoretical underpinning. Experimentalists were navigating without a map. Although Mandelbrot coined the word "fractals," what were they? Could there be a precise definition for them in the usual way of mathematics? In the beginning, Mandelbrot didn't want to do this. He didn't want to destroy the magic of the experience by honing a sharp definition which might be inadequate and limiting. He felt the notion of a fractal, "like a good wine — demanded a bit of aging before being 'bottled.'"

The Mandelbrot set

Mandelbrot and his colleagues were not being particularly abstruse mathematicians. They were playing with the simplest of formulae. The whole idea is based on iteration — the practice of applying a formula time and time again. The formula which generated the Mandelbrot set was simply $x^2 + c$.

The first thing we do is choose a value of c. Let's choose $c = 0.5$. Starting with $x = 0$ we substitute into the formula $x^2 + 0.5$. This first calculation gives 0.5 again. We now use this as x, substituting it into $x^2 + 0.5$ to give a second calculation: $(0.5)^2 + 0.5 = 0.75$. We keep going, and at the third stage this will be $(0.75)^2 + 0.5 = 1.0625$. All these calculations can be done on a handheld calculator. Carrying on, we find that the answer gets bigger and bigger.

Let's try another value of c, this time $c = -0.5$. As before we start at $x = 0$ and substitute it into $x^2 - 0.5$ to give -0.5. Carrying on, we get -0.25, but this time the values do not become bigger and bigger but, after some oscillations, settle down to a figure near $-0.3660\ldots$

So by choosing $c = 0.5$ the sequence starting at $x = 0$ zooms off to infinity, but by choosing $c = -0.5$ we find that the sequence starting at $x = 0$ actually converges to a value near -0.3660. The Mandelbrot set consists of all those values of c for which the sequence starting at $x = 0$ does *not* escape to infinity.

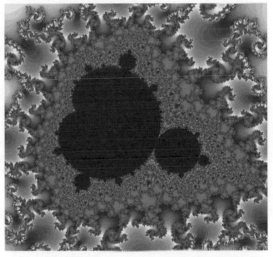

This is not the whole story because so far we have only considered the one-dimensional real numbers — giving a one-dimensional Mandelbrot set, so we wouldn't see much. What needs to be considered is the same formula $z^2 + c$ but with z and c as two-dimensional complex numbers (see Chapter 71). This will give us a two-dimensional Mandelbrot set.

For some values of c in the Mandelbrot set, the sequence of zs may do all sorts of strange things, like dance between a number of points, but they will not escape to infinity. In the Mandelbrot

The Mandelbrot set

set we see another key property of fractals, that of self-similarity. If you zoom into the set you will not be sure of the level of magnification because you will just see more Mandelbrot sets.

Before Mandelbrot

Like most things in mathematics, discoveries are rarely brand new. Looking into the history, Mandelbrot found that mathematicians such as Henri Poincaré and Arthur Cayley had brief glimmerings of the idea a hundred years before him. Unfortunately they did not have the computing power to investigate matters further.

The shapes discovered by the first wave of fractal theorists included crinkly curves and the "monster curves" that had previously been dismissed as pathological examples of curves. As they were so pathological they had been locked up in the mathematician's cupboard and given little attention. What was wanted then were the more normal "smooth" curves which could be dealt with by differential calculus. With the popularity of fractals, other mathematicians whose work was resurrected were Gaston Julia and Pierre Fatou, who worked on fractal-like structures in the complex plane in the years following the First World War. Their curves were not called fractals, of course, and they did not have the technological equipment to see their shapes.

Other famous fractals

The famous Koch curve is named after the Swedish mathematician Niels Fabian Helge von Koch. The snowflake curve is practically the first fractal curve. It is generated from the side of the triangle treated as an element, splitting it into three parts each of length ⅓ and adding a triangle in the middle position.

The curious property of the Koch curve is that it has a finite area, because it always stays within a circle, but at each stage of its generation its length increases. It is a curve which encloses a finite area but has an "infinite" circumference!

Another famous fractal is named after the Polish mathematician Wacław Sierpiński. It is found by subtracting triangles from an equilateral triangle; and by continuing this process, we get the Sierpiński gasket (generated by a different process in Chapter 76).

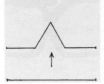

The generating element of the Koch snowflake

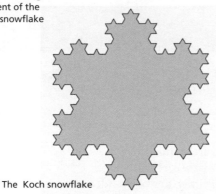

The Koch snowflake

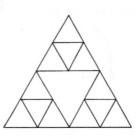

The Sierpiński gasket

Fractional dimension

The way Felix Hausdorff looked at dimension was innovative. It has to do with scale. If a line is scaled up by a factor of 3 it is 3 times longer than it was previously. Since this $3 = 3^1$ a line is said to have dimension 1. If a solid square is scaled up by a factor of 3 its area is 9 times its previous value or 3^2 and so the dimension is 2. If a cube is scaled up by this factor its volume is 27 or 3^3 times its previous value, so its dimension is 3. These values of the Hausdorff dimension all coincide with our expectations for a line, square, or cube.

If the basic unit of the Koch curve is scaled up by 3, it becomes 4 times longer than it was before. Following the scheme described, the Hausdorff dimension is the value of D for which $4 = 3^D$. An alternative calculation is that

$$D = \frac{\log 4}{\log 3}$$

which means that D for the Koch curve is approximately 1.262. With fractals it is frequently the case that the Hausdorff dimension is greater than the ordinary dimension, which is 1 in the case of the Koch curve.

The Hausdorff dimension informed Mandelbrot's definition of a fractal — a set of points whose value of D is not a whole number. Fractional dimension became the key property of fractals.

The applications of fractals

The potential for the applications of fractals is wide. Fractals could well be the mathematical medium which models such natural objects as plant growth, or cloud formation.

Fractals have already been applied to the growth of marine organisms such as corals and sponges. The spread of modern cities has been shown to have a similarity with fractal growth. In medicine they have found application in the modeling of brain activity. And the fractal nature of movements of stocks and shares and the foreign exchange markets has also been investigated. Mandelbrot's work opened up a new vista and there is much still to be discovered.

86 **Chaos**

How is it possible to have a theory of chaos? Surely chaos happens in the absence of theory? The story goes back to 1812. While Napoleon was advancing on Moscow, his compatriot the Marquis Pierre-Simon de Laplace published an essay on the deterministic universe: if at one particular instant, the positions and velocities of all objects in the universe were known, and the forces acting on them, then these quantities could be calculated exactly for all future times. The universe and all objects in it would be completely determined. Chaos theory shows us that the world is more intricate that that.

In the real world we cannot know all the positions, velocities and forces exactly, but the corollary to Laplace's belief was that if we knew approximate values at one instant, the universe would not be much different anyway. This was reasonable, for surely sprinters who started a tenth of a second after the gun had fired would break the tape only a tenth of a second off their usual time. The belief was that small discrepancies in initial conditions meant small discrepancies in outcomes. Chaos theory exploded this idea.

Pinboard box experiment

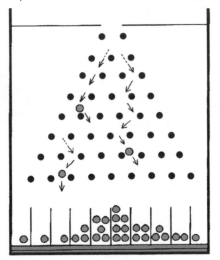

The butterfly effect

The butterfly effect shows how initial conditions slightly different from the given ones can produce an actual result very different from the predictions. If fine weather is predicted for a day in Europe but a butterfly flaps its wings in South America then this could actually presage storms on the other side of the world — because the flapping of the wings changes the air pressure very slightly causing a weather pattern completely different from the one originally forecast.

We can illustrate the idea with a simple mechanical experiment. If you drop a ball-bearing through the opening in the top of a pinboard box it will progress downwards, being deflected one way or the other by the different pins it encounters on route until it reaches a finishing slot at the bottom. You might then attempt to let another identical ball-bearing go from the very same position with exactly the same velocity. If you could do this *exactly* then the Marquis de Laplace would be correct and the path followed by the ball would be exactly the same. If the first ball dropped into the third slot from the right, then so would the second ball.

But of course you cannot let the ball go from exactly the same position with exactly the same velocity and force. In reality, there will be a very slight difference which you might not even be able to measure. The result is that the ball-bearing may take a very different route to the bottom and probably end up in a different slot.

A simple pendulum

The free pendulum is one of the simplest mechanical systems to analyze. As the pendulum swings back and forth, it gradually loses energy. The displacement from the vertical and the (angular) velocity of the bob decrease until it is eventually stationary.

The movement of the bob can be plotted in a phase diagram. On the horizontal axis the (angular) displacement is measured and on the vertical axis the velocity is measured. The point of release is plotted at the point A on the positive horizontal axis. At A the displacement is at a maximum and the velocity is zero. As the bob moves through the vertical axis (where the displacement is zero) the velocity is at a maximum, and this is plotted on the phase diagram at B. At C, when the bob is at the other extremity of its swing, the displacement is negative and the velocity is zero. The bob then swings back through D (where it is moving in the opposite direction so its velocity is negative) and completes one swing at E. In the phase diagram this is represented by a rotation through 360 degrees, but because the swing is reduced the point E is shown *inside* A. As the pendulum swings less and less this phase portrait spirals into the origin until eventually the pendulum comes to rest. This is not the case for the double pendulum in which the bob is at the end of a jointed pair of rods. If the displacement is small the motion of the double pendulum is similar to the simple pendulum, but if the displacement is large the bob swings, rotates and lurches about and the displacement about the intermediate joint is seemingly random. If the motion is not forced, the bob will also come to rest but the curve that describes its motion is far from the well-behaved spiral of the single pendulum.

The free pendulum

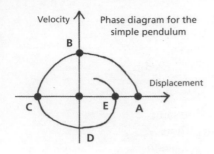

Velocity

Phase diagram for the simple pendulum

B

Displacement

C   E   A

D

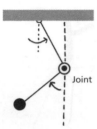

Movement of double pendulum

Joint

Chaotic motion

The characteristic of chaos is that a deterministic system may appear to generate random behavior. Let's look at another example, the repeating, or iterative, formula $a \times p \times (1 - p)$ where p stands for the population, measured as a proportion on a scale from 0 to 1. The value of a must be somewhere between 0 and 4 to guarantee that the value of p stays in the range from 0 to 1.

Let's model the population when $a = 2$. If we pick a starting value of, say, $p = 0.3$ at *time* = 0, then to find the population at *time* = 1, we feed $p = 0.3$ into $a \times p \times (1 - p)$ to give 0.42. Using only a handheld calculator we can repeat this operation, this time with $p = 0.42$, to give us the next figure (0.4872). Progressing in this way, we find the population at later times. In this case, the population quickly settles down to $p = 0.5$. This settling down always takes place for values of a less than 3.

If we now choose $a = 3.9$, a value near the maximum permissible, and use the same initial population $p = 0.3$, the population does *not* settle down but oscillates wildly. This is because the value of a is in the "chaotic region," that is, a is a number greater than 3.57. Moreover, if we choose a different initial population, $p = 0.29$, a value close to 0.3, the population growth shadows the previous growth pattern for the first few steps but then starts to diverge from it completely. This is the behavior experienced by Edward Lorenz in 1961 (see box).

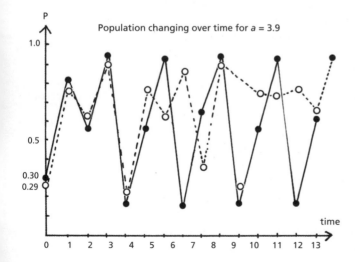

P

Population changing over time for a = 3.9

1.0

0.5

0.30
0.29

time

0   1   2   3   4   5   6   7   8   9   10   11   12   13

Forecasting the weather

Even with very powerful computers we all know that we cannot forecast the weather more than a few days in advance. Over just a few days forecasting the weather still gives us nasty surprises. This is because the equations which govern the weather are nonlinear — they involve the variables multiplied together, not just the variables themselves.

The theory behind the mathematics of weather forecasting was worked out independently by the French engineer Claude Navier

From meteorology to mathematics

The discovery of the butterfly effect happened by chance around 1961. When meteorologist Edward Lorenz at MIT went to have a cup of coffee and left his ancient computer plotting away he came back to something unexpected. He had been aiming to recapture some interesting weather plots but found the new graph unrecognizable. This was strange for he had entered in the same initial values and the same picture should have been drawn out. Was it time to trade in his old computer and get something more reliable?

After some thought he did spot a difference in the way he had entered the initial values: before he had used six decimal places but on the rerun he only bothered with three. To explain the disparity he coined the term "butterfly effect." After this discovery his intellectual interests migrated to mathematics.

in 1821 and the British mathematical physicist George Gabriel Stokes in 1845. The Navier–Stokes equations that resulted are of intense interest to scientists. The Clay Mathematics Institute in Cambridge, Massachusetts has offered a million-dollar prize to whoever makes substantial progress towards a mathematical theory that unlocks their secrets. Applied to the problem of fluid flow, much is known about the steady movements of the upper atmosphere. But air flow near the surface of the Earth creates turbulence and chaos results, with the subsequent behavior largely unknown.

While a lot is known about the theory of linear systems of equations, the Navier–Stokes equations contain nonlinear terms which make them intractable. Practically the only way of solving them is to do so numerically by using powerful computers.

Strange attractors

Dynamic systems can be thought of possessing "attractors" in their phase diagrams. In the case of the simple pendulum the attractor is the single point at the origin that the motion is directed towards. With the double pendulum it's more complicated, but even here the phase portrait will display some regularity and be attracted to a set of points in the phase diagram. For systems like this the set of points may form a fractal (see Chapter 85) which is called a "strange" attractor that will have a definite mathematical structure. So all is not lost. In the new chaos theory, it is not so much "chaotic" chaos that results as "regular" chaos.

87 The parallel postulate

This dramatic story begins with a simple geometric scenario. Imagine a line *l* and a point *P* not on the line. How many lines can we draw through *P* parallel to the line *l*? It appears obvious that there is exactly one line through *P* which will never meet *l* no matter how far it is extended in either direction. This seems self-evident and in perfect agreement with common sense. Euclid of Alexandria included a variant of it as one of his postulates in that foundation of geometry, the *Elements*.

P
•

————————————————— *l*

Common sense is not always a reliable guide. We shall see whether Euclid's assumption makes mathematical sense.

Euclid's *Elements*

Euclid's geometry is set out in the 13 books of the *Elements*, written around 300 B.C. One of the most influential mathematics texts ever written, Greek mathematicians constantly referred to it as the first systematic codification of geometry. Later scholars studied and translated it from extant manuscripts and it was handed down and universally praised as the very model of what geometry should be.

The *Elements* percolated down to school level and readings from the "sacred book" became the way geometry was taught. It proved unsuitable for the youngest pupils, however. As the poet A.C. Hilton quipped: "though they wrote it all by rote, they did not write it right." You might say Euclid was written for men not boys. In English schools, it reached the zenith of its influence as a subject in the curriculum during the 19th century but it remains a touchstone for mathematicians today.

Euclid's postulates

One of the characteristics of mathematics is that a few assumptions can generate extensive theories. Euclid's postulates are an excellent example, and one that set the model for later axiomatic systems. His five postulates are:
1. A straight line can be drawn from any point to any point.
2. A finite straight line can be extended continuously in a straight line.
3. A circle can be constructed with any center and any radius.
4. All right angles are equal to each other.
5. If a straight line falling on two straight lines makes the interior angles on the same side less than two right angles, the two straight lines, if extended indefinitely, meet on that side on which the angles are less than two right angles.

It is the style of Euclid's *Elements* that makes it noteworthy – its achievement is the presentation of geometry as a sequence of proven propositions. Sherlock Holmes would have admired its deductive system, which advanced logically from the clearly stated postulates, and may have castigated Dr. Watson for not seeing it as a "cold unemotional system."

While the edifice of Euclid's geometry rests on the postulates (what are now called axioms; see box), these were not enough. Euclid added "definitions" and "common notions." The definitions include such declarations as "a *point* is that which has no part" and "a *line* is breadthless length." Common notions include such items as "the whole is greater than the part" and "things which are equal to the same thing are also equal to one another." It was only towards the end of the 19th century that it was recognized that Euclid had made tacit assumptions.

The fifth postulate

It is Euclid's fifth postulate that caused controversy over 2,000 years after the *Elements* first appeared. In style alone, it looks out of place through its wordiness and clumsiness. Euclid himself was unhappy with it but he needed it to prove propositions and had to include it. He tried to prove it from the other postulates but failed.

Later mathematicians either tried to prove it or replace it by a simpler postulate. In 1795, John Playfair stated it in a form which gained popularity: for a line *l* and a point *P* not on the line *l* there is a *unique* line passing through *P* parallel to *l*. Around the same time, Adrien Marie Legendre substituted another equivalent version when he asserted the existence of a triangle whose angles add up to 180 degrees. These new forms of the fifth postulate went some way to meet the objection of artificiality. They were more acceptable than the cumbersome version given by Euclid.

Another line of attack was to search for the elusive proof of the fifth postulate. This exerted a powerful attraction on its adherents. If a proof could be found, the postulate would become a theorem and it could retire from the firing line. Unfortunately attempts to do this turned out to be excellent examples of circular reasoning, arguments which assume the very thing they are trying to prove.

Non-Euclidean geometry

A breakthrough came through the work of Carl Friedrich Gauss, János Bolyai and Nikolai Ivanovich Lobachevsky. Gauss did not publish his work, but it seems clear he reached his conclusions in 1817. Bolyai published in 1831 and Lobachevsky, independently, in 1829, causing a priority dispute between these two. There is no doubting the brilliance of all these men. They effectively showed that the fifth postulate was independent of the other four postulates. By adding its negation to the other four postulates, they showed a consistent system was possible.

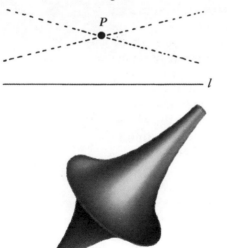

Bolyai and Lobachevsky constructed a new geometry by allowing there to be more than one line through *P* that does not meet the line *l*. How can this be? Surely the dotted lines meet *l*. If we accept this we are unconsciously falling in with Euclid's view. The diagram is therefore a confidence trick, for what Bolyai and Lobachevsky were proposing was a new sort of geometry which does not conform to the commonsense one of Euclid. In fact, their non-Euclidean geometry can be thought of as the geometry on the curved surface of what is known as a "pseudosphere".

The shortest paths between the points on a pseudosphere play the same role as straight lines in Euclid's geometry. One of the curiosities of this non-Euclidean geometry is that the sum of the angles in a triangle is less than 180 degrees. This geometry is called hyperbolic geometry.

Another alternative to the fifth postulate states that every line through *P* meets the line *l*. Put a different way, there are no lines through *P* which are "parallel" to *l*. This geometry is different from Bolyai's and Lobachevsky's, but it is a genuine geometry nevertheless. One model for it is the geometry on the surface of a sphere. Here the great circles (those circles that have the same circumference as the sphere itself) play the role of straight lines in Euclidean geometry. In this non-Euclidean geometry the sum of the angles in a triangle is greater than 180 degrees. This is called elliptic geometry and is associated with the German mathematician Benhard Riemann who investigated it in the 1850s.

The geometry of Euclid which had been thought to be the one true geometry — according to Immanuel Kant, the geometry "hard-wired into man" — had been knocked off its pedestal. Euclidean geometry was now one of many systems, sandwiched between hyperbolic and elliptic geometry. The different versions were unified under one umbrella by Felix Klein in 1872. The advent of non-Euclidean geometry was an earth-shaking event in mathematics and paved the way to the geometry of Einstein's general relativity (see Chapter 58). It is the general theory of relativity which demands a new kind of geometry — the geometry of curved space–time, or Riemannian geometry. It was this non-Euclidean geometry which now explained why things fall down, and not Newton's attractive gravitational force between objects. The presence of massive objects in space, like the Earth and the Sun cause space–time to be curved. A marble on a sheet of thin rubber will cause a small indentation, but try placing a bowling ball on it and a great warp will result.

This curvature measured by Riemannian geometry predicts how light beams bend in the presence of massive space objects. Ordinary Euclidean space, with time as an independent component, will not suffice for general relativity. One reason is that Euclidean space is flat — there is no curvature. Think of a sheet of paper lying on a table; we can say that at any point on the paper the curvature is zero. Underlying Riemannian space–time is a concept of curvature which varies continuously — just as the curvature of a rumpled piece of cloth varies from point to point. It's like looking in a bendy fairground mirror — the image you see depends on where you look in the mirror.

No wonder that Gauss was so impressed by young Riemann in the 1850s and even suggested then that the "metaphysics" of space would be revolutionized by his insights.

88 Graphs

There are two types of graphs in mathematics. At school we draw curves which show the relationship between variables *x* and *y*. In the other more recent sort, dots are joined up by wiggly lines.

Königsberg, the former capital of East Prussia, is famous for the seven bridges which cross the River Pregel. Home to the illustrious philosopher Immanuel Kant, the city and its bridges are also linked with the famous mathematician Leonhard Euler.

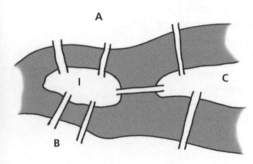

In the 18th century a curious question was posed: was it possible to set off and walk around Königsberg crossing each bridge exactly once? The walk does not require us to finish where we started — only that we cross each bridge once.

In 1735, Euler presented his solution to the Russian Academy, a solution which is now seen as the beginning of modern graph theory. In our semi-abstract diagram, the island in the middle of the river is labeled *I* and the banks of the river by *A*, *B* and *C*. Can you plan a walk for a Sunday afternoon that crosses each bridge just once? Pick up a pencil and try it. The key step is to peel away the semi-abstractness and progress to complete abstraction. In so doing a graph of points and lines is obtained. The land is represented by "points" and the bridges joining them are represented by "lines." We don't care that the lines are not straight or that they differ in length. These things are unimportant. It is only the connections that matter.

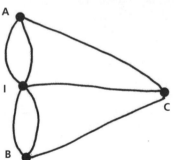

Euler made an observation about a successful walk. Apart from the beginning and the end of the walk, every time a bridge is crossed onto a piece of land it must be possible to leave it on a bridge not previously walked over. Translating this thought into the abstract picture, we may say that lines meeting at a point must occur in pairs. Apart from two points representing the start and finish of the walk, the bridges can be traversed if and only if each point has an even number of lines incident on it.

The number of lines meeting at a point is called the "degree" of the point.

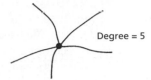

Degree = 5

Euler's theorem states that

The bridges of a town or city may be traversed exactly once if, apart from at most two, all points have even degree.

Looking at the graph representing Königsberg, every point is of odd degree. This means that a walk crossing each bridge only once is not possible in Königsberg. If the bridge set up were changed then such a walk may become possible. If an extra bridge were built between the island I and C the degrees at I and C would both be even. This means we could begin a walk on A and end on B having walked over every bridge exactly once. If yet another bridge were built, this time between A and B (shown right), we could start anywhere and finish at the *same* place because every point would have even degree in this case.

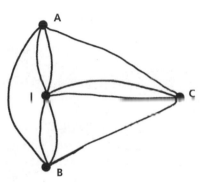

The hand-shaking theorem

If we were asked to draw a graph that contained three points of odd degree, we would have a problem. Try it. It cannot be done because

In any graph the number of points with odd degree must be an even number.

This is the hand-shaking theorem — the first theorem of graph theory. In any graph every line has a beginning and an end, or in other words it takes two people to shake hands. If we add up the degrees of every point for the whole graph we must get an even number, say N. Next we say there are x points with odd degree and y points with even degree. Adding all the degrees of the odd points together we'll have N_x and adding all the degrees of the even points will give us N_y, which is even. So we have $N_x + N_y = N$, and therefore $N_x = N - N_y$. It follows that N_x is even. But x itself cannot be odd because the addition of an odd number of odd degrees would be an odd number. So it follows that x must be even.

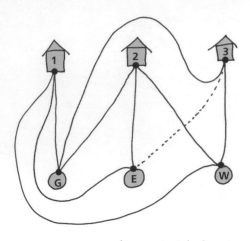

Non-planar graphs

The utilities problem is an old puzzle. Imagine three houses and three utilities — gas, electricity and water. We have to connect each of the houses to each of the utilities, but there's a catch — the connections must not cross.

In fact this cannot be done — but you might try it out on your unsuspecting friends. The graph described by connecting three points to another three points in all possible ways (with only nine lines) cannot be drawn in the plane without crossings. Such a graph is called non-planar. This utilities graph, along with the graph made by all lines connecting five points, has a special place in graph theory. In 1930, the Polish mathematician Kazimierz Kuratowski proved the startling theorem that a graph is planar if and only if it does not contain either one of these two as a subgraph, a smaller graph contained within the main one.

Trees

A "tree" is a particular kind of graph, very different from the utilities graph or the Königsberg graph. In the Königsberg bridge problem there were opportunities for starting at a point and returning to it via a different route. Such a route from a point and back to itself is called a cycle. A tree is a graph which has no cycles.

A familiar example of a tree graph is the way directories are arranged in computers. They are arranged in a hierarchy with a root directory and subdirectories leading off it. Because there are no cycles there is no way to cross from one branch other than through the root directory — a familiar maneuver for computer users.

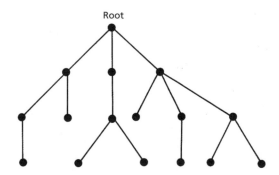

Counting trees

How many different trees can be made from a specific number of points? The problem of counting trees was tackled by the 19th century British mathematician Arthur Cayley. For example, there are exactly three different tree types with five points:

Cayley was able to count the number of different tree types for small numbers of points. He was able to go as far as trees with fewer than 14 points before the sheer computational complexity was too much for a man without a computer. Since then the calculations have advanced as far as trees with as many as 22 points. There are millions of possible types for these.

Even in its own time, Cayley's research had practical applications. Counting trees is relevant in chemistry, where the distinctiveness of some compounds depends on the way atoms are arranged in their molecules. Compounds with the same number of atoms but with different arrangements have different chemical properties. Using his analysis it was possible to predict the existence of chemicals "at the tip of his pen" that were subsequently found in the laboratory.

89 Probability

What is the chance of it snowing tomorrow? What is the likelihood that I will catch the early train? What is the probability of you winning the lottery? Probability, likelihood, and chance are all words we use every day when we want to know the answers. They are also the words of the mathematical theory of probability.

Probability theory is important. It has a bearing on uncertainty and is an essential ingredient in evaluating risk. But how can a theory involving uncertainty be quantified? After all, isn't mathematics an exact science?

The real problem is to *quantify* probability.

Suppose we take the simplest example on the planet, the tossing of a coin. What is the probability of getting a head? We might rush in and say the answer is ½ (sometimes expressed as 0.5 or 50 percent). Looking at the coin we make the assumption it is a fair coin, which means that the chance of getting a head equals the chance of getting a tail, and therefore the probability of a head is ½.

Situations involving coins, balls in boxes, and "mechanical" examples are relatively straightforward. There are two main theories in the assignment of probabilities. Looking at the two-sided symmetry of the coin provides one approach. Another is the relative frequency approach, where we conduct the experiment a large number of times and count the number of heads. But how large is large? It is easy to believe that the number of heads relative to the number of tails is roughly 50:50 but it might be that this proportion would change if we continued the experiment.

But what about coming to a sensible measure of the probability of it snowing tomorrow? There will again be two outcomes: either it snows or it does not snow, but it is not at all clear that they are equally likely as it was for the coin. An evaluation of the probability of it snowing tomorrow will have to take into account the weather conditions at the time and a host of other factors. But even then it is not possible to pinpoint an exact number for this probability. Though we may not come to an actual number, we can usefully ascribe a "degree of belief" that the probability will be low, medium or high.

In mathematics, probability is measured on a scale from 0 to 1. The probability of an impossible event is 0 and a certainty is 1. A probability of 0.1 would mean a low probability while 0.9 would signify a high probability.

Origins of probability

The mathematical theory of probability came to the fore in the 17th century with discussions on gambling problems between Blaise Pascal, Pierre de Fermat and Antoine Gombaud (also known as the Chevalier de Méré). They found a simple game puzzling. The Chevalier de Méré's question is this: which is more likely, rolling a "six" on four throws of a dice, or rolling a "double six" on 24 throws with two dice? Which option would you put your shirt on?

The prevailing wisdom of the time thought the better option was to bet on the double six because of the many more throws allowed. This view was shattered when the probabilities were analyzed. Here is how the calculations go:

Throw one dice: the probability of not getting a six on a single throw is $5/6$, and in four throws the probability of this would be $5/6 \times 5/6 \times 5/6 \times 5/6$ which is $(5/6)^4$. Because the results of the throws do not affect each other, they are "independent" and we can multiply the probabilities. The probability of at least one six is therefore

$$1 - (5/6)^4 = 0.517746 \ldots$$

Throw two dice: the probability of not getting a double six in one throw is $35/36$ and in 24 throws this has the probability $(35/36)^{24}$.

The probability of at least one double six is therefore

$$1 - (35/36)^{24} = 0.491404 \ldots$$

We can take this example a little further.

Playing craps

The two dice example is the basis of the modern game of craps played in casinos and online betting. When two distinguishable dice (red and blue) are thrown there are 36 possible outcomes and these may be recorded as pairs (x,y) and displayed as 36 dots against a set of x/y axes — this is called the "sample space."

Let's consider the "event" A of getting the sum of the dice to add up to 7. There are 6 combinations that each add up to 7, so we can describe the event by

$$A = \{(1,6), (2,5), (3,4), (4,3), (5,2), (6,1)\}$$

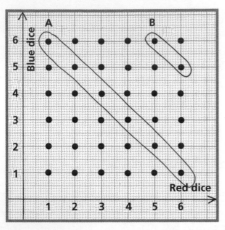

Sample space
(for 2 dice)

and ring it on the diagram. The probability of A is 6 chances in 36, which can be written $\Pr(A) = 6/36 = 1/6$. If we let B be the event of getting the sum on the dice equal to 11 we have the event $B = \{(5,6), (6,5)\}$ and $\Pr(B) = 2/36 = 1/18$.

In the dice game craps, in which two dice are thrown on a table, you can win or lose at the first stage, but for some scores all is not lost and you can go onto a second stage. You win at the first throw if either the event A or B occurs — this is called a "natural." The probability of a natural is obtained by adding the individual probabilities, $6/36 + 2/36 = 8/36$. You lose at the first stage if you throw a 2, 3 or a 12 (this is called "craps"). A calculation like the one above gives the probability of losing at the first stage as $4/36$. If a sum of either 4, 5, 6, 8, 9 or 10 is thrown, you go onto a second stage and the probability of doing this is $24/36 = 2/3$.

In the gaming world of casinos the probabilities are written as odds. In craps, for every 36 games you play, on average you will win at the first throw 8 times and not win 28 times so the odds against winning on the first throw are 28 to 8, which is the same as 3.5 to 1.

The monkey on a typewriter

Alfred is a monkey who lives in the local zoo. He has a battered old typewriter with 26 keys for the letters of the alphabet, a key for a full stop, one for a comma, one for a question mark and one for a space — 30 keys in all. He sits in a corner filled with literary ambition, but his method of writing is curious — he hits the keys at random.

Any sequence of letters typed will have a nonzero chance of occurring, so there is a chance he will type out the plays of Shakespeare word perfect. More than this, there is a chance (albeit smaller) he will follow this with a translation into French, and then Spanish, and then German. For good measure we could allow for the possibility of him continuing on with the poems of William Wordsworth. The chance of all this is minute, but it is certainly not zero. This is the key point.

Let's see how long he will take to type the soliloquy in *Hamlet*, starting off with the opening "To be or." We imagine 8 boxes which will hold the 8 letters including the spaces.

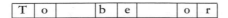

T	o		b	e		o	r

The number of possibilities for the first position is 30, for the second is 30, and so on. So the number of ways of filling out the 8 boxes is $30 \times 30 \times 30 \times 30 \times 30 \times 30 \times 30 \times 30$. The chance of Alfred getting as far as "To be or" is 1 chance in 6.561×10^{11}. If Alfred hits the typewriter once every second there is an expectation he will have typed "To be or" in about 20,000 years, and proved himself a particularly long-lived primate. So don't hold your breath waiting for the whole works of Shakespeare. Alfred will produce nonsense like "xo,h?yt?" for a great deal of the time.

How has the theory developed?

When probability theory is applied the results can be controversial, but at least the mathematical underpinnings are reasonably secure. In 1933, Andrey Nikolaevich Kolmogorov was instrumental in defining probability on an axiomatic basis — much like the way the principles of geometry were defined two millennia before.

Probability is defined by the following axioms:

1. the probability of all occurrences is 1.
2. probability has a value which is greater than or equal to zero.
3. when occurrences cannot coincide their probabilities can be added.

From these axioms, dressed in technical language, the mathematical properties of probability can be deduced. The concept of probability can be widely applied. Much of modern life cannot do without it. Risk analysis, sport, sociology, psychology, engineering design, finance and so on — the list is endless. Who'd have thought the gambling problems that kick-started these ideas in the 17th century would spawn such an enormous discipline? What were the chances of that?

90 **Distributions**

Ladislaus J. Bortkiewicz was fascinated by mortality tables. Not for him a gloomy topic, they were a field of enduring scientific enquiry. He famously counted the number of cavalrymen in the Prussian army that had been killed by horse-kicks. Then there was Frank Benford, an electrical engineer who counted the first digits of different types of numerical data to see how many were ones, twos and so on. And George Kingsley Zipf, who taught German at Harvard, had an interest in philology and analyzed the occurrences of words in pieces of text.

All these examples involve measuring the probabilities of events. What are the probabilities of x cavalrymen in a year receiving a lethal kick from a horse? Listing the probabilities for each value of x is called a distribution of probabilities, or for short, a probability distribution. It is also a *discrete* distribution because the values of x only take isolated values — there are gaps between the values of interest. You can have three or four Prussian cavalrymen struck down by a lethal horse-kick but not $3\frac{1}{2}$. As we'll see, in the case of the Benford distribution we are only interested in the appearance of digits 1, 2, 3, . . . and, for the Zipf distribution, you may have the word "it" ranked eighth in the list of leading words, but not at position, say 8.23.

Life and death in the Prussian army

Bortkiewicz collected records for ten corps over a 20-year period giving him data for 200 corps-years. He looked at the number of deaths (this was what mathematicians call the variable) and the number of corps-years when this number of deaths occurred. For example, there were 109 corps-years when no deaths occurred, while in one corps-year, there were four deaths. At the barracks, Corp C (say) in one particular year experienced four deaths.

How is the number of deaths distributed? Collecting this information is one side of the statistician's job — being out in the field recording results. Bortkiewicz obtained the following data:

Number of deaths	0	1	2	3	4
Frequency	109	65	22	3	1

$$e^{-\lambda} \lambda^x / x !$$

The Poisson formula

Thankfully, being killed by a horse-kick is a rare event. The most suitable theoretical technique for modeling how often rare events occur is to use something called the Poisson distribution. With this technique, could Bortkiewicz have predicted the results without visiting the stables? The theoretical Poisson distribution says that the probability that the number of deaths (which we'll call X) has the value x is given by the Poisson formula, where e is the special number discussed earlier that's associated with growth and the exclamation mark means the factorial, the number multiplied by all the other whole numbers between it and 1 (see Chapter 69). The Greek letter lambda, written λ, is the average number of deaths. We need to find this average over our 200 corps-years so we multiply 0 deaths by 109 corps-years (giving 0), 1 death by 65 corps-years (giving 65), 2 deaths by 22 corps-years (giving 44), 3 deaths by 3 corps-years (giving 9) and 4 deaths by 1 corps-year (giving 4) and then we add all of these together (giving 122) and divide by 200. So our average number of deaths per corps-year is 122/200 = 0.61.

The theoretical probabilities (which we'll call p) can be found by substituting the values r = 0, 1, 2, 3 and 4 into the Poisson formula. The results are:

Number of deaths	0	1	2	3	4
Probabilities, p	0.543	0.331	0.101	0.020	0.003
Expected number of deaths, $200 \times p$	108.6	66.2	20.2	4.0	0.6

It looks as though the theoretical distribution is a good fit for the experimental data gathered by Bortkiewicz.

First numbers

If we analyze the last digits of telephone numbers in a column of the telephone directory we would expect to find 0, 1, 2, . . . , 9 to be uniformly distributed. They appear at random and any number has an equal chance of turning up. In 1938 the electrical engineer Frank Benford found that this was not true for the first digits of some sets of data. In fact he rediscovered a law first observed by the astronomer Simon Newcomb in 1881.

Yesterday I conducted a little experiment. I looked through the foreign currency exchange data in a national newspaper. There were exchange rates like 2.119 to mean you will need (U.S. dollar) $2.119 to buy £1 sterling. Likewise, you will need (Euro) €1.59 to buy £1 sterling and (Hong Kong dollar) HK $15.390 to buy £1. Reviewing the results of the data and recording the number of appearances by first digit, gave the following table:

First digit	1	2	3	4	5	6	7	8	9	Total
Number of occurrences	18	10	3	1	3	5	7	2	1	50
Percentage, %	36	20	6	2	6	10	14	4	2	100

These results support Benford's law, which says that for some classes of data, the number 1 appears as the first digit in about 30 percent of the data, the number 2 in 18 percent of the data and so on. It is certainly not the uniform distribution that occurs in the last digit of the telephone numbers.

It is not obvious why so many data sets do follow Benford's law. In the 19th century when Simon Newcomb observed it in the use of mathematical tables he could hardly have guessed it would be so widespread.

Instances where Benford's distribution can be detected include scores in sporting events, stock market data, house numbers, populations of countries, and the lengths of rivers. The measurement units are unimportant — it does not matter if the lengths of rivers are measured in meters or miles. Benford's law has practical applications. Once it was recognized that accounting information followed this law, it became easier to detect false information and uncover fraud.

Words

One of G.K. Zipf's wide interests was the unusual practice of counting words. It turns out that the ten most popular words appearing in the English language are the tiny words ranked as shown:

Rank	1	2	3	4	5	6	7	8	9	10
Word	the	of	and	to	a	in	that	it	is	was

This was found by taking a large sample across a wide range of written work and just counting words. The most common word was given rank 1, the next rank 2, and so on. There might be small differences in the popularity stakes if a range of texts were analyzed, but it will not vary much.

It is not surprising that "the" is the most common, and "of" is second. The list continues and you might want to know that "among" is in 500th position and "neck" is ranked 1,000. We shall only consider the top ten words. If you pick up a text at random and count these words you will get more or less the same words in rank order. The surprising fact is that the ranks have a bearing on the actual number of appearances of the words in a text. The word "the" will occur twice as often as "of" and three times more frequently than "and," and so on. The actual number is given by a well-known formula. This is an experimental law and was discovered by Zipf from data. The theoretical Zipf's law says that the percentage of occurrences of the word ranked r is given by

$$\frac{k}{r} \times 100$$

where the number k depends only on the size of the author's vocabulary. If an author had command of all the words in the English language, of which there are around a million by some estimates, the value of k would be about 0.0694. In the formula for Zipf's law the word "the" would then account for about 6.94 percent of all words in a text. In the same way "of" would account for half of this, or about 3.47 percent of the words. An essay of 3,000 words by such a talented author would therefore contain 208 appearances of "the" and 104 appearances of the word "of."

For writers with only 20,000 words at their command, the value of k rises to 0.0954, so there would be 286 appearances of "the" and 143 appearances of the word "of." The smaller the vocabulary, the more often you will see "the" appearing.

Crystal ball gazing

Whether Poisson, Benford or Zipf, all these distributions allow us to make predictions. We may not be able to predict a dead cert but knowing how the probabilities distribute themselves is much better than taking a shot in the dark. Add to these three, other distributions like the binomial, the negative binomial, the geometric, the hypergeometric, and many more, the statistician has an effective array of tools for analyzing a vast range of human activity.

91 The normal curve

The "normal" curve plays a pivotal role in statistics. It has been called the equivalent of the straight line in mathematics. It certainly has important mathematical properties but if we set to work analyzing a block of raw data we would rarely find that it followed a normal curve exactly.

The normal curve is prescribed by a specific mathematical formula which creates a bell-shaped curve; a curve with one hump and which tails away on either side. The significance of the normal curve lies less in nature and more in theory, and in this it has a long pedigree. In 1733 Abraham de Moivre, a French Huguenot who fled to England to escape religious persecution, introduced it in connection with his analysis of chance. Pierre Simon Laplace published results about it and Carl Friedrich Gauss used it in astronomy, where it is sometimes referred to as the Gaussian law of error.

Adolphe Quetelet used the normal curve in his sociological studies published in 1835, in which he measured the divergence from the "average man" by the normal curve. In other experiments he measured the heights of French conscripts and the chest measurements of Scottish soldiers and assumed these followed the normal curve. In those days there was a strong belief that most phenomena were "normal" in this sense.

The cocktail party

Let's suppose that Georgina went to a cocktail party and the host, Sebastian, asked her if she had come far? She realized afterwards it was a very useful question for cocktail parties — it applies to everyone and invites a response. It is not taxing and it starts the ball rolling if conversation is difficult.

The next day, slightly hungover, Georgina traveled to the office wondering if her colleagues had come far to work. In the staff canteen she learned that some lived around the corner and some lived 50 miles away — there was a great deal of variability. She took advantage of the fact that she was the Human Resources Manager of a very large company to tack a question on the end of her annual employee questionnaire: "how far have you traveled to work today?" She wanted to work out the average distance of travel of the company's staff. When Georgina drew a histogram of results the distribution showed no particular form, but at least she could calculate the average distance traveled.

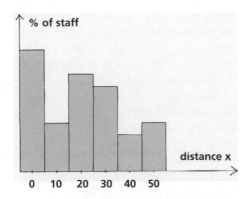

Georgina's histogram of distance traveled by her colleagues to work

This average turned out to be 20 miles. Mathematicians denote this by the Greek letter mu, written μ, and so here $\mu = 20$. The variability in the population is denoted by the Greek letter sigma, written σ, which is sometimes called the standard deviation. If the standard deviation is small the data is close together and has little variability, but if it is large, the data is spread out. The company's marketing analyst, who had trained as a statistician, showed Georgina that she might have got around the same value of 20 by sampling. There was no need to ask all the employees. This estimation technique depends on the Central Limit Theorem.

Take a random sample of staff from all of the company's workforce. The larger the sample the better, but 30 employees will do nicely. In selecting this sample at random it is likely there will be people who live around the corner and some long-distance travelers as well. When we calculate the average distance for our sample, the effect of the longer distances will average out the shorter distances. Mathematicians write the average of the sample as $\bar{x}$, which is read as "x bar." In Georgina's case, it is most likely that the value of $\bar{x}$ will be near 20, the average of the population. Though it is certainly possible, it is unlikely that the average of the sample will be very small or very large.

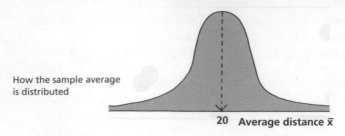

How the sample average
is distributed

20   Average distance $\bar{x}$

The Central Limit Theorem is one reason why the normal curve is important to statisticians. It states that the actual distribution of the sample averages $\bar{x}$ approximates to a normal curve whatever the distribution of x. What does this mean? In Georgina's case, x represents the distance from the workplace and $\bar{x}$ is the average of a sample. The distribution of x in Georgina's histogram is nothing like a bell-shaped curve, but the distribution of $\bar{x}$ is, and it is centered on $\mu = 20$.

This is why we can use the average of a sample $\bar{x}$ as an estimate of the population average μ. That the variability of the sample averages $\bar{x}$ is an added bonus. If the variability of the x values is the standard deviation σ, the variability of $\bar{x}$ is $\sigma/\sqrt{n}$ where n is the size of the sample we select. The larger the sample size, the narrower will be the normal curve, and the better will be the estimate of μ.

Other normal curves

Let's do a simple experiment. We'll toss a coin four times. The chance of throwing a head each time is $p = \frac{1}{2}$. The result for the four throws can be recorded using H for heads and T for tails, arranged in the order in which they occur. Altogether there are 16 possible outcomes. For example, we might obtain three heads in the outcome $THHH$. There are in fact four possible outcomes giving three heads (the others are $HTHH, HHTH, HHHT$) so the probability of three heads is $4/16 = 0.25$.

With a small number of throws, the probabilities are easily calculated and placed in a table, and we can also calculate how the probabilities are distributed. The number of combinations row can be found from Pascal's triangle (see Chapter 76):

Number of heads	0	1	2	3	4
Number of combinations	1	4	6	4	1
Probability	0.0625	0.25	0.375	0.25	0.0625
	(= 1/16)	(= 4/16)	(= 6/16)	(= 4/16)	(= 1/16)

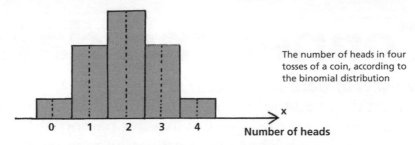

The number of heads in four tosses of a coin, according to the binomial distribution

Number of heads

This is called a binomial distribution of probabilities, which occurs where there are two possible outcomes (here a head or a tail). These probabilities may be represented by a diagram in which both the heights and areas describe them.

Tossing the coin four times is a bit restrictive. What happens if we throw it a large number, say 100, times? The binomial distribution of probabilities can be applied where $n = 100$, but it can usefully be approximated by the normal bell-shaped curve with mean $\mu = 50$ (as we would expect 50 heads when tossing a coin 100 times) and variability (standard deviation) of $\sigma = 5$. This is what de Moivre discovered in the 16th century.

For large values of n, the variable x which measures the number of successes fits the normal curve increasingly well. The larger the value of n the better the approximation and tossing the coin 100 times qualifies as large. Now let's say we want to know the probability of throwing between 40 and 60 heads. The area A shows the region we're interested in and gives us the probability of tossing between 40 and 60 heads which we write as $prob(40 \leq x \leq 60)$. To find the actual numerical value we need to use pre-calculated mathematical tables, and once this has been done, we find $prob(40 \leq x \leq 60) = 0.9545$. This shows that getting between 40 and 60 heads in 100 tosses of a coin is 95.45 percent, which means that this is very likely.

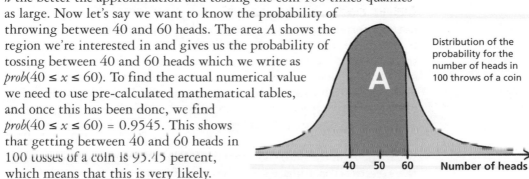

Distribution of the probability for the number of heads in 100 throws of a coin

Number of heads

The area left over is $1 - 0.9545$ which is a mere 0.0455. As the normal curve is symmetric about its middle, half of this will give the probability of getting more than 60 heads in 100 tosses of the coin. This is just 2.275 percent and represents a very slim chance indeed. If you visit Las Vegas this would be a bet to leave well alone.

92 Connecting data

How are two sets of data connected? Statisticians of a hundred years ago thought they had the answer. Correlation and regression go together like a horse and carriage, but like this pairing, they are different and have their own jobs to do. Correlation measures how well two quantities such as weight and height are related to each other. Regression can be used to predict the values of one property (say weight) from the other (in this case, height).

Pearson's correlation

The term correlation was introduced by Francis Galton in the 1880s. He originally termed it "co-relation," a better word for explaining its meaning. Galton, a Victorian gentleman of science, had a desire to measure everything and applied correlation to his investigations into pairs of variables: the wing length and tail length of birds, for instance. The Pearson correlation coefficient, named after Galton's biographer and protégé Karl Pearson, is measured on a scale between minus one and plus one. If its numerical value is high, say +0.9, there is said to be a strong correlation between the variables. The correlation coefficient measures the tendency for data to lie along a straight line. If it is near to zero the correlation is practically non-existent.

We frequently wish to work out the correlation between two variables to see how strongly they are connected. Let's take the example of the sales of sunglasses and see how this relates to the sales of ice creams. San Francisco would be a good place in which to conduct our study and we shall gather data each month in that city. If we plot points on a graph where the x (horizontal) coordinate represents sales of sunglasses and the y (vertical) coordinate gives the sales of ice creams, each month we will have a data point (x, y) representing both pieces of data. For example, the point $(3, 4)$ could mean the May sales of sunglasses were $30,000 while sales of ice creams in the city were $40,000 in that same month. We can plot the monthly data points (x, y) for a whole year on a scatter diagram. For this example, the value of the Pearson correlation coefficient would be around +0.9 indicating a strong correlation. The data has a tendency to follow a straight line. It is positive because the straight line has a positive gradient — it is pointing in a northeasterly direction.

Cause and correlation

Finding a strong correlation between two variables is
not sufficient to claim that one causes the other.
There may be a cause and effect relation between the
two variables but this cannot be claimed on the basis
of numerical evidence alone. On the cause/correlation
issue it is customary to use the word "association"
and wise to be wary of claiming more than this.

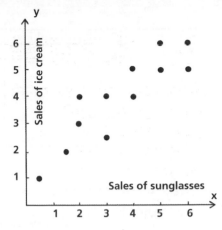

Scatter diagram

In the sunglasses and ice cream example, there is
a strong correlation between the sales of sunglasses
and that of ice cream. As the sales of sunglasses
increase, the number of ice creams sold tends to
increase. It would be ludicrous to claim that the
expenditure on sunglasses caused more ice creams to be sold. With
correlation there may be a hidden intermediary variable at work. For
example, the expenditure on sunglasses and on ice creams is linked
together as a result of seasonal effects (hot weather in the summer
months, cool weather in the winter). There is another danger in using
correlation. There may be a high correlation between variables but no
logical or scientific connection at all. There could be a high correlation
between house numbers and the combined ages of the house's occupants
but reading any significance into this would be unfortunate.

Spearman's correlation

Correlation can be put to other uses. The correlation coefficient can be
adapted to treat ordered data — data where we want to know first,
second, third and so on, but not necessarily other numerical values.

Occasionally we have only the ranks as data. Let's look at Albert
and Zac, two strong-minded ice skating judges at a competition who
have to evaluate skaters on artistic merit. It will be a subjective
evaluation. Albert and Zac have both won Olympic medals and are
called on to judge the final group which has been narrowed down to
five competitors: Ann, Beth, Charlotte, Dorothy and Ellie. If Albert
and Zac ranked them in exactly the same way, that would be fine but
life is not like that. On the other hand we would not expect Albert to
rank them in one way and Zac to rank them in the very reverse order.
The reality is that the rankings would be in between these two
extremes. Albert ranked them 1 to 5 with Ann (the best) followed by
Ellie, Beth, Charlotte and finally Dorothy in fifth position. Zac rated
Ellie the best, followed by Beth, Ann, Dorothy and Charlotte. These
rankings can be summarized in a table.

Skater	Albert's rankings	Zac's rankings	Difference in ranks, d	d^2
Ann	1	3	-2	4
Ellie	2	1	1	1
Beth	3	2	1	1
Charlotte	4	5	-1	1
Dorothy	5	4	1	1
$n = 5$			Sum	8

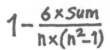

$$1 - \frac{6 \times Sum}{n \times (n^2 - 1)}$$

Spearman's formula

How can we measure the level of agreement between the judges? Spearman's correlation coefficient is the instrument mathematicians use to do this for ordered data. Its value here is $+0.6$ which indicates a limited measure of agreement between Albert and Zac. If we treat the pairs of ranks as points we can plot them on a graph to obtain a visual representation of how closely the two judges agree.

The formula for this correlation coefficient was developed in 1904 by the psychologist Charles Spearman who, like Pearson, was influenced by Francis Galton.

Regression lines

Are you shorter or taller than both your parents or do you fall between their heights? If we were all taller than our parents, and this happened at each generation, then one day the population might be composed of ten-footers and upwards, and surely this cannot be. If we were all shorter than our parents then the population would gradually diminish in height and this is equally unlikely. The truth lies elsewhere.

Francis Galton conducted experiments in the 1880s in which he compared the heights of mature young adults with the heights of their parents. For each value of the x variable measuring parents' height (actually combining height of mother and father into a "mid-parent" height) he observed the heights of their offspring. We are talking about a practical scientist here, so out came the pencils and sheets of paper divided into squares on which he plotted the data. For 205 mid-parents and 928 offspring he found the average height of both sets to be $68\frac{1}{4}$ inches or 5 feet $8\frac{1}{4}$ inches (173.4 cm) which value he called the mediocrity. He found that children of very tall mid-parents were generally taller than this mediocrity but not as

Measuring agreement between two judges

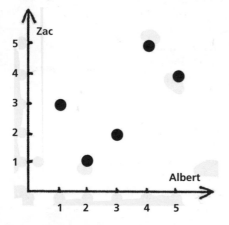

tall as their mid-parents, while shorter children were taller than their mid-parents but shorter than the mediocrity. In other words, the children's heights regressed towards the mediocrity. It's a bit like top-class batter Alex Rodriguez's performances for the New York Yankees. His batting average in an exceptional season is likely to be followed by an inferior average in the next, yet overall would still be better than the average for all players in the league. We say his batting average has regressed to the average (or mean).

Regression is a powerful technique and is widely applicable. Let's suppose that, for a survey, the operational research team of a popular retail chain chooses five of its stores, from small outlets (with 1,000 customers a month) through to mega-stores (with 10,000 customers a month). The research team observes the number of staff employed in each. They plan to use regression to estimate how many staff they will need for their other stores.

Number of customers (1000s)	1	4	6	9	10
Number of staff	24	30	46	47	53

Let's plot this on a graph, where we'll make the x coordinate the number of customers (we call this the explanatory variable) while the number of staff is plotted as the y coordinate (called the response variable). It is the number of customers that explains the number of staff needed and not the other way around. The average number of customers in the stores is plotted as 6 (i.e. 6,000 customers) and the average number of staff in the stores is 40. The regression line always passes through the "average point," here (6, 40).

There are formulae for calculating the regression line, the line which best fits the data (also known as the line of least squares). In our case the line is $\hat{y} = 20.8 + 3.2x$ so the slope is 3.2 and is positive (going up from left to right). The line crosses the vertical y axis at the point 20.8. The term $\hat{y}$ is the estimate of the y value obtained from the line. So if we want to know how many staff should be employed in a store that receives 5,000 customers a month we could substitute the value $x = 5$ into the regression equation and obtain the estimate $\hat{y} = 37$ staff showing how regression has a very practical purpose.

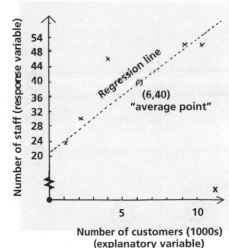

93 **Groups**

Evariste Galois died in a duel aged 20, but left behind enough ideas to keep mathematicians busy for centuries. These involved the theory of groups, mathematical constructs that can be used to quantify symmetry. Apart from its artistic appeal, symmetry is the essential ingredient for scientists who dream of a future theory of everything. Group theory is the glue which binds the "everything" together.

Symmetry is all around us. Greek vases have it, snow crystals have it, buildings often have it and some letters of our alphabet have it. There are several sorts of symmetry: chief among them are mirror symmetry and rotational symmetry. We'll just look at two-dimensional symmetry — all our objects of study live on the flat surface of this page.

Mirror symmetry

Can we set up a mirror so that an object looks the same in front of the mirror as in the mirror? The word MUM has mirror symmetry, but HAM does not; MUM in front of the mirror is the same as MUM in the mirror while HAM becomes MAH. A tripod has mirror symmetry, but the triskelion (tripod with feet) does not. The triskelion as the object before the mirror is right-handed but its mirror image in what is called the image plane is left-handed.

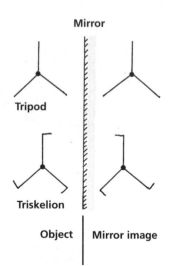

Mirror

Tripod

Triskelion

Object | **Mirror image**

Rotational symmetry

We can also ask whether there is an axis perpendicular to the page so that the object can be rotated in the page through an angle and be brought back to its original position. Both the tripod and the triskelion have rotational symmetry. The triskelion, meaning "three legs," is an interesting shape. The right-handed version is a figure which appears as the symbol of the Isle of Man and also on the flag of Sicily.

If we rotate it through 120 degrees or 240 degrees the rotated figure will coincide with itself; if you closed your eyes before rotating it you would see the same triskelion when you opened them again after rotation.

The curious thing about the three-legged figure is that no amount of rotation keeping in the plane will ever convert a right-handed triskelion into a left-handed one. Objects for which the image in the mirror is distinct from the object in front of the mirror are called chiral — they look similar but are not the same. The molecular structure of some chemical compounds may exist in both right-handed and left-handed forms in three dimensions and are examples of chiral objects. This is the case with the compound limosene which in one form tastes like lemons and in the other like oranges. The drug thalidomide in one form is an effective cure of morning sickness in pregnancy but in the other form has tragic consequences.

The Isle of Man triskelion

Measuring symmetry

In the case of our triskelion the basic symmetry operations are the (clockwise) rotations R through 120 degrees and S through 240 degrees. The transformation I is the one that rotates the triangle through 360 degrees or, alternatively, does nothing at all. We can create a table based on the combinations of these rotations, in the same way we might create a multiplication table.

This table is like an ordinary multiplication table with numbers except we are "multiplying" symbols. According to the most widely used convention, the multiplication $R \circ S$ means first rotate the triskelion clockwise through 240 degrees with S and then by 120 degrees with R, the result being a rotation by 360 degrees, as if you did nothing at all. This can be expressed as $R \circ S = I$, the result found at the junction of the last but one row and the last column of the table.

The symmetry group of the triskelion is made up of I, R and S and the multiplication table of how to combine them. Because the group contains three elements its size (or "order") is three. The table is also called a Cayley table (named after the mathematician Arthur Cayley, distant cousin to Sir George Cayley, a pioneer of flight).

∘	I	R	S
I	I	R	S
R	R	S	I
S	S	I	R

Cayley table for the symmetry group of the triskelion

Like the triskelion, the tripod without feet has rotational symmetry. But it also has mirror symmetry and therefore has a larger symmetry group. We'll call U, V and W the reflections in the three mirror axes.

The larger symmetry group of the tripod, which is of order six, is composed of the six transformations I, R, S, U, V and W and has the multiplication table shown in the margin.

An interesting transformation is achieved by combining two reflections in different axes, such as $U \circ W$ (where the reflection W is applied first and is followed by the reflection U). This is actually a

○	I	R	S	U	V	W
I	I	R	S	U	V	W
R	R	S	I	V	W	U
S	S	I	R	W	U	V
U	U	W	V	I	S	R
V	V	U	W	R	I	S
W	W	V	U	S	R	I

Caley table for the symmetry group of the tripod

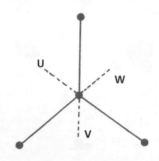

Reflections of a tripod

rotation of the tripod through 120 degrees, in symbols $U \circ W = R$. Combining the reflections the other way around $W \circ U = S$, gives a rotation through 240 degrees. In particular $U \circ W \neq W \circ U$. This is a major difference between a multiplication table for a group and an ordinary multiplication table with numbers.

A group in which the order of combining the elements is immaterial is called an "abelian group", named after the Norwegian mathematician Niels Abel. The symmetry group of the tripod is the smallest group which is not abelian.

Abstract groups

The trend in algebra in the 20th century had been towards abstract algebra, in which a group is defined by some basic rules known as axioms. With this viewpoint the symmetry group of the triangle becomes just one example of an abstract system. There are systems in algebra that are more basic than a group and require fewer axioms; other systems that are more complex require more axioms. However the concept of a group is just right and is the most important algebraic system of all. It is remarkable that from so few axioms such a large body of knowledge has emerged. The advantage of the abstract method is that general theorems can be deduced for all groups and applied, if need be, to specific ones.

A feature of group theory is that there may be small groups sitting inside bigger ones. The symmetry group of the triskelion of order three is a subgroup of the symmetry group of the tripod of order six. J.L. Lagrange proved a basic fact about subgroups. Lagrange's theorem states that the order of a subgroup must always divide exactly the order of the group. So we automatically know the symmetry group of the tripod has no subgroups of order four or five.

Classifying groups

There has been an extensive programme to classify all the possible finite groups. There is no need to list them all because some groups are built up from basic ones, and it is the basic ones that are needed. The principle of classification is much the same as in chemistry where interest is focused on the basic chemical elements and not the compounds which can be made from them. The symmetry group of the tripod of six elements is a "compound" being built up from the group of rotations (of order three) and reflections (of order two).

Axioms for a group

A collection of elements G with "multiplication" ∘ is called a group if

1. There is an element 1 in G so that $1 \circ a = a \circ 1 = a$ for all elements a in the group G (the special element 1 is called the identity element).

2. For each element a in G there is an element $\bar{a}$ in G with $\bar{a} \circ a = a \circ \bar{a} = 1$ (the element $\bar{a}$ is called the inverse element of a).

3. For all elements a, b and c in G it is true that $a \circ (b \circ c) = (a \circ b) \circ c$ (this is called the associative law).

Nearly all basic groups can be classified into known classes. The complete classification, called "the enormous theorem," was announced by Daniel Gorenstein in 1983 and was arrived at through the accumulated work of 30 years' worth of research and publications by mathematicians. It is an atlas of all known groups. The basic groups fall into one of four main types, yet 26 groups have been found that do not fall into any category. These are known as the "sporadic groups."

The sporadic groups are mavericks and are typically of large order. Five of the smallest were known to Emile Mathieu in the 1860s but much of the modern activity took place between 1965 and 1975. The smallest sporadic group is of order $7{,}920 = 2^4 \times 3^2 \times 5 \times 11$ but at the upper end are the "baby monster" and the plain "monster" which has order $2^{46} \times 3^{20} \times 5^9 \times 7^6 \times 11^2 \times 13^3 \times 17 \times 19 \times 23 \times 29 \times 31 \times 41 \times 47 \times 59 \times 71$ which in decimal speak is around 8×10^{53} or, if you like, 8 with 53 trailing zeros — a very large number indeed. It can be shown that 20 of the 26 sporadic groups are represented as subgroups inside the "monster" — the six groups that defy all classificatory systems are known as the "six pariahs."

Although snappy proofs and shortness are much sought after in mathematics, the proof of the classification of finite groups is something like 10,000 pages of closely argued symbolics. Mathematical progress is not always due to the work of a single outstanding genius.

94 **Matrices**

This is the story of "extraordinary algebra" — a revolution in mathematics which took place in the middle of the 19th century. Mathematicians had played with blocks of numbers for centuries, but the idea of treating blocks as a single number took off 150 years ago with a small group of mathematicians who recognized its potential.

Ordinary algebra is the traditional algebra in which symbols such as a, b, c, x and y represent single numbers. Many people find this difficult to understand, but for mathematicians it was a great step forward. In comparison, "extraordinary algebra" generated a seismic shift. For sophisticated applications this progress from a one-dimensional algebra to a multiple dimensional algebra would prove incredibly powerful.

Multiple dimensioned numbers

In ordinary algebra a might represent a number such as 7, and we would write $a = 7$, but in matrix theory a *matrix A* would be a "multiple dimensioned number," for example the block

$$A = \begin{pmatrix} 7 & 5 & 0 & 1 \\ 0 & 4 & 3 & 7 \\ 3 & 2 & 0 & 2 \end{pmatrix}$$

This matrix has three rows and four columns (it's a "3 by 4" matrix), but in principle we can have matrices with any number of rows and columns — even a giant "100 by 200" matrix with 100 rows and 200 columns. A critical advantage of matrix algebra is that we can think of vast arrays of numbers, such as a data set in statistics, as a single entity. More than this, we can manipulate these blocks of numbers simply and efficiently. If we want to add or multiply together all the numbers in two data sets, each consisting of 1,000 numbers, we don't have to perform 1,000 calculations — we just have to perform one (adding or multiplying the two matrices together).

A practical example

Suppose the matrix A represents the output of the AJAX company in one week. The AJAX company has three factories located in different parts of the country and their output is measured in units (say 1,000s of items) of the four products it produces. In our example, the quantities, tallying with matrix A opposite, are:

	product 1	product 2	product 3	product 4
factory 1	7	5	0	1
factory 2	0	4	3	7
factory 3	3	2	0	2

In the next week the production schedule might be different, but it could be written as another matrix B. For example B might be given by

$$B = \begin{pmatrix} 9 & 4 & 1 & 0 \\ 0 & 5 & 1 & 8 \\ 4 & 1 & 1 & 0 \end{pmatrix}$$

What is the total production for both weeks? The matrix theorist says it is the matrix $A + B$ where corresponding numbers are added together,

$$A+B = \begin{pmatrix} 7+9 & 5+4 & 0+1 & 1+0 \\ 0+0 & 4+5 & 3+1 & 7+8 \\ 3+4 & 2+1 & 0+1 & 2+0 \end{pmatrix} = \begin{pmatrix} 16 & 9 & 1 & 1 \\ 0 & 9 & 4 & 15 \\ 7 & 3 & 1 & 2 \end{pmatrix}$$

Easy enough. Sadly, matrix multiplication is less obvious. Returning to the AJAX company, suppose the unit profit of its four products are 3, 9, 8, 2. We can certainly compute the overall profit for Factory 1 with outputs 7, 5, 0, 1 of its four products. It works out as $7 \times 3 + 5 \times 9 + 0 \times 8 + 1 \times 2 = 68$.

But instead of dealing with just one factory we can just as easily compute the total profits T for *all* the factories:

$$T = \begin{pmatrix} 7 & 5 & 0 & 1 \\ 0 & 4 & 3 & 7 \\ 3 & 2 & 0 & 2 \end{pmatrix} \times \begin{pmatrix} 3 \\ 9 \\ 8 \\ 2 \end{pmatrix} = \begin{pmatrix} 7\times3+5\times9+0\times8+1\times2 \\ 0\times3+4\times9+3\times8+7\times2 \\ 3\times3+2\times9+0\times8+2\times2 \end{pmatrix} = \begin{pmatrix} 68 \\ 74 \\ 31 \end{pmatrix}$$

Look carefully and you'll see the *row* by *column* multiplication, an essential feature of matrix multiplication. If in addition to the unit profits we are given the unit *volumes* 7, 4, 1, 5 of each unit of the products, in one fell swoop we can calculate the profits *and* storage requirements for the three factories by the single matrix multiplication:

$$\begin{pmatrix} 7 & 5 & 0 & 1 \\ 0 & 4 & 3 & 7 \\ 3 & 2 & 0 & 2 \end{pmatrix} \times \begin{pmatrix} 3 & 7 \\ 9 & 4 \\ 8 & 1 \\ 2 & 5 \end{pmatrix} = \begin{pmatrix} 68 & 74 \\ 74 & 54 \\ 31 & 39 \end{pmatrix}$$

The total storage is provided by the second column of the resulting matrix, that is 74, 54 and 39. Matrix theory is very powerful. Imagine the situation of a company with hundreds of factories, thousands of products, and different unit profits and storage requirements in different weeks. With matrix algebra the calculations, and our understanding, are fairly immediate, without having to worry about the details which are all taken care of.

Matrix algebra vs ordinary algebra

There are many parallels to be drawn between matrix algebra and ordinary algebra. The most celebrated difference occurs in the multiplication of matrices. If we multiply matrix A with matrix B and then try it the other way round:

$$A \times B = \begin{pmatrix} 3 & 5 \\ 2 & 1 \end{pmatrix} \times \begin{pmatrix} 7 & 6 \\ 4 & 8 \end{pmatrix} = \begin{pmatrix} 3 \times 7 + 5 \times 4 & 3 \times 6 + 5 \times 8 \\ 2 \times 7 + 1 \times 4 & 2 \times 6 + 1 \times 8 \end{pmatrix} = \begin{pmatrix} 41 & 58 \\ 18 & 20 \end{pmatrix}$$

$$B \times A = \begin{pmatrix} 7 & 6 \\ 4 & 8 \end{pmatrix} \times \begin{pmatrix} 3 & 5 \\ 2 & 1 \end{pmatrix} = \begin{pmatrix} 7 \times 3 + 6 \times 2 & 7 \times 5 + 6 \times 1 \\ 4 \times 3 + 8 \times 2 & 4 \times 5 + 8 \times 1 \end{pmatrix} = \begin{pmatrix} 33 & 41 \\ 28 & 28 \end{pmatrix}$$

So in matrix algebra we may have $A \times B$ and $B \times A$ being different, a situation which does not arise in ordinary algebra where the order of multiplying two numbers together makes no difference to the answer.

Another difference occurs with inverses. In ordinary algebra inverses are easy to calculate. If $a = 7$ its inverse is $\frac{1}{7}$ because it has the property that $\frac{1}{7} \times 7 = 1$. We sometimes write this inverse as $a^{-1} = \frac{1}{7}$ and we have $a^{-1} \times a = 1$.

An example in matrix theory is $A = \begin{pmatrix} 1 & 2 \\ 3 & 7 \end{pmatrix}$ and we can verify that $A^{-1} = \begin{pmatrix} 7 & -2 \\ -3 & 1 \end{pmatrix}$ because $A^{-1} \times A = \begin{pmatrix} 7 & -2 \\ -3 & 1 \end{pmatrix} \times \begin{pmatrix} 1 & 2 \\ 3 & 7 \end{pmatrix} = \begin{pmatrix} 1 & 0 \\ 0 & 1 \end{pmatrix}$

where $I = \begin{pmatrix} 1 & 0 \\ 0 & 1 \end{pmatrix}$ is called the identity matrix and is the matrix counterpart

of 1 in ordinary algebra. In ordinary algebra, only 0 does not have an inverse but in matrix algebra many matrices do not have inverses.

Travel plans

Another example of using matrices is in the analysis of a flight network for airlines. This will involve both airport hubs and smaller airports. In practice this may involve hundreds of destinations — here we'll look at a small example: the hubs London (**L**) and Paris (**P**), and smaller airports Edinburgh (**E**), Bordeaux (**B**), and Toulouse (**T**) and the network showing possible *direct* flights. To use a computer to analyze such networks, they are first coded using matrices. If there is a direct flight between airports a 1 is recorded at the intersection of the row and column labeled by these airports (like from London to Edinburgh). The "connectivity" matrix which describes the network above is A.

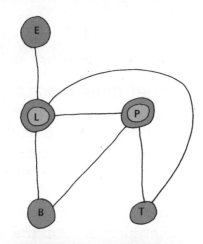

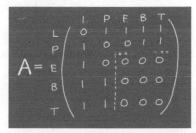

The lower submatrix (marked out by the dotted lines) shows there are no direct links between the three smaller airports. The matrix product $A \times A = A^2$ of this matrix with itself can be interpreted as giving the number of possible journeys between two airports *with exactly one stopover*. So, for example, there are three possible roundtrips to Paris via other cities but no trips from London to Edinburgh which involve stopovers. The number of routes which are either direct or involve one stopover are the elements of the matrix $A + A^2$. This is another example of the ability of matrices to capture the essence of a vast amount of data under the umbrella of a single calculation.

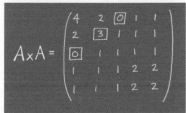

When a small group of mathematicians created the theory of matrices in the 1850s they did so to solve problems in pure mathematics. From an applied perspective, matrix theory was very much a "solution looking for a problem." As so often happens "problems" did arise which needed the nascent theory. An early application occurred in the 1920s when Werner Heisenberg investigated "matrix mechanics," a part of quantum theory. Another pioneer was Olga Taussky-Todd, who worked for a period in aircraft design and used matrix algebra. When asked how she discovered the subject she replied that it was the other way around, matrix theory had found her. Such is the mathematical game.

95 Codes

What does Julius Caesar have in common with the transmission of modern digital signals? The short answer is codes and coding. To send digital signals to a computer or a digital television set, the coding of pictures and speech into a stream of zeros and ones — a binary code — is essential for it is the only language these devices understand. Caesar used codes to communicate with his generals and kept his messages secret by changing around the letters of his message according to a key which only he and they knew.

Accuracy was essential for Caesar and it is also required for the effective transmission of digital signals. Caesar also wanted to keep his codes to himself as do the cable and satellite broadcasting television companies who only want paying subscribers to be able to make sense of their signals.

Let's look at accuracy first. Human error or "noise along the line" can always occur, and must be dealt with. Mathematical thinking allows us to construct coding systems that automatically detect errors and even make corrections.

Error detection and correction

One of the first binary coding systems was the Morse code, which makes use of two symbols, dots • and dashes –. The American inventor Samuel F.B. Morse sent the first intercity message using his code from Washington to Baltimore in 1844. It was a code designed for the electric telegraph of the mid-19th century with little thought to an efficient design. In Morse code, the letter A is coded as • –, B as — ●●●, C as — • — • and other letters as different sequences of dots and dashes. A telegraph operator sending "CAB" would send the string — • — • / • — / — ●●●. Whatever its merits, Morse code is not very good at error detection let alone correction. If the Morse code operator wished to send "CAB," but mistyped a dot for a dash in C, forgot the dash in A and noise on the wire substituted a dash for a dot in B, the receiver getting ●● — • / • / — — ●●, would see nothing wrong and interpret it as "FEZ."

At a more primitive level we could look at a coding system consisting of just 0 and 1 where 0 represents one word and 1 another. Suppose an army commander has to transmit a message to his troops which is either "invade" or "do not invade." The "invade" instruction is coded by "1" and the "do not invade" instruction by "0." If a 1 or a 0 was incorrectly transmitted the receiver would never know — and the wrong instruction would be given, with disastrous consequences.

We can improve matters by using code words of length two. If this time we code the "invade" instruction by 11 and the "do not invade" by 00, this is better. An error in one digit would result in 01 or 10 being received. As only 11 or 00 are legitimate code words, the receiver would certainly know that an error had been made. The advantage of this system is that an error would be detectable, but we still would not know how to correct it. If 01 were received, how would we know whether 00 or 11 should have been sent?

The way to a better system is to combine design with longer code words. If we code the "invade" instruction by 111 and the "do not invade" by 000 an error in one digit could certainly be detected, as before. If we knew that at most one error could be made (a reasonable assumption since the chance of two errors in one code word is small), the correction could actually be made by the receiver. For example, if 110 were received then the correct message would have been 111. With our rules, it could not be 000 since this code word is two errors away from 110. In this system there are only two code words 000 and 111 but they are far enough apart to make error detection and correction possible.

The same principle is used when word processing is in autocorrect mode. If we type "animul" the word processor detects the error and corrects it by taking the nearest word, "animal." The English language is not fully correcting though, because if we type "lomp" there is no unique nearest word; the words lamp, limp, lump, pomp and romp are all equidistant in terms of single errors from lomp.

A modern binary code consists of code words that are blocks consisting of zeros and ones. By choosing the legitimate code words far enough apart, both detection and correction are possible. The code words of Morse code are too close together but the modern code systems used to transmit data from satellites can always be set in autocorrect mode. Long code words with high performance in terms of error correction take longer to transmit so there is a tradeoff between length and speed of transmission. Voyages into deep space by NASA have used codes that are three-error correcting and these have proved satisfactory in combating noise on the line.

Making messages secret

Julius Caesar kept his messages secret by changing around the letters of his message according to a key that only he and his generals knew. If the key fell into the wrong hands his messages could be deciphered by his enemies. In the 16th century, Mary Queen of Scots sent secret messages in code from her prison cell. Mary had in mind the overthrow of her cousin, Queen Elizabeth, but her coded messages were intercepted. More sophisticated than the Roman method of rotating all letters by a key, her codes were based on substitutions but ones whose key could be uncovered by analyzing the frequency of letters and symbols used. During the Second World War the German Enigma code was cracked by the discovery of its key. In this case it was a formidable challenge but the code was always vulnerable because the key was transmitted as part of the message.

A startling development in encryption of messages was discovered in the 1970s. Running counter to everything that had been previously believed, it said that the secret key could be broadcast to all and yet the message could remain entirely safe. This is called public key cryptography. The method depends on a 200-year-old theorem in a branch of mathematics glorified for being the most useless of all.

Public key encryption

Mr. John Sender, a secret agent known in the spying fraternity as "J," has just arrived in town and wants to send his minder Dr. Rodney Receiver a secret message to announce his arrival. What he does next is rather curious. He goes to the public library, takes a town directory off the shelf and looks up Dr. R. Receiver. In the directory he finds two numbers alongside Receiver's name — a long one, which is 247, and a short one, 5. This information is available to all and sundry, and it is all the information John Sender requires to encrypt his message, which for simplicity is his calling card, J. This letter is number 74 in a list of words, again publicly available.

Sender encrypts 74 by calculating 74^5 (modulo 247), that is, he wants to know the remainder on dividing 74^5 by 247. Working out 74^5 is just about possible on a handheld calculator, but it has to be done exactly:

$$74^5 = 74 \times 74 \times 74 \times 74 \times 74 = 2{,}219{,}006{,}624$$

and

$$2{,}219{,}006{,}624 = 8{,}983{,}832 \times 247 + 120$$

so dividing his huge number by 247 he gets the remainder 120. Sender's encrypted message is 120 and he transmits this to Receiver. Because the numbers 247 and 5 were publicly available anyone could encrypt a message. But not everyone could decrypt it. Dr. R. Receiver has more information up his sleeve. He made up his personal number 247 by multiplying together two prime numbers. In this case he obtained the number 247 by multiplying $p = 13$ and $q = 19$, but only he knows this.

This is where the ancient theorem due to Leonhard Euler is taken out and dusted down. Dr. R. Receiver uses the knowledge of $p = 13$ and $q = 19$ to find a value of a where $5 \times a \equiv 1$ modulo $(p - 1)(q - 1)$ where the symbol $\equiv$ means equals in modular arithmetic. What is a so that dividing $5 \times a$ by $12 \times 18 = 216$ leaves remainder 1? Skipping the actual calculation he finds $a = 173$.

Because he is the only one who knows the prime numbers p and q, Dr. Receiver is the only one who can calculate the number 173. With it he works out the remainder when he divides the huge number 120^{173} by 247. This is outside the capacity of a handheld calculator but is easily found by using a computer. The answer is 74, as Euler knew 200 years ago. With this information, Receiver looks up word 74 and sees that J is back in town.

You might say, surely a hacker could discover the fact that $247 = 13 \times 19$ and the code could be cracked. You would be correct. But the encryption and decryption principle is the same if Dr. Receiver had used another number instead of 247. He could choose two very big prime numbers and multiply them together to get a much larger number than 247.

Finding the two prime factors of a very large number is virtually impossible — what are the factors of 24,812,789,922,307 for example? But numbers much larger than this could also be chosen. The public key system is secure and if the might of supercomputers joined together are successful in factoring an encryption number, all Dr. Receiver has to do is increase its size still further. In the end it is considerably easier for Dr. Receiver to "mix boxes of black sand and white sand together" than for any hacker to unmix them.

96 Latin squares

For a few years the world has been Sudoku mad. Across the land, pens and pencils are chewed waiting for the right inspiration for the number to put in that box. Is it 4 or is it 5? Maybe it's 9. Commuters emerge from their trains in the mornings having expended more mental effort than they will for the rest of the day. In the evening the dinner burns in the oven. Is it 5, 4, or maybe 7? All these people are playing with Latin squares — they are being mathematicians.

	4		8		3			
		7						3
		9	7			2	6	
3				1		7		9
			6	9	8			
1		5		2				6
	2	3			6	5		
6						1		
		5		2		8		

Sudoku unlocked

In Sudoku we are given a 9×9 grid with some numbers filled in. The object is to fill in the rest using the given numbers as clues. Each row and each column should contain exactly one of the digits 1, 2, 3, …, 9, as do the small constituent 3×3 squares.

It is believed that Sudoku (meaning "single digits") was invented in the late 1970s. It gained popularity in Japan in the 1980s before sweeping to mass popularity by 2005. The appeal of the puzzle is that, unlike crosswords, you don't have to be widely read to attempt them but, like crosswords, they can be compelling. Addicts of both forms of self-torture have much in common.

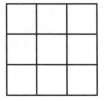

3×3 Latin squares

A square array containing exactly one symbol in each row and each column is called a Latin square. The number of symbols equals the size of the square and is called its "order." Can we fill out a blank 3×3 grid so that each row and column contains exactly one of the symbols *a*, *b* and *c*? If we can, this would be a Latin square of order 3.

In introducing the concept of a Latin square, Leonhard Euler called it a "new kind of magic square." Unlike magic squares, however, Latin squares are not concerned with arithmetic and the symbols do not have to be numbers. The reason for the name is simply that the symbols used to form them are taken from the Latin alphabet, while Euler used Greek with other squares.

A 3×3 Latin square can be easily written down.

If we think of *a*, *b* and *c* as the days of the week Monday, Wednesday and Friday, the square could be used to schedule meetings between two teams of people. Team One is made up of **Larry**, **Mary** and **Nancy** and Team Two of **Ross**, **Sophie** and **Tom**.

a	*b*	*c*
b	*c*	*a*
c	*a*	*b*

	R	S	T
L	*a*	*b*	*c*
M	*b*	*c*	*a*
N	*c*	*a*	*b*

For example, **Mary** from Team One, has a meeting with **Tom** from Team Two on Monday (the intersection of the **M** row with the **T** column is *a* = Monday). The Latin square arrangement ensures a meeting takes place between each pair of team members and there is no clash of dates.

This is not the only possible 3×3 Latin square. If we interpret *A*, *B* and *C* as topics discussed at the meetings between Team One and Team Two, we can produce a Latin square which ensures each person discusses a different topic with a member of the other team.

	R	S	T
L	*A*	*B*	*C*
M	*C*	*A*	*B*
N	*B*	*C*	*A*

So **Mary** from Team One discusses topic *C* with **Ross**, topic *A* with **Sophie** and topic *B* with **Tom**.

But *when* should the discussions take place, between *whom*, and on *what* topic? What would be the schedule for this complex organization? Fortunately the two Latin squares can be combined symbol by symbol to produce a composite Latin square in which each of the possible nine pairs of days and topics occurs in exactly one position.

	R	S	T
L	*a,A*	*b,B*	*c,C*
M	*b,C*	*c,A*	*a,B*
N	*c,B*	*a,C*	*b,A*

Another interpretation for the square is the historical "nine officers problem" in which nine officers belonging to three regiments *a*, *b* and *c* and of three ranks *A*, *B* and *C* are placed on the parade ground so that each row and column contains an officer of each regiment and rank. Latin squares which combine in this way are called "orthogonal." The 3×3 case is straightforward but finding pairs of orthogonal Latin squares for some larger ones is far from easy. This is something Euler discovered.

In the case of a 4×4 Latin square, a "16 officers problem" would be to arrange the 16 court cards in a pack of cards in a square in such a way that there is one rank (Ace, King, Queen or Jack) and one suit (spades, clubs, hearts or diamonds) in each row and column. In 1782 Euler posed the same problem for "36 officers." In essence he was looking for two orthogonal squares of order 6. He couldn't find them and conjectured there were no pairs of orthogonal Latin squares of orders 6, 10, 14, 18, 22 … Could this be proved?

Along came Gaston Tarry, an amateur mathematician who worked as a civil servant in Algeria. He scrutinized examples and by 1900 had verified Euler's conjecture in one case: there is no pair of orthogonal Latin squares of order 6. Mathematicians naturally assumed Euler was correct in the other cases 10, 14, 18, 22 …

In 1960, the combined efforts of three mathematicians stunned the mathematical world by proving Euler wrong in *all* the other cases. Raj Bose, Ernest Parker and Sharadchandra Shrikhande proved there were indeed pairs of orthogonal Latin squares of orders 10, 14, 18, 22, … The *only* case where Latin squares do not exist (apart from trivial ones of orders 1 and 2) is order 6.

We've seen that there are two mutually orthogonal order 3 Latin squares. For order 4 we can produce three squares which are mutually orthogonal to each other. It can be shown that there are never more than *n* – 1 mutually orthogonal Latin squares of order *n*, so for *n* = 10, for example, there cannot be more than nine mutually orthogonal squares. But finding them is a different story. To date, no one has been able even to produce three Latin squares of order 10 that are mutually orthogonal to each other.

Are Latin squares useful?

R.A. Fisher, an eminent statistician, saw the practical use of Latin squares. He used them to revolutionize agricultural methods during his time at Rothamsted Research Station in Hertfordshire, UK.

Fisher's objective was to investigate the effectiveness of fertilizers on crop yield. Ideally we would want to plant crops in identical soil conditions so that soil quality wasn't an unwanted factor influencing crop yield. We could then apply the different fertilizers safe in the knowledge that the "nuisance" of soil quality was eliminated. The only way of ensuring identical soil conditions would be to use the *same* soil — but it is impractical to keep digging up and replanting crops. Even if this were possible different weather conditions could become a new nuisance.

A way round this is to use Latin squares. Let's look at testing four treatments. If we mark out a square field into 16 plots we can envisage the Latin square as a description of the field where the soil quality varies "vertically" and "horizontally."

The four fertilizers are then applied at random in the scheme labeled *a*, *b*, *c* and *d*, so exactly one is applied in each row and column in an attempt to remove the variation of soil quality. If we suspect another factor might influence crop yield, we could deal with this too. Suppose we think that the time of day of applying treatment is a factor. Label four time zones during the day as *A*, *B*, *C* and *D* and use orthogonal Latin squares as the design for a scheme to gather data. This ensures each treatment and time zone is applied in one of the plots. The design for the experiment would be:

a, time *A*	*b*, time *B*	*c*, time *C*	*d*, time *D*
b, time *C*	*a*, time *D*	*d*, time *A*	*c*, time *B*
c, time *D*	*d*, time *C*	*a*, time *B*	*b*, time *A*
d, time *B*	*c*, time *A*	*b*, time *D*	*a*, time *C*

Other factors can be screened out by going on to create even more elaborate Latin square designs. Euler could not have dreamt of the solution to his officers' problem being applied to agricultural experiments.

97 The diet problem

Tanya Smith takes her athletics very seriously. She goes to the gym every day and monitors her diet closely. Tanya makes her way in the world by taking part-time jobs and has to watch where the money goes. It is crucial that she takes the right amount of minerals and vitamins each month to stay fit and healthy. The amounts have been determined by her coach. He suggests that future Olympic champions should absorb at least 120 milligrams (mg) of vitamins and at least 880 mg of minerals each month. To make sure she follows this regime Tanya relies on two food supplements. One is in solid form and has the trade name Solido and the other is in liquid form marketed under the name Liquex. Her problem is to decide how much of each she should purchase each month to satisfy her coach.

The classic diet problem is to organize a healthy diet and pay the lowest price for it. It was a prototype for problems in linear programing, a subject developed in the 1940s that is now used in a wide range of applications.

At the beginning of March Tanya takes a trip to the supermarket and checks out Solido and Liquex. On the back of a packet of Solido she finds out it contains 2 mg vitamins and 10 mg minerals, while a carton of Liquex contains 3 mg vitamins and 50 mg minerals. She dutifully fills her trolley with 30 packets of Solido and 5 cartons of Liquex to keep herself going for the month. As she proceeds towards the checkout she wonders if she has the right amount.

	Solido	Liquex	Requirements
Vitamins	2 mg	3 mg	120 mg
Minerals	10 mg	50 mg	880 mg

First she calculates how many vitamins she has in the trolley. In the 30 packets of Solido she has $2 \times 30 = 60$ mg vitamins and in the Liquex, $3 \times 5 = 15$. Altogether she has $2 \times 30 + 3 \times 5 = 75$ mg vitamins. Repeating the calculation for minerals, she has $10 \times 30 + 50 \times 5 = 550$ mg minerals.

As the coach required her to have at least 120 mg vitamins and 880 mg minerals, she needs more packets and cartons in the trolley. Tanya's problem is juggling the right amounts of Solido and Liquex with the vitamin and mineral requirements. She goes back to the health section of the supermarket and puts more packets and cartons into her trolley. She now has 40 packets and 15 cartons. Surely this will be OK? She recalculates and finds she has $2 \times 40 + 3 \times 15 = 125$ mg vitamins and $10 \times 40 + 50 \times 15 = 1,150$ mg minerals. Now Tanya certainly satisfies her coach's recommendation and has even exceeded the required amounts.

Feasible solutions

The combination (40, 15) of foods will enable Tanya to satisfy the diet. This is called a possible combination, or a "feasible" solution. We have seen already that (30, 5) is not a feasible solution so there is a demarcation between the two types of combinations — feasible solutions in which the diet is fulfilled and non-feasible solutions in which it is not.

Tanya has many more options. She could fill her trolley with only Solido. If she did this she would need to buy at least 88 packets. The purchase (88, 0) satisfies both requirements, because this combination would contain $2 \times 88 + 3 \times 0 = 176$ mg vitamins and $10 \times 88 + 50 \times 0 = 880$ mg minerals. If she bought only Liquex she would need at least 40 cartons, the feasible solution (0, 40) satisfies both vitamin and mineral requirements, because $2 \times 0 + 3 \times 40 = 120$ mg vitamins and $10 \times 0 + 50 \times 40 = 2,000$ mg minerals. We may notice that the intake of vitamins and minerals is not met exactly with any of these possible combinations though the coach will certainly be satisfied Tanya is having enough.

Optimum solutions

Money is now brought into the equation. When Tanya gets to the checkout she must pay for the purchases. She notes that the packets and cartons are equally priced at $5 each. Of the feasible combinations we have found so far (40, 15), (88, 0) and (0, 40) the bills would be $275, $440 and $200, respectively so the best solution so far will be to buy no Solido and 40 cartons of Liquex. This will be the least cost purchase and the dietary requirement will be achieved. But how much food to buy has been hit and miss. On the spur of the moment Tanya has tried various combinations of Solido and Liquex and figured out the cost in

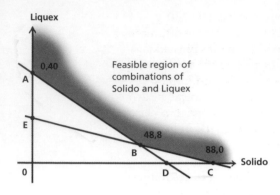

Liquex

0,40

A

Feasible region of
combinations of
Solido and Liquex

E

48,8

B

88,0

0

D

C

Solido

these cases only. Can she do better? Is there a possible combination of Solido and Liquex that will satisfy her coach and at the same time cost her the least? What she would like to do is to go home and analyze the problem with a pencil and paper.

Linear programing problems

Tanya's always been coached to visualize her goals. If she can apply this to winning Olympic gold, why not to mathematics? So she draws a picture of the feasible region. This is possible because she is only considering two foods. The line AD represents the combinations of Solido and Liquex that contain exactly 120 mg vitamins. The combinations above this line have more than 120 mg vitamins. The line EC represents the combinations that contain exactly 880 mg minerals. The combinations of foods that are above both these lines is the feasible region and represents all the feasible combinations Tanya could buy.

Problems with the framework of the diet problem are called linear programing problems. The word "programing" means a procedure (its usage before it became synonymous with computers) while "linear" refers to the use of straight lines. To solve Tanya's problem with linear programing, mathematicians have shown that all we need to do is to work out the size of the food bill at the corner points on Tanya's graph.Tanya has discovered a new feasible solution at the point B with coordinates (48, 8) which means that she could purchase 48 packets of Solido and 8 cartons of Liquex. If she did this she would satisfy her diet exactly because in this combination there is 120 mg of vitamins and 880 mg of minerals. At $5 for both a packet and a carton this combination would cost her $280. So the optimum purchase will remain as it was before, that is, she should purchase no Solido at all and 40 cartons of Liquex at a total cost of $200, even though she will have 1,120 mg of vitamins in excess of the 880 mg required.

The optimum combination ultimately depends on the relative costs of the supplements. If the cost per packet of Solido went down to $2 and Liquex went up to $7 then the bills for the corner point combinations A (0, 40), B (48, 8) and C (88, 0) would be respectively $280, $152 and $176.

The best purchase for Tanya with these prices is 48 packets of Solido and 8 cartons of Liquex, with a bill of $152.

History

In 1947 the American mathematician George Dantzig, then working for the U.S. Air Force, formulated a method for solving linear programing problems called the "simplex method". It was so successful that Dantzig became known in the West as the father of linear programing. In Soviet Russia, cut off during the Cold War, Leonid Kantorovich independently formulated a theory of linear programing. In 1975, Kantorovich and the Dutch mathematician Tjalling Koopmans were awarded the Nobel Prize for Economics for work on the allocation of resources, which included linear programing techniques.

Tanya considered only two foods — two variables — but nowadays problems involving thousands of variables are commonplace. When Dantzig found his method there were few computers but there was the Mathematical Tables Project — a decade-long job creation scheme which began in New York in 1938. It took a team of some ten human calculators working for 12 days with hand calculators to solve a diet problem in nine "vitamin" requirements and 77 variables.

While the simplex method and its variants have been phenomenally successful, other methods have also been tried. In 1984 the Indian mathematician Narendra Karmarkar derived a new algorithm of practical significance, and the Russian Leonid Khachiyan proposed one of chiefly theoretical importance.

The basic linear programing model has been applied to many situations other than choosing a diet. One type of problem, the transportation problem, concerns itself with transporting goods from factories to warehouses. It is of a special structure and has become a field in its own right. The objective in this case is to minimize the cost of transportation. In some linear programing problems the objective is to maximize (like maximizing profit). In other problems the variables only take integer values or just two values 0 or 1, but these problems are quite different and require their own solution procedures.

It remains to be seen whether Tanya Smith wins her gold medal at the Olympic Games. If so, it will be another triumph for linear programing.

Some said Johnny was the smartest person alive. John von Neumann was a child prodigy who became a legend in the mathematical world. When people heard that he arrived at a meeting in a taxi having just scribbled out his "minimax theorem" in game theory, they just nodded. It was exactly the sort of thing von Neumann did. He made contributions to quantum mechanics, logic, algebra, so why should game theory escape his eye? It didn't — with Oskar Morgenstern he coauthored the influential *Theory of Games and Economic Behavior*. In its widest sense game theory is an ancient subject, but von Neumann was key to sharpening the theory of the "two-person zero-sum game."

Two-person zero-sum games

It sounds complicated, but a two-person zero-sum game is simply one "played" by two people, companies, or teams, in which one side wins what the other loses. If A wins $200 then B loses that $200; that's what zero-sum means. There is no point in A cooperating with B — it is pure competition with only winners and losers. In "win–win" language A wins $200 and B wins –$200 and the sum is 200 + (–200) = 0. This is the origin of the term "zero-sum."

Let's imagine two TV companies ATV and BTV are bidding to operate an extra news service in either Scotland or England. Each company must make a bid for one country only and they will base their decision on the projected increased size of their viewing audiences. Media analysts have estimated the increased audiences and both companies have access to their research. These are conveniently set down in a "payoff table" and measured in units of a million viewers.

If both ATV and BTV decide to operate in Scotland then ATV will gain 5 million viewers, but BTV will *lose* 5 million viewers. The

		BTV	
		Scotland	England
ATV	Scotland	+5	–3
	England	+2	+4

meaning of the minus sign, as in the payoff −3, is that ATV will lose an audience of 3 million. The + payoffs are good for ATV and the − payoffs are good for BTV.

We'll assume the companies make their *one-off* decisions on the basis of the payoff table and that they make their bids simultaneously by sealed bids. Obviously both companies act in their own best interests.

If ATV chooses Scotland the worst that could happen would be a loss of 3 million; if it bids for England, the worst would be a gain of 2 million. The obvious strategy for ATV would be to choose England (row 2). It couldn't do worse than gain 2 million viewers whatever BTV chooses. Looking at it numerically, ATV works out −3 and 2 (the row minimums) and chooses the row corresponding to the maximum of these.

BTV is in a weaker position but it can still work out a strategy that limits its potential losses and hope for a better payoff table next year. If BTV chooses Scotland (column 1) the worst that could happen would be a loss of 5 million; if it chooses England, the worst would be loss of 4 million. The safest strategy for BTV would be to choose England (column 2) for it would rather lose an audience of 4 million than 5 million. It couldn't do worse than lose 4 million viewers whatever ATV decides.

These would be the safest strategies for each player and, if followed, ATV would gain 4 million extra viewers while BTV loses them.

When is a game determined?

The following year, the two TV companies have an added option — to operate in Wales. Because circumstances have changed there is a new payoff table.

	BTV			
	Wales	Scotland	England	row minimum
ATV Wales	+3	+2	+1	+1
Scotland	+4	−1	0	−1
England	−3	+5	−2	−3
column maximum	+4	+5	+1	

As before, the safe strategy for ATV is to choose the row which maximizes the worst that can happen. The maximum from {+1, −1, −3} is to choose Wales (row 1). The safe strategy for BTV is to choose the column which minimizes from {+4, +5, +1}. That is England (column 3).

By choosing Wales, ATV can *guarantee* to win no less than 1 million viewers whatever BTV does, and by choosing England (column 3), BTV can *guarantee* to lose no more than 1 million viewers whatever ATV does. These choices therefore represent the *best* strategies for each company, and in this sense the game is determined (but it is still unfair to BTV). In this game the

$$\text{maximum of } \{+1, -1, -3\} = \text{minimum of } \{+4, +5, +1\}$$

and both sides of the equation have the common value of +1. Unlike the first game, this version has a "saddle-point" equilibrium of +1.

Repetitive games

The iconic repetitive game is the traditional game of "paper, scissors, stone." Unlike the TV company game which was a one-off, this game is usually played half a dozen times, or a few hundred times by competitors in the annual World Championships.

	paper	scissors	stone	row minimum
paper	draw = 0	lose = −1	win = +1	−1
scissors	win = +1	draw = 0	lose = −1	−1
stone	lose = −1	win = +1	draw = 0	−1
column maximum	+1	+1	+1	

In "paper, scissors, stone," two players show either a hand, two fingers or a fist, each symbolizing paper, scissors or stone. They play simultaneously on the count of three: paper draws with paper, is defeated by scissors (since scissors can cut paper), but defeats stone (because it can wrap stone). If playing "paper" the payoffs are therefore 0, –1, +1, which is the top row of our completed payoff table.

There is no saddle point for this game and no obvious *pure* strategy to adopt. If a player always chooses the same action, say paper, the opponent will detect this and simply choose scissors to win every time. By von Neumann's "minimax theorem" there is a "mixed strategy" or a way of choosing different actions based on probability.

According to the mathematics, players should choose randomly but overall the choices of paper, scissors, stone should each be made a third of the time. "Blind" randomness may not always be the best course, however, as world champions have ways of choosing their strategy with a little "psychological" spin. They are good at second-guessing their opponents.

When is a game *not* zero-sum?

Not every game is zero-sum — each player sometimes has their own separate payoff table. A famous example is the "prisoner's dilemma" designed by A.W. Tucker.

Two people, **Andrew** and **Bertie**, are picked up by the police on suspicion of highway robbery and held in separate cells so they cannot confer with each other. The payoffs, in this case jail sentences, not only depend on their individual responses to police questioning but on how they jointly respond. If **A** confesses and **B** doesn't then **A** gets only a one-year sentence (from **A**'s payoff table) but **B** is sentenced to ten years (from **B**'s payoff table). If **A** doesn't confess but **B** does, the sentences go the other way around. If both confess they get four years each but if neither confesses and they both maintain their innocence they get off scot-free!

A		B	
		confess	not confess
A	confess	+4	+1
	not confess	+10	0

B		B	
		confess	not confess
A	confess	+4	+10
	not confess	+1	0

If the prisoners could cooperate they would take the optimum course of action and not confess — this would be the "win–win" situation.

99 Fermat's last theorem

We can add two square numbers together to make a third square. For instance, $5^2 + 12^2 = 13^2$. But can we add two cubed numbers together to make another cube? What about higher powers? Remarkably, we cannot. Fermat's last theorem says that for any four whole numbers, x, y, z and n, there are no solutions to the equation $x^n + y^n = z^n$ when n is bigger than 2. Fermat claimed he'd found a "wonderful proof," tantalizing the generations of mathematicians that followed, including a ten-year-old boy who read about this mathematical treasure hunt one day in his local library.

Fermat's last theorem is about a Diophantine equation, the kind of equation which poses the stiffest of all challenges. These equations demand that their solutions be whole numbers. They are named after Diophantus of Alexandria whose *Arithmetica* became a milestone in the theory of numbers. Pierre de Fermat was a 17th-century lawyer and government official in Toulouse in France. A versatile mathematician, he enjoyed a high reputation in the theory of numbers, and is most notably remembered for the statement of the last theorem, his final contribution to mathematics. Fermat proved it, or thought he had, and he wrote in his copy of Diophantus' *Arithmetica* "I have discovered a truly wonderful proof, but the margin is too small to contain it."

Fermat solved many outstanding problems, but it seems that Fermat's last theorem was not one of them. The theorem has occupied legions of mathematicians for 300 years, and has only recently been proved. This proof could not be written in any margin and the modern techniques required to generate it throw extreme doubt on Fermat's claim.

The equation $x + y = z$

How can we solve this equation in three variables x, y and z? In an equation we usually have one unknown, x, but here we have three.

Actually this makes the equation $x + y = z$ quite easy to solve. We can choose the values of x and y any way we wish, add them to get z and these three values will give a solution. It is as simple as that.

For example, if we choose $x = 3$ and $y = 7$, the values $x = 3$, $y = 7$ and $z = 10$ make a solution of the equation. We can also see that some values of x, y and z are not solutions of the equation. For example, $x = 3$, $y = 7$ and $z = 9$ is not a solution because these values do not make the left-hand side $x + y$ equal the right hand side z.

The equation $x^2 + y^2 = z^2$

We'll now think about squares. The square of a number is that number multiplied by itself, the number which we write as x^2. If $x = 3$ then $x^2 = 3 \times 3 = 9$. The equation we are thinking of now is not $x + y = z$, but

$$x^2 + y^2 = z^2$$

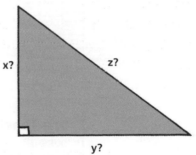

Can we solve this as before, by choosing values for x and y and computing z? With the values $x = 3$ and $y = 7$, for example, the left-hand side of the equation is $3^2 +7^2$ which is $9 + 49 = 58$. For this z would have to be the square root of 58 ($z = \sqrt{58}$) which is approximately 7.6158. We are certainly entitled to claim that $x = 3$, $y = 7$ and $z = \sqrt{58}$ is a solution of $x^2 + y^2 = z^2$ but unfortunately Diophantine equations are primarily concerned with whole number solutions. As $\sqrt{58}$ is not a whole number, the solution $x = 3$, $y = 7$ and $z = \sqrt{58}$ will not do.

The equation $x^2 + y^2 = z^2$ is connected with triangles. If x, y and z represent the lengths of the three sides of a right-angled triangle they satisfy this equation. Conversely, if x, y and z satisfy the equation then the angle between x and y is a right angle. Because of the connections with Pythagoras' theorem solutions for x, y and z are called Pythagorean triples.

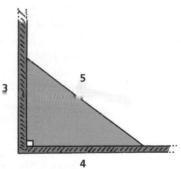

How can we find Pythagorean triples? This is where the local builder comes to the rescue. Part of the builder's equipment is the ubiquitous 3–4–5 triangle. The values $x = 3$, $y = 4$ and $z = 5$ turn out to be a solution of the kind we are looking for because $3^2 + 4^2 = 9 + 16 = 5^2$. From the converse, a triangle with dimensions 3, 4 and 5, must include a right angle. This is the mathematical fact that the builder uses to build his walls at right angles.

In this case we can break up a 3×3 square, and wrap it around a 4×4 square to make a 5×5 square.

There are other whole number solutions $x^2 + y^2 = z^v$. For example $x = 5, y = 12$ and $z = 13$ is another solution because $5^2 + 12^2 = 13^2$ and in fact there are an infinite number of solutions to the equation. The builder's solution $x = 3, y = 4$ and $z = 5$ holds pride of place since it is the smallest solution, and is the only solution composed of consecutive whole numbers. There are many solutions where two numbers are consecutive, such as $x = 20, y = 21$ and $z = 29$ as well as $x = 9, y = 40$ and $z = 41$, but no others with all three.

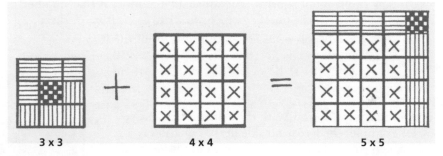

3 x 3   4 x 4   5 x 5

From feast to famine

It looks like a small step to go from $x^2 + y^2 = z^2$ to $x^3 + y^3 = z^3$. So, following the idea of reassembling one square around another to form a third square, can we pull off the same trick with a cube? Can we reassemble one cube around another to make a third? It turns out this can't be done. The equation $x^2 + y^2 = z^2$ has an infinite number of different solutions but Fermat was unable to find even one whole number example of $x^3 + y^3 = z^3$. Worse was to follow, and Leonhard Euler's lack of findings led him to phrase the last theorem:

> *There is no solution in whole numbers to the equation $x^n + y^n = z^n$ for all values of n higher than 2.*

One way to approach the problem of proving this is to start on the low values of n and move forward. This was the way Fermat went to work. The case $n = 4$ is actually simpler than $n = 3$ and it is likely Fermat had a proof in this case. In the 18th and 19th centuries, Euler filled in the case $n = 3$, Adrien-Marie Legendre completed the case $n = 5$ and Gabriel Lamé proved the case $n = 7$. Lamé initially thought he had a proof of the general theorem but was unfortunately mistaken.

Ernst Kummer was a major contributor and in 1843 submitted a manuscript claiming he had proved the theorem in general, but Dirichlet pointed out a gap in the argument. The French Academy of Sciences offered a prize of 3,000 francs for a valid proof, eventually awarding it to Kummer for his worthy attempt. Kummer proved the theorem for all primes less than 100 (and other values) but excluding the irregular primes 37, 59 and 67. For example, he could not prove there were no whole numbers which satisfied $x^{67} + y^{67} = z^{67}$. His failure to prove the theorem generally opened up valuable techniques in abstract algebra. This was perhaps a greater contribution to mathematics than settling the question itself.

Ferdinand von Lindemann, who did prove that the circle could not be squared (see Chapter 68) claimed to have proved the theorem in 1907 but was found to be in error. In 1908 Paul Wolfskehl bequeathed a 100,000 marks prize to be awarded to the first provider of a proof, a prize made available for 100 years. Over the years, something like 5,000 proofs have been submitted, checked and all returned to the hopefuls as being false.

The proof

While the link with Pythagoras' theorem only applies for $n = 2$, the link with geometry proved the key to its eventual proof. The connection was made with the theory of curves and a conjecture put forward by two Japanese mathematicians Yutaka Taniyama and Goro Shimura. In 1993 Andrew Wiles gave a lecture on this theory at Cambridge, England, and included his proof of Fermat's theorem. Unhappily this proof was wrong.

The similarly named French mathematician André Weil dismissed such attempts. He likened proving the theorem with climbing Everest and added that if a man falls short by 100 yards he has not climbed Everest. The pressure was on. Wiles cut himself off and worked on the problem incessantly. Many thought Wiles would join that throng of the nearly people.

With the help of colleagues, however, Wiles was able to excise the error and substitute a correct argument. This time he convinced the experts and proved the theorem. His proof was published in 1995 and he claimed the Wolfskehl prize just inside the qualifying period to become a mathematical celebrity. The ten-year-old boy sitting in a Cambridge public library reading about the problem years before had come a long way.

100 The Riemann hypothesis

The Riemann hypothesis represents one of the stiffest challenges in pure mathematics. The Poincaré conjecture and Fermat's last theorem have been conquered but not the Riemann hypothesis. Once decided, one way or the other, elusive questions about the distribution of prime numbers will be settled and a range of new questions will be opened up for mathematicians to ponder.

The story starts with the addition of fractions of the kind

$$1 + \frac{1}{2} + \frac{1}{3}$$

The answer is $^{15}/_6$ (approximately 1.83). But what happens if we keep adding smaller and smaller fractions, say up to ten of them?

$$1 + \frac{1}{2} + \frac{1}{3} + \frac{1}{4} + \frac{1}{5} + \frac{1}{6} + \frac{1}{7} + \frac{1}{8} + \frac{1}{9} + \frac{1}{10}$$

Using only a handheld calculator, these fractions add up to approximately 2.9 in decimals. A table shows how the total grows as more and more terms are added.

Number of terms	Total (approximate)
1	1
10	2.9
100	5.2
1,000	7.5
10,000	9.8
100,000	12.1
1,000,000	14.4
1,000,000,000	21.3

The series of numbers

$$1 + \frac{1}{2} + \frac{1}{3} + \frac{1}{4} + \frac{1}{5} + \frac{1}{6} + \cdots$$

is called the harmonic series. The harmonic label originates with the Pythagoreans who believed that a musical string divided by a half, a third, a quarter, gave the musical notes essential for harmony.

In the harmonic series, smaller and smaller fractions are being added but what happens to the total? Does it grow beyond all numbers, or is there a barrier somewhere, a limit that it never rises above? To answer this, the trick is to group the terms, doubling the runs as we go. If we add the first 8 terms (recognizing that $8 = 2 \times 2 \times 2 = 2^3$), for example

$$S_{2^3} = 1 + \frac{1}{2} + \left(\frac{1}{3} + \frac{1}{4}\right) + \left(\frac{1}{5} + \frac{1}{6} + \frac{1}{7} + \frac{1}{8}\right)$$

(where S stands for sum) and, because $\frac{1}{3}$ is bigger than $\frac{1}{4}$ and $\frac{1}{5}$ is bigger than $\frac{1}{8}$ (and so on), this is greater than

$$1 + \frac{1}{2} + \left(\frac{1}{4} + \frac{1}{4}\right) + \left(\frac{1}{8} + \frac{1}{8} + \frac{1}{8} + \frac{1}{8}\right) = 1 + \frac{1}{2} + \frac{1}{2} + \frac{1}{2}$$

So we can say

$$S_{2^3} > 1 + \frac{3}{2}$$

and more generally

$$S_{2^k} > 1 + \frac{k}{2}$$

If we take $k = 20$, so that $n = 2^{20} = 1{,}048{,}576$ (more than a million terms), the sum of the series will only have exceeded 11 (see table). It is increasing in an excruciatingly slow way — but, a value of k can be chosen to make the series total beyond *any* preassigned number, however large. The series is said to diverge to infinity. By contrast, this does not happen with the series of squared terms

$$1 + \frac{1}{2^2} + \frac{1}{3^2} + \frac{1}{4^2} + \frac{1}{5^2} + \frac{1}{6^2} + \cdots$$

We are still using the same process: adding smaller and smaller numbers together, but this time a limit is reached, and this limit is less than 2. Quite dramatically the series converges to $\pi^2/6 = 1.64493\ldots$

In this last series the power of the terms is 2. In the harmonic series the power of the denominators is silently equal to 1 and this value is critical. If the power increases by a minuscule amount to a number just above 1 the series converges, but if the power decreases by a minuscule amount to a value just below 1, the series diverges. The harmonic series sits on the boundary between convergence and divergence.

The Riemann zeta function

The celebrated Riemann zeta function $\zeta(s)$ was actually known to Euler in the 18th century but Bernhard Riemann recognized its full importance. The ζ is the Greek letter zeta, while the function is written as:

$$\zeta(s) = 1 + \frac{1}{2^s} + \frac{1}{3^s} + \frac{1}{4^s} + \frac{1}{5^s} + \cdots$$

Various values of the zeta function have been computed, most prominently, $\zeta(1) = \infty$ because $\zeta(1)$ is the harmonic series. The value of $\zeta(2)$ is $\pi^2/6$, the result discovered by Euler. It has been shown that the values of $\zeta(s)$ all involve π when s is an even number while the theory of $\zeta(s)$ for odd values of s is far more difficult. Roger Apéry proved the important result that $\zeta(3)$ is an irrational number but his method did not extend to $\zeta(5)$, $\zeta(7)$, $\zeta(9)$, and so on.

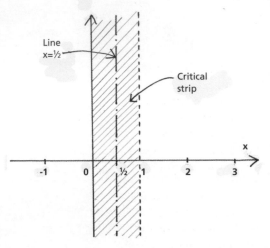

Line x=½

Critical strip

x

-1 0 ½ 1 2 3

The Riemann hypothesis

The variable s in the Riemann zeta function represents a real variable but this can be extended to represent a complex number (see Chapter 71). This enables the powerful techniques of complex analysis to be applied to it.

The Riemann zeta function has an infinity of zeros, that is, an infinity of values of s for which $\zeta(s) = 0$. In a paper presented to the Berlin Academy of Sciences in 1859, Riemann showed all the important zeros were complex numbers that lay in the critical strip bounded by $x = 0$ and $x = 1$. He also made his famous hypothesis:

All the zeros of the Riemann zeta function $\zeta(s)$ lie on the line $x = \frac{1}{2}$; the line along the middle of the critical strip.

The first real step towards settling this hypothesis was made independently by Charles de la Vallée-Poussin and Jacques Hadamard in 1896. They showed that the zeros must lie on the interior of the strip (so x could not equal 0 or 1). In 1914, the English mathematician G.H. Hardy proved that an infinity of zeros lie along the line $x = \frac{1}{2}$ though this does not prevent there being an infinity of zeros lying off it.

As far as numerical results go, the non-trivial zeros calculated by 1986 (1,500,000,000 of them) do lie on the line $x = \frac{1}{2}$ while up-to-date calculations have verified this is also true for the first 100 billion zeros. While these experimental results suggest that the conjecture is reasonable, there is still the possibility that it may be false. The conjecture is that all zeros lie on this critical line, but this awaits proof or disproof.

Why is the Riemann hypothesis important?

There is an unexpected connection between the Riemann zeta function $\zeta(s)$ and the theory of prime numbers (see Chapter 72). The prime numbers are 2, 3, 5, 7, 11 and so on, the numbers only divisible by 1 and themselves. Using primes, we can form the expression

$$\left(1-\frac{1}{2^s}\right)\times\left(1-\frac{1}{3^s}\right)\times\left(1-\frac{1}{5^3}\right)\times\cdots$$

and this turns out to be another way of writing $\zeta(s)$, the Riemann zeta function. This tells us that knowledge of the Riemann zeta function will throw light on the distribution of prime numbers and enhance our understanding of the basic building blocks of mathematics.

In 1900, David Hilbert set out his famous 23 problems for mathematicians to solve. He said of his eighth problem, "if I were to awaken after having slept for 500 years, my first question would be: Has the Riemann hypothesis been proven?"

Hardy used the Riemann hypothesis as insurance when crossing the North Sea after his summer visit to his friend Harald Bohr in Denmark. Before leaving the port he would send his friend a postcard with the claim that he had just proved the Riemann hypothesis. It was a clever each way bet. If the boat sank he would have the posthumous honor of solving the great problem. On the other hand, if God did exist he would not let an atheist like Hardy have that honor and would therefore prevent the boat from sinking.

The person who can rigorously resolve the issue will win a prize of a million dollars offered by the Clay Mathematics Institute. But money is not the driving force — most mathematicians would settle for achieving the result and a very high position in the pantheon of great mathematicians.

GLOSSARY

GENETICS

Allele Alternative variant of a gene. Individuals will generally have two alleles of each gene, which may vary.

Amino acids Molecules from which proteins are built. All life uses 20 different amino acids, the instructions of which are carried by codons or triplets of DNA and RNA.

Chromosome Strand of DNA that carries genes and other genetic information. Humans have 22 – 46 pairs of autosomes, and 2 sex chromosomes.

Codon (Triplet) A string of three DNA or RNA bases that codes for production of an amino acid.

DNA Deoxyribonucleic acid, the molecule that holds the genetic instructions for most forms of life. Its structure is a double helix.

Dominant Allele that is always expressed, even when it differs from the other allele as in heterozygotes.

Exons The units within genes that contain protein-coding information. They are broken up by introns.

Gamete Reproductive cells, which contain only half the usual complement of chromosomes. In humans, these are sperm and eggs, which each contain 23 unpaired chromosomes.

Gene The fundamental unit of inheritance. Usually taken to mean a section of DNA that contains the code for a protein, but the definition is being widened to include DNA that carries other genetic instructions. The full genetic code of an organism is the genome.

Haplotype A sub-section of a chromosome that tends to remain intact during recombination. Haplotype blocks are responsible for genetic linkage.

Heterozygote (Heterozygous adj) An individual with two different alleles of a particular gene or DNA sequence.

Homozygote (Homozygous adj) An individual with two identical alleles of a particular gene or DNA sequence.

Imprinted gene A gene marked so it is expressed according to maternal or paternal origin.

Introns Non-coding passages of DNA that break up the coding regions, or exons, of genes.

In-vitro fertilization (IVF) Assisted reproduction procedure, by which eggs are fertilized by sperm in the laboratory, and resulting embryos are then transferred to the womb.

Junk DNA DNA that does not code for protein. Much of it, however, is transcribed into RNA and regulates gene expression.

Meiosis Cell division process by which germ cells create gametes. Cells produced by meiosis have only one set of chromosomes instead of the usual two. Recombination occurs during meiosis.

Methylation Process by which DNA is chemically modified, often associated with gene silencing. Important to epigenetics and imprinting.

Mitochondria Cellular structures outside the nucleus that generate energy, which contain DNA. Mitochondria are always inherited from the mother, and mitochondrial DNA is useful for tracing maternal ancestry.

Mitosis Normal process of cell division, by which a cell copies its genetic material and splits. The resulting daughter cells carry the same DNA as their parents, save for any random mutations.

Mutation Process by which the DNA sequence is altered. This can involve substitution of one base for another. It can occur randomly, as a result of copying errors, or through damage by radiation or chemicals.

Nucleotide (base) A DNA or RNA "letter," in which the genetic code is spelled. DNA's nucleotides are adenine (A), cytosine (C), guanine (G) and thymine (T). In RNA, uracil (U) is used in place of thymine.

Nucleus Cellular structure containing the chromosomes and most of an organism's DNA. Organisms with nuclei are known as *eukaryotes*.

Phenotype An observed characteristic of an organism, which can be influenced by either inheritance or the environment.

Protein Large organic compound made from a long chain of amino acids. Many proteins are enzymes that catalyze chemical reactions in cells. Others are structural, like collagen.

Recessive Allele that is expressed only when two copies are present, in homozygotes.

Recombination (crossing-over) Process that occurs during meiosis, by which chromosomes exchange chunks of genetic material.

RNA Ribonucleic acid, a chemical cousin of DNA, usually single-stranded, which carries genetic messages inside cells. A key type is messenger RNA, which translates DNA into protein.

Single nucleotide polymorphism (SNP) Point at which the genetic code commonly differs by one base between individuals of the same species. Standard form of genetic variation.

Transcription Process by which DNA is copied into RNA to make proteins and regulate gene expression.

Translation Process by which messenger RNA is used to manufacture protein, in structures called ribosomes.

PHYSICS

Acceleration The change in something's speed in a given time.

Atom The smallest unit of matter that can exist independently. Atoms contain a hard central nucleus made up of (positively charged) protons and (uncharged) neutrons surrounded by clouds of (negatively charged) electrons.

Black-body radiation Light glow emitted by a black object at a specific temperature, which has a characteristic spectrum.

Cosmic microwave background radiation A faint microwave glow that fills the sky. It is the afterglow of the big bang that has since cooled and been redshifted to a temperature of 3 kelvins.

Electricity The flow of electric charge. It has some voltage (energy), may cause a current (a flow) and can be slowed or blocked by resistance.

Energy A property of something that dictates its potential for change. It is conserved overall but can be exchanged between many different types.

Entanglement In quantum theory, the idea that particles that are related at one point in time carry away information with them thereafter and can be used for instantaneous signaling.

Entropy A measure of disorder. The more ordered something is, the lower its entropy.

Fields A means of transmitting a force at a distance. Electricity and magnetism are fields, as is gravity.

Force A lift, pull or push, causing the motion of something to change. Newton's second law defines a force as being proportional to the acceleration it produces.

Frequency The rate at which wave crests pass some point.

Gravity A fundamental force through which masses attract one another. Gravity is described by Einstein's theory of general relativity.

Inertia *see* Mass

Interference The combining of waves of different phases that may produce reinforcement (if in phase) or cancellation (if out of phase).

Many-worlds hypothesis In quantum theory and cosmology, the idea that there are many parallel universes that branch off as events occur, and that we are at any time in one branch.

Mass A property that is equivalent to the number of atoms or amount of energy that something contains. Inertia is a similar idea that describes mass in terms of its resistance to movement, such that a heavy (more massive) object is harder to move.

Momentum The product of mass and velocity that expresses how hard it is to stop something once moving.

Nucleus The hard central core of the atom, made of protons and neutrons held together by the strong nuclear force.

Observer In quantum theory, an observer is someone who performs an experiment and measures the outcome.

Phase The relative shift between one wave and another measured in wavelength fractions. One whole wavelength shift is 360 degrees; if the relative shift is 180 degrees, the two waves are exactly out of phase (*see also* Interference).

Photon Light manifesting as a particle.

Pressure Defined as the force per unit area. The pressure of a gas is the force exerted by its atoms or molecules on the inside surface of its container.

Quanta The smallest sub-units of energy, as used in quantum theory.

Quark A fundamental particle, three of which combine to make up protons and neutrons. Forms of matter made of quarks are called hadrons.

Redshift The shift in wavelength of light from a receding object, due to the Doppler effect or cosmological expansion. In astronomy it is a way of measuring distances to far away stars and galaxies.

Space–time metric Geometric space combined with time into one mathematical function in general relativity. It is often visualized as a rubber sheet.

Spectrum The sequence of electromagnetic waves, from radio waves through visible light to x-rays and gamma rays.

Universe All of space and time. By definition it includes everything, but some physicists talk of parallel universes separate from our own. Our universe is about 14 billion years old, determined from its rate of expansion and the ages of the stars.

Vacuum A space that contains no atoms is a vacuum. None exists in nature — even outer space has a few atoms per cubic centimeter — but physicists come close in the laboratory.

Velocity Velocity is speed in a particular direction. It is the distance in that direction by which something moves in a given time.

Wave function In quantum theory, a mathematical function that describes all the characteristics of some particle or body, including the probability that it has certain properties or is in some location.

Wavelength The distance from one wave crest to the next adjacent one.

Wave–particle duality Behavior, particularly of light, that is sometimes wave-like and at other times like a particle.

MATHEMATICS

Algebra Dealing with letters instead of numbers so as to extend arithmetic, algebra is now a general method applicable to all mathematics and its applications. The word "algebra" derives from "al-jabr" used in an Arabic text of the ninth century A.D.

Argand diagram A visual method for displaying the two-dimensional plane of complex numbers.

Axiom A statement, for which no justification is sought, that is used to define a system. The term "postulate" served the same purpose for the ancient Greeks but for them it was a self-evident truth.

Cardinality The number of objects in a set. The cardinality of the set $\{a, b, c, d, e\}$ is 5, but cardinality can also be given meaning in the case of infinite sets.

Chaos theory The theory of dynamical systems that appear random but have underlying regularity.

Conic section The collective name for the classical family of curves which includes circles, straight lines, ellipses, parabolas and hyperbolas. Each of these curves is found as cross-sections of a cone.

Counterexample A single example that disproves a statement. The statement "All swans are white" is shown to be false by producing a black swan as a counterexample.

Differentiation A basic operation in Calculus which produces the derivative or rate of change. For an expression describing how distance depends on time, for example, the derivative represents the velocity. The derivative of the expression for velocity represents acceleration.

Diophantine equation An equation in which solutions have to be whole numbers or perhaps fractions. Named after the ancient Greek mathematician Diophantus of Alexandria (c.A.D. 250).

Discrete A term used in opposition to "continuous." There are gaps between discrete values, such as the gaps between the whole numbers 1, 2, 3, 4, . . .

Distribution The range of probabilities of events that occur in an experiment or situation. For example, the Poisson distribution gives the probabilities of x occurrences of a rare event happening for each value of x.

Divisor A whole number that divides into another whole number exactly. The number 2 is a divisor of 6 because $6 \div 2 = 3$. So 3 is another because $6 \div 3 = 2$.

Exponent A notation used in arithmetic. Multiplying a number by itself, 5×5 is written 5^2 with the exponent 2. The expression $5 \times 5 \times 5$ is written 5^3, and so on. The notation may be extended: for example, the number $5^{1/2}$ means the square root of 5. Equivalent terms are power and index.

Geometry Dealing with the properties of lines, shapes and spaces, the subject was formalized in Euclid's *Elements* in the third century B.C. Geometry pervades all of mathematics and has now lost its restricted historical meaning.

Hypothesis A tentative statement awaiting either proof or disproof. It has the same mathematical status as a conjecture.

Imaginary ("numbers") Numbers involving the "imaginary" $i = \sqrt{-1}$. They help form the complex numbers when combined with ordinary (or "real") numbers.

Integration A basic operation in Calculus that measures area. It can be shown to be the inverse operation of differentiation.

Irrational numbers Numbers which cannot be expressed as a fraction (e.g. the square root of 2).

Matrix An array of numbers or symbols arranged in a square or rectangle. The arrays can be added together and multiplied and they form an algebraic system.

One-to-one correspondence The nature of the relationship when each object in one set corresponds to exactly one object in another set, and vice versa.

Polyhedron A solid shape with many faces. For example, a tetrahedron has four triangular faces and a cube has six square faces.

Prime number A whole number that has only itself and 1 as divisors. For example, 7 is a prime number but 6 is not (because $6 \div 2 = 3$). It is customary to begin the prime number sequence with 2.

Pythagoras' theorem If the sides of a right-angled triangle have lengths x, y and z then $x^2 + y^2 = z^2$ where z is the length of the longest side (the hypotenuse) opposite the right angle.

Quaternions Four-dimensional imaginary numbers discovered by Sir W.R. Hamilton.

Set A collection of objects: for example, the set of some items of furniture could be $F = \{$chair, table, sofa, stool, cupboard$\}$.

Symmetry The regularity of a shape. If a shape can be rotated so that it fills its original imprint it is said to have rotational symmetry. A figure has mirror symmetry if its reflection fits its original imprint.

Theorem A term reserved for an established fact of some consequence.

Twin primes Two prime numbers separated by at most one number. For example, the twins 11 and 13. It is not known whether there is an infinity of these twins.

Venn diagram A pictorial method (balloon diagram) used in set theory.

x–y axes The idea due to René Descartes of plotting points having an x-coordinate (horizontal axis) and y-coordinate (vertical axis).

INDEX

A FIREFLY BOOK

Published by Firefly Books Ltd. 2011

First printing

Publisher Cataloging-in-Publication Data (U.S.)

A CIP record of this book is available from Library of Congress (US)

Library and Archives Canada Cataloguing in Publication

Henderson, Mark, 1974-
100 most important science ideas : key concepts from genetics, physics and
mathematics / Mark Henderson, Joanne Baker, and Tony Crilly.

Includes index.
ISBN 978-1-55407-948-3

1. Genetics--Miscellanea. 2. Physics--Miscellanea. 3. Mathematics--
Miscellanea. I. Baker, Joanne II. Crilly, A. J. III. Title. IV. Title: One
hundred most important science ideas.

Q126.H46 2011   500   C2011-900742-8

Published in the United States by
Firefly Books (U.S.) Inc.
P.O. Box 1338, Ellicott Station
Buffalo, New York 14205

Published in Canada by
Firefly Books Ltd.
66 Leek Crescent
Richmond Hill, Ontario L4B 1H1

Printed in China

Picture credits:
Illustrations by Patrick Nugent
Mathematics illustrations by Patrick Nugent and Tony Crilly
Title page: www.istockphoto.com: DNA© Andrey Prokhorov;
Plasma ball © Siniša Botaš; X-ray hand © Dirk Freder;
Human brain © Mads Abildgaard
p. 363: iStockphotos